MODERN METALLOGRAPHIC TECHNIQUES AND THEIR APPLICATIONS

Modern Metallographic Techniques and Their Applications

VICTOR A. PHILLIPS

General Electric Company
Corporate Research and Development
Schenectady, New York

WILEY-INTERSCIENCE, a Division of John Wiley & Sons, Inc.
New York · London · Sydney · Toronto

PREFACE

The science of metallurgy may be said to have originated in the discovery by Sorby that metals could be polished and etched to reveal significant structure when viewed through a crude optical microscope. Metallography, as it came to be called, rapidly became the backbone of metallurgy and has remained so to this day. The advent of the electron microscope as a reliable tool during the last decade and the birth of the field of transmission electron microscopy, extended the resolution of the light optical microscope by four orders of magnitude, and has given tremendous impetus to the field of metallography. The demands of modern industrial civilization and the birth of nuclear and space engineering have greatly diversified the range of materials in everyday use. Thus the term "metallography" has come to embrace ceramics, polymers, and inorganic substances, as well as metals, and might more aptly be called "materiallography." The contents of this book are, therefore, addressed to material scientists, engineers, metallurgists, in fact to all scientists concerned with the relation between structure and properties or behavior. A number of the techniques discussed are equally applicable to such diverse fields as mineralogy, petrography, forensic science, and the study of particulate matter in the pollution field. The real division these days is perhaps between nonbiological and biological materials; even here there is a good deal of common interest in techniques for the examination of structure.

Although the light optical microscope retains a central place in metallography and, as will be shown, has become a versatile and sophisticated tool for studying materials, the tools of the metallographer have become much more diversified, and now include x-ray diffraction,

v

electron diffraction, electron microscopy, scanning electron microscopy, electron microprobe analysis, fluorescent x-ray analysis, ion microscopy, and microhardness. All of these techniques, with the exception of ion microscopy which is still essentially a research tool, will be found in a large modern metallography laboratory. The staff of such a laboratory comprises a team of specialists, since it is difficult work and takes many years for any one individual to master even several of the available metallographic techniques. It is thus not surprising that the individual scientist or engineer lacking a knowledge of the principles of modern metallography is unable to visualize an approach to his problem. Once the material is specified and the type of information required defined, there is usually a choice of several alternate techniques and a path of action can be decided based on such factors as speed, cost, and the probability of success.

In this book I have endeavored to collect a reasonably comprehensive account of the techniques employed in modern metallography. In view of the scope, emphasis is placed on principles rather than detail wherever reference can be made to other sources of detailed information. I have deliberately excluded x-ray diffraction, since there are many excellent and comprehensive texts available; furthermore, the principles have much in common with electron diffraction which is discussed.

An elementary knowledge of metallography and crystallography on the part of the reader is assumed. The book is addressed mainly to graduates, but it is expected that undergraduates, and practical metallographers, many of whom have only a two-year college background, will profit from reading selected sections. Some emphasis will be placed on the quantitative aspects of metallography, since this appears to me to be one of the most significant trends of recent years, one that is accelerating with the advent of automated and computerized techniques.

It is often overlooked that metallography gives information on composition and properties, as well as structure. Thus the skilled light-metallographer will note the relative hardness of the phases in an alloy from their behavior during grinding and polishing, will note their degree of optical anisotropy and their chemical behavior with various etchants, and will finally record their structure. From these observations the approximate composition of the alloy can often be deduced, and some information on the properties, as well as the structure, obtained. Considerable information on the processing history can now be deduced. With the advent of instruments such as the scanning electron microprobe analyzer, the point made above is somewhat obvious. Indeed, but for the fact that the instrument is giving chemical information from microareas in which the previous metallographic preparation and the microstructural features may be of vital significance, it might be placed in a chemical analysis, rather than metallographic, laboratory. To me, one of the most fascinating and rewarding features of modern metallography is the interweaving of new and old techniques and the interplay of them. It is indeed dangerous to become too specialized, since there is often a simple and effective way of getting required information at a fraction of the time and cost involved in using elaborate and complex tools such as the electron microprobe. The latter should be reserved for applications in which it can give unique information.

I am very grateful to the General Electric Company and to the many individuals and instrument makers acknowledged in the text, who generously made available information and numerous outstanding micrographs. I have been very fortunate in being able to draw freely on the mechanical services at the General Electric Research and Development Center, whose management encouraged the writing of this book. I would like to thank the many colleagues who have assisted directly and indirectly in its preparation. I would also like to express my gratitude to my wife, Mavis, and daughter, Jennifer, for their patience and encouragement through the long and often difficult spare-time hours that went into the making of this book.

V. A. PHILLIPS

Schenectady, New York
February 1971

CONTENTS

1. Metallographic Methods of Revealing Structure   1

2. Microindentation Hardness Testing   27

3. Optical Microscopy   53

4. Low Energy Electron Diffraction and Auger Electron Analysis 111

5. Conventional Electron Microscopy   145

6. Specialized Electron Microscopy   281

7. X-Ray Microscopy   373

8. Scanning Electron Microscopy   422

9. Electron Microprobe Analysis   461

    Index   517

MODERN METALLOGRAPHIC TECHNIQUES AND THEIR APPLICATIONS

1

METALLOGRAPHIC METHODS OF REVEALING STRUCTURE

I. Polishing 2

II. Relief Polishing 2

III. Sulfur Prints 3

IV. Chemicomechanical Polishing 4

V. Electromechanical Polishing 4

VI. Etching 4
    A. Chemical Etching 5
        1. Filming Etchants 6
    B. Electrolytic Etching (and Anodization) 7
    C. Heat Tinting 8
    D. Ionic Bombardment (Cathodic) Etching 9
        1. Cathodic Vacuum Oxidation 9
    E. Thermal Etching 9

VII. Vacuum-Deposited Dielectric Coatings 10

VIII. Etch Pitting 11
    A. Dislocation Etch Pitting 12
        1. Gilman and Johnston Etch-Pitting Technique 12
    B. Tests of Correspondence between Etch Pits and Dislocations . . . 14
    C. Other Methods for Dislocation Density Determination 16

IX. Optical Anisotropy Produced by Etching 16

X. Surface Markings 17

XI. Taper Sectioning 17

XII. Reflectivity 19

XIII. Polarized Light 19
          A. Metallurgical Applications 19
                1. Oxide Coating Thickness 20
                2. Internal Stress Analysis 20
                3. Ferromagnetic Domains 20
                4. Identification of Alloy Phases and Inclusions 20
          B. Optical Properties 22
                1. Immersion Method of Determining Refractive Indices 22
                2. Interference Figures 23
                3. Pleochroism 23
                4. Extinction 24

XIV. References 24

I. POLISHING

Familiarity is assumed with the various methods of metallographic polishing, namely, mechanical,[1-5] electrolytic[5-6] (for theories see Refs.[7-10]), and chemical[5] (for list of chemical polishing agents see Refs.[10-11]). Brief reference is made to the newer methods of chemicomechanical and electromechanical polishing. Where large areas of surface have to be prepared, machined surfaces may sometimes be macroetched to reveal coarser features such as the grain size. The use of a sharp, for example, diamond-tipped tool and a fine finishing cut is necessary to avoid undue distortion of the structure. The introduction of diamond knife microtomes has revived interest in the possibility of preparing undistorted surfaces of softer, readily cuttable materials for microscopic examination. Fine porosity in copper wires of <1-mm diameter may be very readily revealed in this manner. For larger samples, the cost of diamond knives is prohibitive; but for some materials such as wood, plastic, and paper, knives made of glass, cemented carbide, or steel may be quite suitable.

Diamond abrasives have, to a substantial extent, replaced other abrasives for the wet mechanical polishing of metals and alloys. The procedures employed lend themselves with but slight modification to the polishing of most other materials such as ceramics, minerals, plastics, paper, and so forth. In the case of porous materials it is necessary first to fill the pores, which is commonly done by vacuum impregnation with plastic, as in the case of porous metals. Automatic vibratory polishing machines have found considerable application.[2-4]

II. RELIEF POLISHING

When constituents of widely differing "hardness" are present in a sample, it is common to see some structure revealed in the as-polished, or even in the finely ground condition, as a result of level differences due to the different rates of removal of the material. A criterion of the metallographer's skill is his ability to devise and carry out a procedure resulting in a relief-free surface. Conversely, the skilled metallographer will be able to detect even small differences in composition due to "coring" in solid solutions, as a result of their tendency to give relief polishing. In a similar way the grains may be discernible on the machined surface of a polycrystalline sample, as a result of machining differences. Observation of as-polished surfaces will often reveal hard inclusions, for example, oxides and sulphides, and other second-phase particles, standing out in relief, while soft phases are depressed. A sample should always be carefully examined in the as-polished condition before etching.

III. SULFUR PRINTS

Sulfur prints can be obtained by applying photographic paper soaked in dilute (3–5%) sulfuric acid to the finely ground surface of a steel[12]. Hydrogen sulfide liberated from sulfur-rich inclusions reacts with the emulsion, producing dark spots of silver sulfide. The print is rinsed to remove excess acid and fixed in the usual way. This is essentially a macro-technique and can be applied to slices of large ingots.

A similar approach has been used[13] to study phosphorus segregation in steel, using cellophane as the print carrier. The test was based on the formation of molybdenum blue in the reduction of phosphomolybdate by a solution of stannous chloride in hydrochloric acid. In this case, moist cellophane film was pressed onto the horizontal sample surface, and the top film surface wetted with etchant which was allowed to diffuse through the cellophane film for a few minutes. The etchant was rinsed off, and the film was transferred to a white porcelain dish with the side that was in contact with the sample downward. A thin film of reducing solution was then applied and allowed to diffuse through the cellophane to form colored ions. The cellophane film was washed and transferred onto the gelatin surface of a photographic plate to prevent shrinkage during drying. It was claimed that a few hundredths of a percent of phosphorus could be detected. The technique was also applied to copper and phosphor bronze.

In principle the method is a general one and could be applied to study segregations of any element for which a specific reagent giving colored ions is available. A range of impregnated surface test papers is commercially available[14] for both anion and cation tests. Anions that may be checked include chromium, cobalt, copper, molybdenum, nickel, and lead.

The surface test papers are intended for testing for pores in electro-deposited coatings and for qualitative determination of the components of an alloy on a coated surface.

Contact printing has also been applied for qualitative analysis of ore specimens.[15] In one method the photographic paper or hard filter paper was impregnated with a selective attacking reagent, and after application and removal from the sample was "developed" in a reagent that was specific for the ion in question, producing a distinct coloration. In other cases both the specific reagent and the attacking reagent were placed on the paper, giving a "direct print." Since slow attack resulted in blurring of the print as a result of the diffusion of ions through the paper, an *electrographic* method was developed for electrically conducting minerals, in which a metal plate was pressed on top of the impregnated photographic paper and the sample attacked anodically.[15,16]

IV. CHEMICOMECHANICAL POLISHING

Chemicomechanical polishing, previously known as polish-attack or polish-etch, involves, as the name implies, the simultaneous use of mechanical polishing and chemical etching.[12] It is particularly effective for revealing the true structure of soft, easily flowed metals such as pure copper and lead, and also prevents staining. In the case of copper, the final polishing pad may be moistened with dilute ammonia. Samples polished by this technique are inevitably somewhat etched.

V. ELECTROMECHANICAL POLISHING

A more recent technique involves simultaneous mechanical and electrolytic polishing.[17] The polishing lap is made of an electrically conducting, corrosion-resistant material such as stainless steel, which serves as the cathode. A positive voltage is applied to the specimen, and the polishing cloth is soaked in an electrolytic polishing solution with the addition of a polishing compound such as alumina. The method has found some application in polishing soft materials that tend to flow. More rapid polishing would be expected than with mechanical polishing alone, and waviness on a macroscale resulting from electrolytic polishing should be avoided by the combined method.

VI. ETCHING

The following methods of etching will be discussed: (*a*) chemical, (*b*) electrolytic, (*c*) heat tinting, (*d*) cathodic, and (*e*) thermal.

Grain contrast may also be produced as a result of etchpitting of the surface on a submicroscopic scale. The tiny etch pits, whose morphology is related to the orientation of the grain in which they form, alter the reflectivity of the grain and can also give color contrast when viewed between crossed nicols using a sensitive tint plate.[18] (See Etch Pitting in Section VIII.)

Submicroscopic detail may sometimes be made visible by etching; for example, particles too small to resolve may give rise to pits or mounds. This is a common effect in precipitation-hardened alloys. Similarly, diffusion fronts and segregation such as coring may be revealed by etching, although there is no sharp phase boundary but rather a gradual variation in composition.

A. Chemical Etching

Some structural features can be made visible under vertical bright field illumination by etching the polished surface of an (opaque) sample with an etchant whose rate of attack varies either with orientation, or with composition, or with both. Some ways in which etching can convert orientation and compositional differences into topographical features that reflect the incident light differently are illustrated in Figure 1. The grains

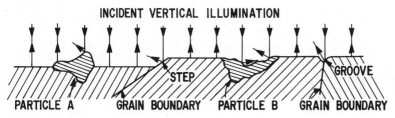

Figure 1 Some origins of contrast on an etched sample under vertical illumination.

are made visible as a result of level differences, that is, steps, grooves, or ridges (not illustrated) at the boundaries. The grains reflect equally so that no contrast would be seen between them. This is essentially a *grain boundary etch*. Particle A was left in relief because it was electropositive relative to the matrix, whereas particle B was electronegative and was preferentially etched.

A different type of etchant would be one in which the rate of attack is strongly dependent on crystal orientation. This can lead to a difference in inclination of the grain surfaces, to grain contrast in vertical illumination. Grain contrast can also arise in other ways which are discussed below. An etch of this kind would be called a *grain contrast etch*.

It should be pointed out that ridging or grooving effects at grain boundaries are usually associated with chains of particles, thin films of second phase, or segregation.

1. Filming Etchants

Compositional or structural variations may be made visible by the use of filming etchants that produce a transparent film of oxide, sulfide, or some other type on the surface of a sample.[18]

In the *interference method*, normal illumination results in interference between light reflected from the surface of the film, which may be amorphous or crystalline, and light transmitted through the film and reflected from the metal-film interface. The thickness of the film is determined by the orientation of the grain; the optical path difference is simply the product of the refractive index of the film and its thickness. Suitable films may be produced chemically, electrolytically, or by the heat-tinting technique described later. The aim is to obtain interference colors, usually of first order. Suitable chemical filming etchants of the sulfiding type have been developed for copper[18,19] and iron,[20] and of the oxiding type for aluminum[21] (see Ref. 18 for other examples).

In the *polarization method*, using polarized light, an otherwise optically isotropic sample is coated with an (epitaxial) oxide, sulfide, or other film that is anisotropic and strongly absorbing. Alternatively, a transparent birefringent film of varying thickness is used to vary the ellipticity of the polarized light. In this way the contrast between differently oriented crystals in the sample is increased. The birefringence of the film can result either from its anisotropic crystal structure or from micropores of which the shape and size determine the birefringence.[18] Colored contrast may be obtained in polarized light between crossed nicols with the aid of a sensitive tint plate. The technique is particularly useful for aluminum and its alloys, employing an oxide film produced by anodic oxidation (see Ref. 18 for references).

Thinner films may be employed with the polarization method than with the interference method. The two methods are complimentary. Two crystals with the surface parallel to the same plane such as (100) show the same interference color, irrespective of the direction of [001] in that plane; whereas different colors would be obtained, between crossed nicols,[18] and could be varied by rotating the stage. Information on the texture is thus obtainable. Other effects that can be studied include twinning and microchemical or micromechanical inhomogeneities. Although thick films still give contrast effects in polarized light, they tend to integrate microinhomogeneities, so that sensitivity is lost.[18]

Compositional or structural variations may sometimes be revealed by another type of etchant which deposits a metallic film on the surface of the metal. This would also be called a filming etchant. Thus cupric chloride solutions of controlled pH can be used on iron[11,12] to reveal phosphorus segregation (copper deposits first on areas lowest in phosphorus—Stead's reagent), subgrains, strain lines, and depth of nitrogen diffusion (Fry's reagent). This type of etchant should be distinguished from other types, such as copper-ammonium-chloride used for macroetching iron (Heyn's and Humfrey's reagents[11,12]), that deposit a loosely adherent film of copper which has to be removed in order to see the structure. Colored chemical films may also be formed which are helpful in identifying phases; for example, in steel, an alkaline potassium ferricyanide solution will blacken cementite, turn pearlite brown, and leave massive nitride unattacked.[11]

Double-etching procedures may sometimes be employed advantageously, when the first etch reveals some features of the structure, for example, outlines a constituent, and is followed by a second etch to reveal some other feature such as the grain size of the matrix.

The reader interested in lists of chemical etchants may refer to other sources.[11,22,23] The mechanism of chemical etching is discussed in Refs.[24-27]

B. Electrolytic Etching (and Anodization)

Following electrolytic polishing, a sample may be etched by reducing the voltage for a few seconds before withdrawing from the solution (Figure 2). Alternatively, the sample may be transferred to a different electrolytic cell for etching; for example, copper can be polished electrolytically, using orthophosphoric acid (Jacquet method[5]), and then transferred to another cell for an electrolytic sulphide filming etch, using the Williams and Rieger method.[28] This etch can also be employed on a mechanically polished sample, but the results are then less satisfactory.

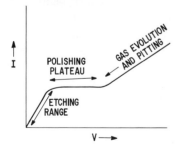

Figure 2   Current (*I*) versus voltage (*V*) diagram, showing region useful in electrolytic etching.

Electrolytic polishing of, for example, aluminum, titanium, and niobium alloys can be followed by *anodization* in a separate anodizing bath to produce a transparent optically active film, whose thickness is a function of substrate orientation and composition, and also of the anodizing voltage. Methods for aluminum have been reviewed by Lacombe and Mouflard.[18] On viewing between crossed polarizers with the addition of a sensitive tint plate, different grains and phases can be made visible by color contrast. By the careful control of procedure, it is possible to obtain very reproducible results, so that the colors can be used in a particular alloy for phase identification. (See Filming Etchants in Section VI.A. 1)

In some cases, etching may be produced after electrolytic polishing by simply short-circuiting the electrodes, and may then be regarded as electrolytic etching caused by the galvanic voltage between the electrodes. This technique was developed by Jacquet[29] for copper and gave contrast as the result of the development of extremely fine etch pits.

Electrolytic etching by the *potentiostatic method* appears to be assuming increased importance. In this method an additional standard reference electrode is employed, with a potentiometric circuit to maintain a selected dissolution voltage on the sample. If the potential versus current curve is known for each phase present, the etching conditions may be chosen to etch selectively a given phase in a multiphase alloy. The characteristic voltage for a phase is a function of composition and will tend to vary in a complex alloy with the processing and heat treatment. It may therefore prove difficult to standardize the etching conditions for a material such as a superalloy, in order to use the technique to identify phases present. The technique has been applied to a variety of materials, as in the review by Lacombe.[5]

C. Heat Tinting

By carefully controlled heating in a suitable atmosphere, many metals and alloys may be induced to form transparent oxide films showing interference colors.[12,18,30] Since the rate of oxidation is sensitive to compositional differences and often to orientation differences, constituents and grains may be revealed. Although now infrequently used, the method can be very useful; thus Westbrook[30] in phase diagram studies on the Ni-Ti-Si system, which contains many phases, was able to use a standardized heat-tinting procedure and identify the phases by their characteristic colors. In the case of high alloy steel, heating for 10 to 60 sec at 1000 to 1400°F darkens ferrite more rapidly than austenite and leaves carbides white.[11] Contrast may also be obtainable in polarized light (see the discussion of filming etchants in Section VI.A. 1).

D. Ionic Bombardment (Cathodic) Etching

Samples, whether metallic or ceramic, may usually be etched by bombardment with heavy gas ions such as argon, readily revealing features such as grains and constituents. The equipment required is relatively simple and may consist of a water-cooled specimen holder, located in a bell jar capable of evacuation to about 10^{-5} torr. Argon is admitted via a controlled leak to maintain a pressure of about 10^{-4} torr. A glow discharge is initiated by applying a potential of about 5 kV between the sample (cathode) and an anode situated 1 to 2 in. above the sample surface. The etching obtained is a function of the pressure, voltage, and time, for a given geometry and gas. Depending on the experimental conditions, either the grain boundaries may be preferentially etched while the grains are leveled, or the grains themselves can be etched. The equipment described in Chapter 5, Section III.B.5 for thinning samples for electron microscopy may also be employed for this kind of etching. The technique has been reviewed by Lacombe.[5] Bombardment with oxygen ions has been used in the case of uranium to produce transparent oxide films capable of giving contrast effects.[18] (See Filming Etchants in Section VI.A. 1)

1. Cathodic Vacuum Oxidation

Some success has been achieved with a filming method of etching in which, following cathodic cleaning or etching by argon ion bombardment, 5×10^{-2} torr of oxygen is admitted and oxygen ion bombardment carried out at 1 to 5 kV for 1 to 10 min.[5] Oxidation is preferably stopped when first order blue and green interference colors are observed, corresponding to a film thickness of 2000 to 2500 Å for a refractive index of 1.5. The method has been successfully applied to iron, uranium, nickel, copper, zirconium-niobium alloys, and alloys of uranium.[5,31] The transparent oxide films give contrast effects in the usual way (see Filming Etchants). However, if the oxidation time was only a few seconds, Aubert[32] showed that grain, subgrain, and twin boundaries in uranium could be decorated and thus made visible in the light microscope, by oxide patches 1 to 2 μ in diameter.

E. Thermal Etching

The rate of evaporation from the surface of a polished crystal heated in high vacuo is a function of the orientation, since this determines the arrangement of atoms and atom planes at that face. Topographical features therefore develop which are related to the structure. This fact is one of the bases of thermal etching. In fact the situation is considerably more complicated, since, even in high vacuo, absorption effects on the sample surface play an important role in modifying the surface energy and hence the topographical features developed. Striations on the grain surfaces related to the orientation, and grain boundary grooves, may develop even in the absence of evaporation. Surface and grain boundary diffusion can redistribute material. Observation of the groove profile is in fact a standard method of relating the surface and grain boundary energies. In principle, grains and constituents may be revealed in both metal and ceramic materials. The method is less useful than ion bombardment etching because it is more difficult to control, and the required structural features tend to be swamped by unwanted effects such as striations and microfacets. Thermal etching has most often proved useful in direct observational research studies on phenomena such as grain growth and phase transformations, where other etching methods cannot readily be used.

Thermal etching may also be carried out in various kinds of atmosphere involving oxidation or other chemical modification of the sample surface. The technique then overlaps heat tinting. The laser has recently been applied to thermal etching.[33] Further information and references are given in Refs. 5 and 34 to 36.

VII. VACUUM-DEPOSITED DIELECTRIC COATINGS

Different phases in a sample are often visible in the as-polished condition under vertical illumination, as a result of differences in reflectivity and the optical phase angle. These effects are often small; however, Pepperhoff[37] showed that they could be enhanced by vacuum depositing a thin uniform transparent layer of a dielectric material on the surface of the sample. The technique has since been extended and developed.[38-47] The contrast is enhanced first as a result of multiple reflection of light between the metal surface and the top of the film, and second, because of phase shifts at the film-metal interface, whose amount depends on the optical constants of the film and substrate.

Since grain boundaries are not revealed except in anisotropic crystals, the sample may be given a grain boundary etch prior to coating. The compounds TiO_2, ZrO_2, ZnS, ZnSe, ZnTe, and SiO have been employed as coating materials, usually deposited at an angle to give a range of coating thickness, so that the optimum region can be selected. The thickness is such that a uniform interference color such as crimson red is observed under white light. The sample can be viewed in the light microscope by white, monochromatic, or polarized light. Color photography is advantageous, but with black and white photography it is possible to obtain high contrast by selecting the wavelength of the illumination and the film thickness, so that one phase appears black.

The technique, which is not yet well known, has the advantage that it may be applied to any material. Even epoxy mounting material in impregnated porous samples may be distinguished.[46] Whereas phase contrast microscopy requires phase differences of at least 6 to 7°, the use of the present technique permits phase differences of about 1° to be detected.[43] Furthermore, a phase contrast image is obtained by making use of purely optical properties. The technique has been applied for the identification of nonmetallic[47] and intermetallic[46] inclusions.

VIII. ETCH PITTING

Randomiy distributed etch pits tend to form on metal surfaces if they are electropolished in either the etching region or in the region of gas evolution (Figure 2). Gas bubbles clinging to the surface can give complex dissolution conditions. If semiprotective reaction product films form on the surface, dissolution tends to be concentrated at flaws in the films. Clearly, pits are not likely to give useful structural information in these cases. They may do so, however, if they are located exclusively at structural "defects" of interest such as submicroscopic precipitate particles.

Another type of pit results from the fact that the dissolution rate of a crystal may differ in different crystallographic directions; thus, if a spherical single crystal is etched, it may develop well-defined crystallographic faces. Alternatively, if for some reason attack is localized at many points on a crystal surface, faceted pits may develop. Crystallographic pits may give information in a number of ways: (a) they may be located exclusively at submicroscopic structural defects such as decorated dislocations, thus revealing their number and distribution. (b) If the etch develops known crystal faces exclusively, for example, {111} or {001}, then the orientation of a grain may be determined by means of an optical goniometer with a precision comparable to that of the usual x-ray technique. (c) "Crystallographic pits" give rise to contrast effects with vertical bright-field illumination, so that regions which differ in orientation are distinguishable; it has proved possible to determine qualitatively the preferred orientation in large sheets by suitable examination in reflected light to determine the directions of minimum reflectivity.

Crystallographic pits are also "phase-detail," since they represent level differences related to orientation, and thus can provide contrast effects due to orientation differences visible in reflected polarized light between crossed polarizers. It is also clear, if the facets are regarded as little mirrors, that the reflectivity of a crystal for plane-polarized light will vary as the crystal surface is rotated on the microscope stage. Optically isotropic materials may thus be rendered "optically active" in reflection by suitable etching.

In order to obtain quantitative structural information by means of etch pits, very careful control of the etching procedure is necessary.

A. Dislocation Etch Pitting

There is a large literature on this subject (see Refs. 48 and 49), mainly because every material requires a different etchant and it is necessary to demonstrate in each case that there is a one to one correlation between dislocations and etch pits.

Gilman and Johnston,[50-53] working on lithium fluoride crystals, pioneered the dislocation etch-pitting technique, culminating in the determination of average dislocation velocities as a function of the applied stress. Their technique merits brief description.

1. Gilman and Johnston Etch-Pitting Technique

The etchants used for production of pits contained Fe^{3+} or Al^{3+} ions, which were thought to be adsorbed on the surface of the lithium fluoride crystal, inhibiting the motion of kinks in surface steps without markedly influencing the nucleation of pits at dislocations.

A freshly cleaved {100} surface of a LiF crystal, when dipped in pure water, became highly polished and dissolved at about 250 Å/sec. If small amounts of FeF_3 were now added to the water, small regularly shaped pits appeared. When the concentration of FeF_3 was increased, then, for the same etching time, the pits increased in sharpness and decreased in size. The rate of dissolution at the center of the pits remained about 250 Å/sec but dissolution of the rest of the surface was almost completely inhibited.

Under optimum conditions the pits were square pyramids, with edges parallel to $<001>$ (Figure 3a). The sides of the pyramids were not simple low index crystal planes and the angle they made with the surface increased to $> 10°$ as the concentration of inhibiting ion was increased. With this neutral etchant, pits at fresh dislocations were much more distinct than those at grown-in and therefore presumably decorated ones, permitting their distinction. On the other hand, an acid etchant, produced by adding FeF_3 to concentrated mixtures of hydrofluoric and acetic acids, gave similar pits at both fresh and grown-in dislocations with edges parallel to $<110>$ (Figure 3b, right). By etching to produce a sharp pit, stressing to move the dislocation responsible for the pit, and then reetching, a second sharp-bottomed pit was produced at the new location; whereas the initial pit became flat bottomed during the second etch (Figure 3b, left). The initial and final positions of the dislocation were thus defined. Gilman and Johnston were able to demonstrate a substantially 1:1 correspondence between emergent dislocations and etch pits. Unlike decoration techniques, etch pitting did not immobilize the dislocations.

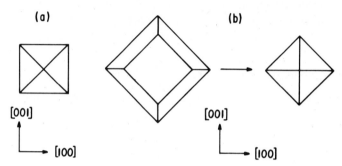

Figure 3   Gilman and Johnson[50-53] etch-pitting technique for lithium fluoride: (a) etched in neutral etchant to produce sharp-bottomed pit at fresh dislocation. (b) Etched in acid etchant, stressed and reetched in acid etchant; flat-bottomed pit (left) represents old site of dislocation, while sharp-bottomed pit (right) represents new site.

Dislocation etch pitting is suitable for determining dislocation densities up to 10^6 to $10^8/\text{cm}^2$; however, the latter figure would require that the pits be < 1 μ square. Using replicas and electron microscopy, the technique can be extended to higher densities ($\sim 10^{10}/\text{cm}^2$) if desired, or one can use transmission electron microscopy which has the advantage of being a direct observational technique. The latter tends to be limited to densities over about $10^6/\text{cm}^2$ due to the fact that the small width of a dislocation image (~ 100 Å) necessitates the use of a magnification of approximately 20,000X. A density of $10^6/\text{cm}^2$ then corresponds to only two dislocations in a 30×30-cm photograph, corresponding to a specimen area of approximately 2.2×10^{-6} cm^2. Dislocation etch pitting is particularly valuable, then, at low dislocation densities, where the optical microscope can be employed.

Sears[54] concluded that dislocation etch pitting can only occur in the presence of adsorbed dissolution poisons that maintain the local undersaturation, for example, of lithium fluoride, below a critical value. Iron evidently acts as a poison in the case of lithium fluoride. Hari Babu, Sirdeshmukh, and Bansigir[55] found that mercuric chloride acted as a poison in the case of the dissolution of sodium chloride in ethyl alcohol, and were able to distinguish between pits formed at decorated and undecorated dislocations. It seems probable that minute amounts of the order of 1 ppm of poison are always required for dislocation etch pitting, although the nature of the poison is often unknown. Cabrera's theory of etching[56] is probably applicable if modified to take the effect of dissolution poisons into account.[54]

In order to use etch-pitting techniques for the determination of dislocation densities, it is necessary to establish a correspondence between pits and dislocations. Some useful criteria will now be discussed.

B. Tests of Correspondence between Etch Pits and Dislocations

The following are some useful criteria for establishing the validity of a dislocation etch-pitting technique.

1. The observed density of pits must be reasonable and must not increase with increase in etching time.

2. The nonuniformity of pit size would indicate that the pits probably nucleated at different times, that is, the number increased with time, or two or more different kinds of sites were involved.

3. A genuine substructure, for example, subboundaries, can generally be distinguished from artifact. By controlled deformation experiments on single crystals with suitably oriented faces, edge dislocations can be distinguished from screw and/or mixed dislocations. The etch patterns on the matching cleavage faces should agree. The identical pattern should be obtained on repolishing to remove completely the previous pits and reetching.

4. Characteristic patterns should be obtained following deformation. If dislocations are being etched, characteristic patterns of slip lines will be observed after deforming in specific ways, for example, by stretching, compressing, bending (pattern across neutral plane), scratching, or impact with hard particles such as silicon carbide. It can be determined by deforming a little and reetching, whether undecorated, as well as decorated, dislocations are revealed by etching.

5. Polygonization structures obtained by suitably bending and annealing should be revealed by etch pitting.

In order to establish a 1:1 correlation between etch pits and dislocations, more quantitative tests are necessary:

6. The misorientation across a subboundary or polygon boundary determined, for instance, by x-ray diffraction, should agree with that calculated from the linear pit density observed at the boundary. Considering the case of a simple tilt boundary, consisting of a set of parallel edge dislocations of Burger's vector **b** arranged in a plane perpendicular to their glide plane, with spacing d, then if θ is the misorientation produced by the boundary, we have

$$\frac{\mathbf{b}}{d} = 2 \sin \frac{\theta}{2} \approx \theta \tag{1}$$

If θ is very small ($< 10^{-3}$ radian), the spacing $\mathbf{b}/\theta$ should be just resolvable by a light microscope.

7. A pileup analysis may be made. The etch-pit spacing is compared with the expected spacing of dislocations piled up in a slip band. A perfect correlation should not be expected, unless the dislocations were pinned before the stress was removed.

8. The pit shape is a useful criterion. If a dislocation is etch-pitted, a sharp-bottomed pit should be obtained. If the crystal is now deformed to move the dislocation, then on reetching, the sharp-bottomed pit will tend to become flat bottomed. This has been demonstrated on a variety of crystals, including LiF_2, MgO, CaF_2, $CaCO_3$, Cu, and NaCl.

9. A 1:1 correlation may be established by examining the identical area by a second technique such as the Berg-Barrett x-ray method or transmission electron microscopy.

It should be noted that, if a dislocation line is not perpendicular to the crystal surface, the pit will be asymmetrical. In order to decide the smallest angle that a dislocation can make with the surface, the most asymmetrical pit is selected and the angle measured. Livingston[57] was able from such measurements to distinguish positive from negative dislocations on a bent rectangular copper crystal with {111} faces. The distinction was possible because edge dislocations on a {111} plane intersecting two opposite {111} faces of a crystal have their compression side in the acute angle between slip plane and specimen surface on one face, whereas those of the opposite sign have their compression side in the obtuse angle. Under vertical illumination the two kinds of pits appeared light and dark, respectively. Confirmation was obtained by the fact that dislocations corresponding to dark and light pits moved in opposite directions under an applied stress.

The smallest *dislocation loops* that are likely to be made visible by etch pitting are those with a diameter greater than the etch-pit size used.

For the effective use of the dislocation etch-pitting technique, it is usually necessary to use high purity material in order to avoid small precipitates, oxide particles, and so on, which cause pits, and to use clean surfaces, since dust and oxide can cause pitting. Pits tend to form at cracks, pores, and at steps such as those usually present on cleaved surfaces. Some of the etches are extremely orientation sensitive and will only work successfully if the orientation of the crystal surface is within 1 to 2° of a certain orientation, as shown by Livingston[57] in the case of copper.

C. Other Methods for Dislocation Density Determination

There are also a number of useful x-ray microscopic techniques,[49,58] either using reflection (Berg Barrett for spacings $> 2 \mu$) or transmission (for spacings $> 5 \mu$) which are discussed later. In the case of transparent crystals, internal decoration techniques, that is, precipitation at dislocations, coupled with light microscopy may be used to reveal the three-dimensional distribution.[48,49,58,59] Even a normally opaque metal silicon is transparent to infrared light; Dash[60-62] made use of this fact to study dislocations in silicon by decorating, for example, with copper precipitates, and then viewing in an infrared microscope. During the last decade the diffraction contrast electron microscope technique has been extensively used for dislocation studies,[48,49,63] as discussed later.

IX. OPTICAL ANISOTROPY PRODUCED BY ETCHING

Cubic metals, although optically isotropic and therefore appearing black when viewed between crossed polaroids, may often be rendered optically active by etching, as first shown by Jones.[64] This is apparently a result of the formation of ridges related to the underlying grain orientation which result in multiple oblique reflection of the light (Figure 4), causing elliptical polarization of incident plane-polarized light, so that complete extinction is no longer obtained with crossed polarizers. The groove direction, and consequently the light intensity observed, are a function of the grain orientation, and complete extinction is only observed when the surface is normal to the incident light. Similar effects can result from the formation of crystallographic etch pits. The cementite plates in pearlite tend to stand in relief and can give similar effects; minima in intensity are then observed on rotating the sample if the plane of the incident light is parallel or perpendicular to the plates. Grooved structures can also result from thermal etching or slip. Thick anodized films sometimes show groove structures that can give contrast.

Optically anisotropic films can be formed on otherwise isotropic materials by filming etchants, as discussed under Filming Etchants in Section VI.A. 1. If the film is also epitaxial, then grain contrast effects may be produced by reflecting polarized light from the top of the film if the plane is rotated by reflection. Other sources of contrast are varying thickness of the film related to the underlying grain orientation, and interference, both discussed under Filming Etchants.

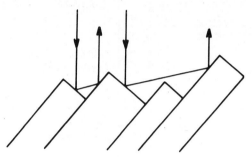

Figure 4   Multiple oblique reflection due to surface ridges.

X. SURFACE MARKINGS

Topographical studies on surfaces polished prior to deformation or transformation have proved very rewarding. Electropolishing of single crystals followed by deformation is a standard procedure in slip line studies and is usually followed by optical, replica, or scanning electron microscopic examination. Both the crystallography and the distribution of slip may be studied. Mechanical twinning and martensitic transformations, since they involve shear and volume changes, are also commonly studied by this technique.

XI. TAPER SECTIONING

In preparing a metallographic sample such as a wire or rod for examination with the light microscope, it is customary to section in a plane normal to the surface and to the wire axis (cross- or transverse section) and normal to the surface containing the axis (longitudinal section). In this way typical variations in structure from surface to center and along the length can be revealed. If either surface layers such as oxide films and coatings or the surface topography is of particular interest, more information can frequently be obtained by taper sectioning.[65-66] The principle is illustrated in Figure 5a.

By the use of taper sectioning, the apparent thickness of a thin surface layer can be increased by a factor of up to approximately 10×, compared with a normal section taken perpendicular to the surface. Thus, whereas

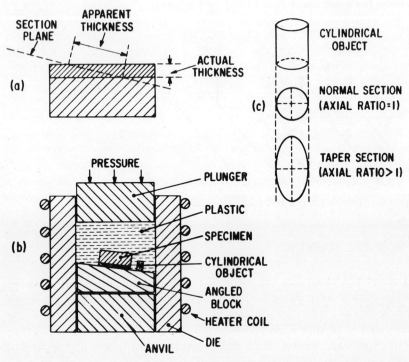

Figure 5   Taper sectioning: (*a*) principle; (*b*) apparatus; and (*c*) calibrating object.

in a normal section the detectable layer thickness is limited by the resolution of the light microscope to about 0.25 μ, in a section taken at the appropriate taper angle, a layer of about 0.025 μ is detectable. It is convenient to place a cylindrical object alongside the sample in the mount (Figure 5*b*) and determine its axial ratio after the preparation is completed (Figure 5*c*). The exact section angle is then readily calculated and the equivalent normal section magnification computed. If a 1500× microscope magnification is employed and the axial ratio observed is 10, this would be 15,000×, but applies, of course, only in one direction in the micrograph. Interpretation is aided if a normal sample section is also prepared for comparison purposes.

In principle, taper sectioning is applicable to transmission electron microscopy using ultramicrotomy. There seems to be no reason why, under ideal conditions, surface films only a few atom layers thick should not be revealed in this way.

XII. REFLECTIVITY

When a light wave falls on a metal, it causes sympathetic oscillation of the conduction electrons, resulting in a strong reflected wave and absorption of transmitted light.[67] Although polished silver can reflect up to about 95% of the incident light, steel reflects only about 60% and the remainder of the light is transmitted and absorbed. The amount reflected depends upon the state of polish. Many noncubic metals such as magnesium, zinc, and cadmium have a different reflectivity, depending on whether the plane of vibration of the light is parallel or perpendicular to the principal (hexagonal) axis. In cadmium this difference amounts to about 8%. Even in the case of an isotropic material, *obliquely* incident light undergoes a change in its state of polarization as a result of reflection. Usually, plane polarized light becomes elliptically polarized on reflection, a phenomenon referred to as *bireflection*. Fortunately, at normal incidence cubic metals do not react to polarized light; however, the higher the numerical aperture of the objective, the more oblique the incidence of the outer cone of rays will be, so that complete extinction is never obtained, even with an optically isotropic sample. Nevertheless, the intensity observed from different grains will be the same, and extinction can be made more complete by stopping down the lens to remove obliquely incident rays.

Tests for establishing the state of polarization of a beam of light are discussed on page 76.

XIII. POLARIZED LIGHT

A. Metallurgical Applications

Elliptical polarization occurs when plane polarized light is reflected from anisotropic crystals because vibrations in different directions are differently affected. This means that grain contrast is seen when a polycrystalline anisotropic material is viewed between crossed nicol prisms. Crystals with a cubic structure are normally isotropic and show no contrast under such conditions; however, other crystal classes such as hexagonal and tetragonal are anisotropic. A polished sample can be tested for optical activity by viewing between crossed nicols and rotating the sample; if the field of view remains black, it is isotropic. It should be noted that even bireflecting grains will appear isotropic if light is incident parallel to the principal axis. If the sample is optically active, grains and twins will be visible as a result of the difference in orientation. Optically active inclusions or constituents such as graphite, silica (crystalline); or titanium, chromium, or zirconium sulfides will be apparent.[68] Deformed grains may show contrast variations due to rotations produced by deformation or mechanical twinning. Careful preparation of a sample is thus important, in order to avoid unwanted deformation.

1. Oxide Coating Thickness

The determination of the thickness of oxide coatings on a metal is sometimes facilitated by the use of polarized light; for example, anodized layers on aluminum can be made to appear bright, while the metal and mounting material appear dark.

2. Internal Stress Analysis

Internal stresses in birefringent transparent materials such as glass cause changes in birefringence. Internal stress patterns may thus be made visible by viewing between crossed nicols. This is one reason why it is important to use good quality lenses in a polarizing microscope. Small differences may be seen more clearly if a sensitive tint plate is employed. This birefringent technique is the basis of photoelastic strain analysis employing plastic models to study problems such as the elastic strain distribution around a hole in a stressed plastic plate.[69,70]

Polycrystalline silver chloride in many ways resembles metals and may be extruded, rolled, annealed, fatigued, and so forth. The material is transparent and optically isotropic in the unstressed state but becomes optically anisotropic when stressed. Nye[71] made use of this fact to study the internal stress distribution resulting from deformation.

3. Ferromagnetic Domains

Ferromagnetic domains in cobalt may be observed as a result of the Kerr effect which is the rotation of the plane of polarization of light reflected from the surface of a magnetic material.[72] Figure 6 shows the results obtainable.

4. Identification of Alloy Phases and Inclusions

Polarized light has been entensively used as an aid in the identification of phases in alloy systems. The literature prior to 1952 has been reviewed by Perryman.[73] Considerable qualitative use has been made of the be-

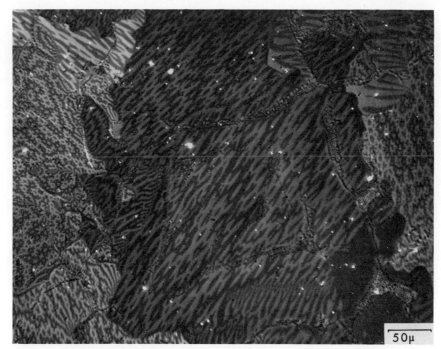

50μ

Figure 6 Polarized light micrograph of as-polished $Co_{17}Sm_2$ sample, illustrating the use of the Kerr effect. The magnetic domain structure revealed within the grains by polarized light indicates that this material has a very large magnetocrystalline anisotropy. Micrograph by A. S. Holik. (Private communication from J. J. Becker, courtesy of General Electric Company.)

havior on rotating a phase between crossed nicols in reflected light for phase identification; for example, Phragmen (see Ref. 73) has listed the behavior of 16 different phases commonly found in commercial aluminum alloys, in addition to their color in ordinary vertical illumination.

The optical constants such as the refractive index and adsorption coefficient of metallic surfaces can be measured in polarized light, using vertical illumination. Such quantitative measurements could, in principle, form the basis of a microscopic phase identification system similar to that used in powder x-ray diffraction analysis. A major practical difficulty is that the illumination is not parallel but consists of a cone of rays impinging on the sample at varying angles of incidence. Other complications arise because of variations in the optical properties with the state of the surface and also relief effects.

Polarized light has proved a valuable aid in the identification of inclusions in metals. The literature prior to 1952 has been reviewed by Mor-rogh.[74] The use of polarized light is mainly qualitative because of the complexities. Morrogh points out that examination under plane-polarized light without the use of an analyzer can be valuable. Vertical illumination in an ordinary metallurgical microscope, in which a glass slip illuminator is employed, actually consists of a mixture of unpolarized light and light polarized in the direction parallel to the reflecting surface of the glass slip. If a polarizer is inserted in the illuminating train and rotated to the position of maximum brightness, using an isotropic sample, most of the ordinary light is excluded.

The use of reflected polarized light in the microscopic examination of ores has been discussed by Cohen.[75] The use is mainly qualitative because of the complexities already mentioned. In the case of thin section petrographic examination by transmitted polarized light, very exact and relatively simple methods are used which are very helpful in identification.

B. Optical Properties

1. Immersion Method of Determining Refractive Indices

The immersion or Becke line method of determining refractive indices is based on optical effects produced when a solid transparent substance is immersed in a liquid of a different refractive index.[76-77] This is a standard method widely used by mineralogists and others. A set of standard immersion liquids is used that span the refractive index range between 1.430 to 1.740 or more, at intervals of 0.004 or 0.005. The interval limits the accuracy obtainable to 0.002 or 0.003, since this is a matching method. Standard immersion liquids are usually calibrated on the Abbé refractometer: range 1.30 to 1.71 or 1.45 to 1.84, accuracy ±0.0002. The cheaper Jelly microrefractometer can be used: range 1.33 to 1.92, accuracy ±0.001.

If the refractive index match is perfect, very little contrast is seen at the border of a crystal immersed in the oil, so that it is nearly invisible. As the mismatch increased in either direction, the edge of the crystal stands out more and more clearly. The question of whether the crystal is higher or lower in refractive index than the liquid is decided by observations on the Becke Lines. These consist of two thin lines (one dark and one bright), parallel to the crystal border, which are visible when the objective is slightly over focus. The bright line is always closest to the material having the high refractive index and always moves toward the medium having the higher refractive index, when the distance above focus is increased. The Becke lines are attributable either to refraction at the edge of the crystal which acts as a converging lens if its refractive index is greater than that of the liquid, or to total reflection at boundaries between the crystal and the oil which are nearly parallel to the illumination.

The refractive index of the oil depends on the wavelength of the light used and is normally calibrated for sodium light. Use may be made of the temperature coefficient of the refractive index for adjusting the value. An immersion liquid with a refractive index n of approximately 1.633 would typically have a coefficient of $-0.0004/°C$. so that n decreases 0.0004 for each 1°C rise. Solids usually show a much smaller change.

Isotropic crystals (cubic) have the same refractive index in all directions, so that unpolarized monochromatic light can be used and the crystal direction ignored. Anisotropic crystals are either uniaxial with two refractive indices or biaxial with three indices. Tetragonal and hexagonal crystals are uniaxial; orthorhombic, monoclinic, and triclinic crystals are biaxial. The *birefringence* is the numerical difference between the highest and lowest of the indices, and is usually between 0.003 and 0.170. Methods of distinguishing and measuring are to be found elsewhere.[76-77]

The refractive indices and birefringence are valuable for identification and may be compared with tabulated complications for known crystals or with actual known crystals.

2. Interference Figures

When viewed in transmission by convergent polarized light, a transparent anisotropic crystal gives an interference figure in the back focal plane of the objective.[67] Uniaxial and biaxial crystals have to be correctly oriented before a characteristic figure can be obtained. The figure observed depends on the system to which the crystal belongs and on its orientation in the viewing position. It may or may not change on rotating the stage.

3. Pleochroism

The absorption of some intensely colored transparent anisotropic crystals varies with their orientation relative to incident plane-polarized white light. On rotating the stage or polarizer, greater or lesser absorption takes place and a color change occurs (no analyzer is necessary). The phenomenon is called pleochroism. The material is referred to as dichroic if two different colors are observed, and as trichroic if three colors are obtained. This property can be of value in the differentiation and recognition of certain inorganic and organic crystals.

4. Extinction

When a transparent anisotropic crystal platelet is rotated between crossed nicols, an extinction position will be reached, where no light is transmitted. The angle between some crystallographic feature such as a long edge of the crystal or a trace of a cleavage plane, and the position of extinction can be measured using a cross-hair eyepiece and graduated circular stage. The angle measured in this way depends upon the crystal system, the orientation of the platelet, and the direction of the feature selected relative to the optic axes. The *extinction angle* is measured relative to an optic axis and lies between 0 and 45°. A triclinic crystal would have three different optic axes, that is, three characteristic extinction angles; but a given platelet could contain only two of them.

XIV. REFERENCES

1. L. E. Samuels, *Metallographic Polishing by Mechanical Methods*, Sir Isaac Pitman & Sons, London, 1968.
2. E. L. Long and R. J. Gray, *Metal Prog.*, **74**, 145 (1958).
3. "Symposium on Methods of Metallographic Specimen Preparation," ASTM Special Technical Publication, No. 285, 1960.
4. J. L. Whitton, *J. Appl. Phys.*, **36**, 3917 (1965).
5. P. Lacombe, "Polishing and Etching Methods and Their Applications to Optical Metallography," in *Metallography, 1963*, Proc. Sorby Centenary Meeting, Sheffield, Special Report **80**, Iron and Steel Institute, London, 1966, pp. 50–120.
6. W. J. McG. Tegart, *Electrolytic and Chemical Polishing of Metals*, 2nd ed., Pergamon Press, London, 1959.
7. J. Mercadié, *Compt. Rend.*, **226**, 1450, 1519 (1948).
8. T. P. Hoar and J. H. S. Mowat, *J. Electrodep. Tech. Soc.*, **26**, 7 (1950).
9. J. Edwards, *J. Electrochem. Soc.*, **100**, 223 (1953).
10. P. Brouillet, *Metaux Corr. Ind.*, **30**, 141, 192, 243 (1955).
11. *Metals Handbook*, 1948 ed., Taylor Lyman, Ed., American Society for Metals, Cleveland, Ohio, p. 396.
12. R. H. Greaves and H. Wrighton, *Practical Microscopical Metallography*, 3rd ed., Chapman and Hall, London, 1939.
13. H. Grubitsch and P. Warbichler, *Arch Eisenhüttenw.*, **16**, 77 (1942).
14. Gallard-Schlesinger Chemical Manufacturing Corp., Carle Place, Long Island, N.Y.
15. E. N. Cameron, "Symposium on Microscopy 1959," ASTM Special Technical Publication, No. 257, p. 39.
16. H. W. Hermance and H. V. Wadlow, in *Physical Methods in Chemical Analysis*, Vol. 2, W. G. Berl, Ed., Academic Press, New York, 1951, pp. 155–228.
17. J. M. Dickinson, *Metal Progr.*, **74** (4), 142 (1958).
18. P. Lacombe and M. Mouflard, *Metaux Corr. Ind.*, **28**, 471 (1953).
19. P. Jacquet, *Rev. Met.*, **42**, 133 (1945).
20. K.-H. Dottinger, *Prakt. Metallogr.*, **3**, 115 (1966).
21. C. Bückle, C. Changarnier, and J. Calvet, *Compt. Rend.*, **235**, 1040 (1952).

22. A. Schrader, *Ätzheft; Verfahren Zur Gefügeentwicklung für die Metallographie*, Gebrüder Borntraeger, Berlin, 1957.
23. H. Hanemann and A. Schrader, *Atlas Metallographicus*, Vols. 1–3, Gebrüder Borntraeger, Berlin, 1941.
24. A. Antonioli and A. Ferri, *Rev. Met.*, **46**, 627 (1949).
25. H. J. Engel, *Arch. Eisenhüttenw.*, **29**, 73 (1958).
26. F. C. Frank, in *Growth and Perfection of Crystals*, R. H. Doremus, B. W. Roberts, and D. Turnbull, Eds., John Wiley & Sons, New York, 1958, p. 411.
27. W. D. Biggs, in *Physical Metallurgy*, R. W. Cahn, Ed., John Wiley & Sons, New York, 1965, p. 541.
28. G. C. Williams and G. Rieger, *Metal Industry*, **56**, 461 (1940).
29. P. Jacquet, *Rev. Met.*, **42**, 133 (1945).
30. J. H. Westbrook, private communication.
31. F. Hilbert and H. Lorenz, *Jena Review*, **5**, 218–223 (1963).
32. H. Aubert, *J. Nucl. Mater.*, **5**, 173 (1962).
33. R. J. Murphy and G. J. Ritter, *Nature*, **210**, 191 (1966).
34. R. Shuttleworth, *Metallurgia*, **38**, 125 (1948).
35. M. J. Olney, *Metal Treatment*, **19**, 347 (1942).
36. B. D. Cuming and A. J. W. Moore, *J. Aust. Inst. Met.*, **3**, 124 (1958).
37. W. Pepperhoff, *Naturwissenschaften*, **16**, 375 (1960).
38. W. Pepperhoff, *Arch. Eisenhüttenw.*, **32**, 269 (1961).
39. W. Pepperhoff, *Arch. Eisenhüttenw.*, **33**, 651 (1962).
40. W. Pepperhoff and H.-E. Bühler, *Arch. Eisenhüttenw.*, **34**, 839 (1963).
41. H.-E. Bühler, W. Pepperhoff, and H.-J. Schüller, *Arch. Eisenhüttenw.*, **36**, 457 (1965).
42. W. Pepperhoff, *Arch. Eisenhüttenw.*, **36**, 941 (1965).
43. L. Habraken and J. L. deBrouwer, *De Ferri Metallographia I*, Presses Académiques Européenes S. C., Brussels, Belgium, 1966, pp. 116–118.
44. H.-E. Bühler, *Radex-Rundschau*, 672–678 (1967).
45. H.-E. Bühler and L. Meyer, "Zeiss Information (Bulletin)," No. 66, 118 (1968).
46. J. B. Buhr, T. M. Kegley, Jr., and R. J. Gray, Proc. 21st AEC Metallographic Group Meeting, 1967 (Report CONF-670533 available from Clearing House for Federal Scientific, and Technical Information, Natl. Bur. Standards, U.S. Dept. of Commerce, Springfield, Va.), p. 38.
47. T. R. Allmand and D. H. Houseman, *Microscope*, **18**, 11 (1970).
48. J. W. Mitchell, in *Direct Observation of Imperfections in Crystals*, J. B. Newkirk and J. H. Wernick, Eds., Interscience Publishers, New York, 1962, pp. 3–27.
49. S. Amelinckx, *The Direct Observation of Dislocations*, Academic Press, New York, 1964.
50. J. J. Gilman and W. G. Johnston, *J. Appl. Phys.*, **27**, 1018 (1956).
51. J. J. Gilman and W. G. Johnston, in *Dislocations and Mechanical Properties of Crystals*, J. Fisher, Ed., John Wiley & Sons, New York, 1957, p. 116.
52. J. J. Gilman and W. G. Johnston, *J. Appl. Phys.*, **30**, 129 (1959).
53. J. J. Gilman and W. G. Johnston, *J. Appl. Phys.*, **31**, 632 (1960).
54. G. W. Sears, *J. Chem. Phys.*, **32**, 1317 (1960).
55. V. Hari Babu, D. B. Sirdeshmukh, and K. G. Bansigir, *Phil. Mag.*, **14**, 1067 (1966).
56. N. Cabrera, *J. Chim. Phys.*, **53**, 675 (1956).
57. J. D. Livingston, in *Direct Observations of Imperfections in Crystals*, J. B. Newkirk and J. H. Wernick, Eds., Interscience Publishers, New York, 1962, p. 115.

58. J. B. Newkirk and J. H. Wernick, Eds., *Direct Observation of Imperfections in Crystals*, Interscience Publishers, New York, 1962.
59. J. M. Hedges and J. W. Mitchell, *Phil. Mag.*, **44**, 223, 357 (1953).
60. W. J. Dash, *J. Appl. Phys.*, **27**, 1193 (1956).
61. W. J. Dash, in *Dislocations and Mechanical Properties of Crystals*, J. Fisher, Ed., John Wiley & Sons, New York, 1957, p. 57.
62. W. J. Dash, *J. Appl. Phys.*, **29**, 705 (1958).
63. P. B. Hirsch, A. Howie, R. B. Nicholson, D. W. Pashley, and M. J. Whelan, *Electron Microscopy of Thin Crystals*, Butterworths, Washington, 1965.
64. O. Jones, *Phil. Mag.*, **48**, 207 (1926).
65. A. J. W. Moore, *Metallurgia*, **38**, 71 (1948).
66. E. C. W. Perryman, *Metal Industry*, **79**, 23 (1951).
67. G. K. T. Conn and F. J. Bradshaw, Eds., *Polarized Light in Metallography*, Academic Press, New York, 1952.
68. H. Morrogh, *J. Iron Steel Inst.* (London), **155**, 21 (1947).
69. M. Frocht, *Photoelasticity*, Vol. 1, John Wiley & Sons, New York, 1941; Vol. 2, John Wiley & Sons, 1948.
70. E. G. Coker and L. N. G. Filon, revised by H. T. Jessop, *Photo-Elasticity*, 2nd ed., Cambridge University Press, New York, 1957.
71. J. F. Nye, *Proc. Roy. Soc., Ser. A*, **198**, 190 (1949).
72. H. J. Williams, F. G. Foster, and E. A. Wood, *Phys. Rev.*, **82**, 119 (1951).
73. E. C. W. Perryman, in *Polarized Light in Metallography*, G. K. T. Conn and F. J. Bradshaw, Eds., Academic Press, New York, 1952, p. 70.
74. H. Morrogh, in *Polarized Light in Metallography*, G. K. T. Conn and F. J. Bradshaw, Eds., Academic Press, New York, 1952, p. 88.
75. E. Cohen, in *Polarized Light in Metallography*, G. K. T. Conn and F. J. Bradshaw, Eds., Academic Press, New York, 1952, p. 105.
76. F. D. Bloss, *An Introduction to the Methods of Optical Crystallography*, Holt, Rinehart and Winston, New York, 1961.
77. G. H. Needham, *The Practical Use of the Microscope*, Charles C Thomas, Springfield, Ill., 1958.

2

MICROINDENTATION HARDNESS
TESTING

I. Introduction   28

II. Indenters   29
    A. Diamond Pyramid Indenter   29
    B. Knoop Indenter   30

III. Sources of Error   30
    A. Load Error   31
    B. Rate of Loading   32
    C. Vibration   33
    D. Duration of Load Application   34
    E. Shape of Indenter   35
    F. Elastic Recovery of Impression   36
    G. Measurement of Impression   36
    H. Poor Impression Shape   37
    I. Hardness Scale Conversion   38
    J. Anvil Effect   38
    K. Load Dependence   38
    L. Surface Preparation   39
    M. Grain Size Effects   39

IV. Interpretation of Diamond Pyramid Hardness   39

V. Microindentation Hardness Testers.   40
    A. Zeiss Microhardness Tester ·   41
    B. Kentron Microhardness Tester   42
    C. BNFMRA Microhardness Tester   .   .   .   43
    D. GE High Temperature Microhardness Tester   .   .   .   44

VI. Fields of Application. 46
    A. Bulk Property Determination 46
        1. Inclusions 46
        2. Kinetic Studies 46
    B. Microproperty Determination 47
        1. Hardness Gradients 47
        2. Thickness of Thin Layers 47
        3. Grain Size Dependence of Strength 47
    C. Crystallographic Studies 47
    D. Creep 48
    E. Anomalous Creep and Effects Due to Adsorbed Water 48
    F. Determination of Dislocation Mobilities 49
    G. Photomechanical and Electromechanical Effects 49

VII. References 49

I. INTRODUCTION

Microindentation hardness testing, otherwise known as low load hardness testing, or simply microhardness testing, is a valuable tool for the metallographer, furnishing a rather unique method of making a mechanical property measurement on microareas with many applications. The literature has been reviewed by Mott,[1] Bückle,[2] Brown and Ineson,[3] and Glazov and Vigorovich.[4] The indentation hardness of metals is primarily associated with the resistance to plastic deformation, not with the elastic properties.

Normal or macroindentation hardness is normally carried out in the load range above about 5 kg, where the surface condition of the sample is not too important. This is outside the range of principle interest to the metallographer and will not be specifically discussed, although some of the discussion here is applicable. The reader is referred to Refs. 5 through 8.

The low load or microindentation hardness range will be defined, somewhat arbitrarily because it really depends on the sample material, as extending from loads of approximately 5 kg down to approximately 20 g. In this range the surface condition of the sample is of considerable importance, but, with proper care and attention to detail, fairly good absolute (true) hardness values are obtainable on most materials, with the exception of very soft materials. In the low load range extending from 10 to 20 g down to approximately 1 g, the influence of surface condition, vibration, and other factors is vital and deviations from true hardness values are common, so that the values obtained vary with the experimental conditions. Microloads of one or a few grams, although fairly extensively used for certain specialized research studies, are of little general utility because of the extreme difficulties. It appears impossible with presently available equipment to obtain true hardness values at the 1-g load level. Reproducibility is difficult if not impossible; nevertheless, comparative studies have been made that claim to show significant effects relating, for example, to grain boundary hardening (see Section VI.B.1).

The use of impressions of small size means that studies can be made of point-to-point hardness variations, thin layers, small regions such as second phase particles and inclusions, and on brittle materials such as ceramics and some intermetallic compounds, which would otherwise shatter. Measurements can be made over a wide temperature range. Nevertheless, it should be born in mind that the properties of a finite volume are being averaged which extends well beyond the physical size of the impression. Slip can propagate over large distances in a soft crystal. Most of the sources of error can be reduced by using as large a load as is compatible with the requirements of the test.

The discussion will apply primarily to static indentation testing using the diamond pyramid indenter which is the most generally useful test and has a reasonable theoretical basis for interpretation.[9]

II. INDENTERS

A. Diamond Pyramid Indenter

The diamond pyramid indenter[1,6,8,9] is a standard square-based pyramid with $136°$ angle α between opposite faces and $146°$ angle between opposite edges. In making a hardness test, the indenter is pressed into the sample under a known load L kg, removed, and the mean diagonal of the indentation d measured in millimeters. The indentation hardness is defined as the load per unit area of the indentation, that is, has the dimensions of a pressure (ML^{-2}). Hence the diamond pyramid hardness (DPH) is given in kg/mm^2 by

$$\text{DPH} = \left(2L \sin \frac{\alpha}{2}\right) d^{-2} \tag{1}$$

$$= 1.8544 \frac{L}{d^2} \tag{2}$$

Computation is usually carried out using tables.[1,9] The terms DPH, DPN (diamond pyramid number), HV, VPH (Vickers pyramid hardness) and VPN are synonomous.

It may be noted that, with the diamond pyramid indenter, the geometry is such that the impression diagonal $d = 6.542h$, where h is the depth of the impression. This relation ignores elastic recovery effects.

B. Knoop Indenter

The Knoop indenter[1,6,9,10] employed in measuring the Knoop hardness is a rhombohedron based diamond pyramid. The included conical angles subtended by the shorter and longer edges are $130°$ and $172°30'$, respectively. The impression shape is a parallelogram in which the long diagonal is about seven times as great as the short diagonal. The impression depth is about one-thirtieth of the length. It is usual to measure only the longer diagonal l (mm) of the indentation and to calculate the *projected* area A from l and the indenter geometry. Then, if L is the load in kg,

$$H_{\text{Knoop}} = \frac{L}{A} = \frac{L}{0.07028l^2} \tag{3}$$

Knoop hardness values are rather close to the diamond pyramid hardness.

A number of advantages over the diamond pyramid indenter have been claimed. It is said that the Knoop impression can be measured more accurately using the long diagonal. In practice, at low loads, it is more difficult to locate precisely the ends of the diagonal because of the shallow angle.[3] It is claimed that there is less tendency to crack brittle materials which results in false readings. In practice, cracking can almost always be avoided with the diamond pyramid indenter if a low load is employed and impact is avoided. The Knoop indenter penetrates only about half as far as the diamond pyramid for a given load and thus samples a thinner layer. The surface preparation of the sample is then more critical. The shape permits closer spacing of indentations and tends to concentrate the elastic recovery in the shorter diagonal which is not used in measurement. It also permits hardness effects due to crystal anisotropy to be studied.

The Knoop indenter is relatively little used, compared with the diamond pyramid indenter. In view of the uncertainties of conversion from one hardness scale to another, there would seem to be a considerable advantage in standardizing the use of the diamond pyramid indenter for scientific work, although the Knoop indenter may still be employed for control purposes.

III. SOURCES OF ERROR

It is important in microindentation hardness testing to minimize errors. We now consider various sources of error.

A. Load Error

The indentation hardness H is the load L per unit area of indentation A, that is, $H = L/A$. Differentiating, $dH/dL = 1/A = H/L$. Thus

$$\frac{dH}{H} = \frac{dL}{L} \tag{4}$$

The percentage error in hardness thus increases directly with the decrease in load, so that the error can be minimized by using as high a load as possible.

Errors in true hardness result if the correct load is not employed. If the load is too high, for example, as a result of inertial effects of vibration, the impression size will be increased, resulting in a low apparent hardness. Conversely, if the load is too low because of, for example, frictional effects in the microtester, a high apparent hardness will result. Arcing effects in electrical contacts can also be a source of load errors.

Since the impressions obtained with a diamond pyramid indenter are geometrically similar, the same hardness number should be obtained, regardless of the load. Since DPH $\propto L/d^2$, where L is load and d is the impression diagonal, a straight line should be obtained on plotting L versus d^2 for a particular microtester, whose slope depends on the material tested. The microtester should only be used over the linear portion of the curve. Deviations from linearity can be caused by factors inherent in the design of the microtester. The careful selection and preparation of a specimen is necessary, since departure from linearity can arise if the specimen has a hardness gradient normal to the surface.

If the applied load is the true load and the correct values of d are read, the linear portion of the L versus d^2 curve should pass through zero. A positive load intercept for zero impression size is common and indicates that the true load is greater than the applied load. Vibration is a common source of a positive deviation of a gram or more. In one microtester, a positive 18-g deviation was observed because of the malfunction of a solenoid switch circuit in the mechanism. When the load is applied through a lever arm, wear can result in positive or negative deviations. Frictional effects are another source of load error and can vary, depending on the state of cleanliness and lubrication of the bearing surfaces.

In selecting the load to be used in testing a new material on a particular microtester, log-log plots of true hardness versus ocular reading (impression diagonal) for constant load are another useful guide (Figure 1). These are constructed from measurements made on a set of, say, 10 homogeneous samples of known different DPH tested at various loads. The materials are selected to cover a wide range of hardness, for example, silicon (~ 1000 DPH), hard steel (~ 500 DPH), and copper (~ 40 DPH).

Deviations at the small impression size could be due to measurement errors. Deviations at the large impression size could be attributable to factors inherent in the machine design. The load to be used on a new material should be such that one is operating in the linear portion of the curve.

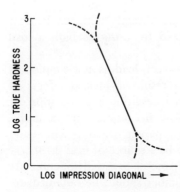

Figure 1   Log-log plot of true DPH hardness versus impression diagonal (ocular reading) on a range of standard materials as a guide to load selection for a new material.

Periodic load calibration is desirable.[7-8] The load on the indenter can be measured directly, using a simple lever balance[11] or a load cell previously calibrated by deadweights. Load cells incorporating resistance strain gauges are commercially available.[12] Such checks should be carried out regularly. Intermediate checks can be carried out more frequently, using standard hardness test blocks available from the manufacturers of hardness testers; or, for small indentations, a well-annealed single crystal of known hardness can be used as a control.

B. Rate of Loading

The rate of loading is normally adjustable and preset, often by means of a dashpot device. Too slow a rate of loading, means that a long time is spent lowering the indenter into contact with the sample and achieving maximum load. The value of the transient load increases directly with the mass of the moving parts and as the square of the loading rate. Inertia effects thus tend to give low DPH values. The DPH value obtained for a given sample may vary with the loading rate even at low rates, particularly if specimen creep occurs, as in soft materials such as lead at room temperature. Standardization of the loading rate is clearly desirable for a given series of tests on a material. Rapid loading may chip the indenter and may tend to cause cracking in hard brittle materials such as ceramics. Cracking leads to low hardness values.

It is clearly desirable to keep the transient load to a very small percentage ($<$ 1%) of the applied load. The mass of moving parts in the microtester should thus be small, but it is also important to use a low loading rate. A rate of 1 mm/min is in general satisfactory for loads of 10 g or more.

C. Vibration

Errors due to vibration are always present, since vibration-free conditions are almost impossible to achieve. Vibration tends to increase the impression size, resulting in low apparent hardness. The smaller the load used for a given material, the bigger are the vibration errors, since it is the amplitude of the vibration relative to the impression dimensions that is important. Similarly, for a given load, the error will be worse for a hard than for a soft material, since the impression size is smaller. On a metal such as copper, vibration effects make it extremely difficult to obtain true hardness values at low loads $< \sim 25$ g.

If an indenter is placed in contact with a testpiece with no applied load for the usual load time, vibration will produce an indentation. From the size of this, one can compute the equivalent load or for a material of known hardness, the equilavent decrease in the hardness number.

The errors are caused by the relative movement of the indenter and the testpiece. This motion is probably substantially less than the combined motion of the two which is much easier to measure. Thus the amplitude and frequency can be measured using a standard vibration tester equipped with a piezoelectric pickup head. Measurements can be made normal to the specimen table, that is, parallel to the indenter motion, and along the other two symmetrical axes lying parallel to the table. Considering the effect of the vibration of a given amplitude on the diagonal length of an indentation made by the diamond pyramid indenter, the vertical motion has about seven times as great an effect as the horizontal motion, due to the geometry of the impression.

There are numerous sources of vibration. Vibration transmitted from the floor or bench may be reduced by the careful choice of location and various types of shock mounting. Vibration transmitted from the operator may be eliminated by automation. Vibration transmitted through the air such as that caused by draughts may be avoided by suitable screening. Vibration generated within the machine is largely a function of design and may be caused by motors, moving parts, or mechanical triggers. Vibration due to a motor can be eliminated by mounting the motor separately from the equipment and using a belt drive. Mechanical triggers can be operated pneumatically instead of by finger control. Resonant frequencies may be present in the machine. For tests < 10 g it is usually advantageous to stand the tester on an inch-thick steel or cast iron plate.

Experience indicates that two types of vibration, namely, continuous and intermittent, are troublesome during testing in the range $< \sim 10$ g. The full effects of continuous, that is, background, vibration are felt in every test, since this occupies several seconds and the frequencies are typically of the order of 60 cps. Thus low apparent hardness values are obtained. Superimposed on this is intermittent (transient) vibration due, for example, to a door's slamming. This is present during one indentation, but not during the rest, and varies in magnitude. Intermittent vibration leads to scatter in a set of data. One would like to be able to discard such impressions, but, unfortunately, experience in a large building indicates that large transients occur without any visible or audible warning to the operator. A possible solution is to attach permanently a vibration meter or recording device to the microhardness tester for use by the operator.

Although these effects can be neglected at loads of a few hundred grams, they are of vital concern in an area of great experimental interest, namely, in the detection and measurement of grain boundary hardening due to solute segregation by making a hardness traverse across the boundary. Here it is necessary on a material such as a copper alloy to use very low loads of 5 g or less, since the width of the hardened region is small. Intermittent vibration could completely mask the effect looked for, whereas continuous vibration would tend to displace the whole curve to anomalously low values (Figure 2).

D. Duration of Load Application

The duration of load application, or load dwell time, should be substantially longer than the time between the first contact of the indenter and the sample and the achievement of maximum load which is typically 1 to 3 sec. With most materials the impression size will not change if the load dwell is further increased. However, with soft materials such as lead, rapid (transient) creep may occur initially, followed by steady state creep at a reduced rate which continues almost indefinitely. The apparent hardness will decrease with increase in the load dwell time, so that the latter should be reported, as well as the hardness and load. A time of 15 sec is commonly used for all materials, regardless of whether or not creep occurs. This is long enough so that the rapid transient creep is completed in a material such as lead.

E. Shape of Indenter

The indenter shape is of vital importance in obtaining true hardness values. The diamond has to be ground to the correct shape with the angles correct within specified tolerances. Reliance is usually placed on

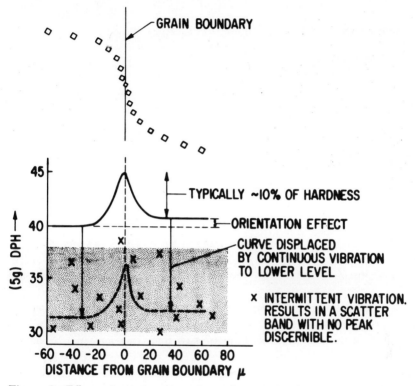

Figure 2 Effect of continuous, and continuous plus intermittent, vibration on microload (5 g) hardness traverse across a segregated grain boundary in an alloy.

the manufacturer; however, these angles can be checked with an optical goniometer. If the apex angle of a diamond pyramid indenter is 137° instead of 136°, and the 136° table of hardness is used, the error is approximately 0.34%. The indenter should have a point, not a chisel tip. This can be checked by microscopic examination of the indenter or by the appearance of indentations made by the indenter. The operator should be alert for signs of wear or damage such as chipping of the indenter. The problem is clearly more acute when small indentations are used.

F. Elastic Recovery of Impression

The sample deformation caused by the indenter is partly elastic. Since the recovered impression is measured, the apparent hardness will be somewhat increased as a result of this effect. The amount of recovery is somewhat controversial, but Mott[1] concludes that the amount increases with increase in load. The effect is small and is usually neglected.

G. Measurement of Impression

The diamond pyramid hardness H for a load L is given by

$$H = kLd^{-2}$$

where k is a constant and d is the mean diagonal length of the impression. Differentiating, we obtain

$$\frac{dH}{dd} = -2kLd^{-3} = -2Hd^{-1}$$

Hence

$$\frac{\Delta H}{H} = \frac{-2\Delta d}{d} \tag{5}$$

Relation 5 shows that the percentage error in DPH is twice the percentage error in reading the mean diagonal length of the impression. The percentage error is also inversely proportional to the impression size.

The resolution of the optical system used to measure the diagonal length is one limiting factor. The resolving power of the objective lens, if defined as the minimum spacing of two lines that can just be distinguished, is about $0.5\lambda/\text{NA}$, where λ is the wavelength ($\sim 0.55\ \mu$ for white light) and NA is the numerical aperture of the objective lens. Thus about $\pm 0.2\ \mu$ resolution is obtainable using a 1.8-mm oil immersion lens of NA 1.30, or slightly better using a monochromatic blue filter. It is more convenient to use an air objective or an objective with the indenter mounted in the front lens, and one may then only get $\pm 0.5\ \mu$ resolution. This is an absolute error, and the same observer may get somewhat better reproducibility on the apparent size of a single impression or the absolute difference in size between two impressions. Even $\pm 0.1\ \mu$ may then be achievable with an oil immersion objective.

If the impression is measured to $\pm 0.5\ \mu$, the approximate error in DPH on annealed copper is $\pm 3/4\%$ for 100-g load, $\pm 2\frac{1}{2}\%$ for a 10-g load, and $\pm 7\frac{1}{2}\%$ for a 1-g load. The measurement error is therefore critical in studying such effects as grain boundary hardening, discussed in Section III.C; it would be highly desirable to use either the scanning electron microscope or replication electron microscopy for such studies.

Impressions are usually measured optically, either employing some kind of micrometer eyepiece or an image projected on a screen. In the latter case it is hard to measure to < 1 mm at $1000\times$ magnification, that is, to $< 1\ \mu$ accuracy. Micrometer backlash can be avoided by making all measurements in one direction. Fringe effects can be avoided by using a blue filter. The accurate calibration of the magnification of the optical system is necessary and can be carried out using a high quality stage micrometer or ruled grating of suitable accuracy. Curvature of the field can be checked during calibration. It is usually preferable to use a flat field objective and to use the center of the field for measurements. Other sources of error are observer bias, poor impression shape (should discard), and poor specimen surface.

It is unfortunate that the optical systems provided on some commercial microhardness testers are, for reasons of economy, inadequate in measurement accuracy at the low load end of the scale, that is, for very small impressions.

H. Poor Impression Shape

The specimen surface should be perpendicular to the direction of travel of the indenter. Tilting results in elongated indentations. This is apparent by comparing the two diagonal lengths of an impression made by a diamond pyramid indenter which should be separately recorded. A tilt of up to $2°$ is tolerable.

Two common effects studied by Tolansky[13] are the piling-up or sinking-in of the material near the indenter, causing departure from a perfect square shape, usually referred to as "barreling" and "pincushioning," respectively (Figure 3). "Barreling" is commonly observed in a work-hardened sample and results because the material, which now has little

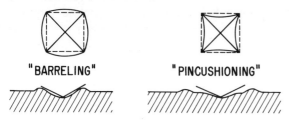

"BARRELING"   "PINCUSHIONING"

Figure 3 "Barreling" and "pincushioning" shape distortions at diamond pyramid hardness impressions.

capacity for further workhardening, rises more at the center of the indenter faces than at the corners. In annealed samples, material tends to workharden near the indenter and push up the softer material further away, resulting in "pincushioning." In practice these effects are usually ignored and the impression diagonals measured as well as possible.

Crystal anisotropy may also lead to an unusual impression shape, the impression being elongated in the direction of the easiest deformation. Hardness anisotropy is most conveniently studied using the Knoop indenter.

I. Hardness Scale Conversion

Hardness conversion from one scale to another is done empirically, using tables or plots of data obtained on the same samples by different hardness measurement techniques.[1,9,14,15] Since there is no sound theoretical basis for conversion, the accuracy is reduced, perhaps seriously.

J. Anvil Effect

The impression size should be much smaller than the thickness of the testpiece, so that the anvil hardness has no effect on the measured hardness. It is usually recommended that the thickness be not less than 10 times the depth of the impression, that is, at least $\frac{3}{2}$ times the mean diagonal length. There should be no visible bulge on the lower surface of the testpiece after making a test. It is good practice to use a glass or hardened steel anvil, in order to give good support when testing thin material. Anvil effect can result in high or low apparent hardness readings, depending on whether the anvil is harder or softer, respectively, than the testpiece.

K. Load Dependence

Since the impressions made with a pyramidal indenter are geometrically similar, unlike those made with a ball indenter, the diamond pyramid hardness should be independent of the load. This has been the subject of much controversy.[1] For a diamond pyramid test the load L is related to the mean impression diagonal d by the relation:

$$L = ad^n \qquad (6)$$

where a and n are constants.

The constants a and n can be determined from a plot of $\log L$ versus $\log d$; $n = 2$ if the diamond pyramid hardness is independent of the load. Deviations from $n = 2$ have often been observed at low loads. If $n < 2$, the apparent hardness is greater; if $n > 2$, the apparent hardness is less than expected. Unfortunately, it is difficult to evaluate and compare previous data, in view of the many sources of error. Brown and Ineson[3] tested the same 7 specimens on 11 different machines. Their data show that the value of the logarithmic index n varied somewhat with the machine. Mott[1] concluded that $n = 2$ if the *unrecovered* impression size was used, so that the hardness was then independent of the load.

L. Surface Preparation

Electrolytic preparation is highly desirable for all microindentation hardness work, although repeated metallographic polishing and etching is usually adequate for loads above about 100 g. Metallographic preparation of an annealed sample, followed by a high temperature anneal, might be thought to provide a suitable surface. However, if low loads (< 10 g) are to be employed, there is a considerable risk that significant differences may arise as a result of changes in the substructure near the surface. Other changes that might occur are carburization, decarburization, internal oxidation, or nitridation. If there is doubt whether or not a worked layer resulting from sawing or grinding has been completely removed by electropolishing, the hardness should be measured, the surface reelectropolished, and the hardness checked to see if it is constant.

It is desirable to use as large a load as possible to minimize the effects of surface preparation.

M. Grain Size Effects

In general a higher hardness will be obtained if the impression size is much larger than the grain size, rather than if it is much smaller. This fact should therefore be reported. The impression should normally be big enough so that the volume tested is typical of the properties of the bulk sample, unless the interest is in the point-to-point variation. In the latter case low load hardness tests have shown that variations in hardness can occur from grain to grain as a function of orientation.[3]

IV. INTERPRETATION OF DIAMOND PYRAMID HARDNESS

We shall briefly consider the interpretation of the diamond pyramid hardness after Tabor[9] in the region where it is independent of load. Although the DPH is simply the load divided by the impression area and has the dimensions of a pressure, it is not a true pressure because the pyramidal area was chosen, not the projected area. The DPH = 0.927P, where P is the vertical pressure.

Because the indenter has a sharp point, fully plastic deformation starts immediately at the smallest load. If we consider an ideally plastic material, which shows no workhardening but deforms plastically at constant yield stress Y, it can be shown[9] that the vertical pressure P is approximately 3.3Y all the time as the load is increased, since the shape of the impression remains constant. That is, the DPH has a value of 2.9 to 3Y.

If workhardening occurs on deformation, as in a typical metal, Y increases as a result of indentation. This increase in yield stress is about that caused by a uniaxial strain of 8 to 10%, independent of the size of the indentation. Thus the hardness number is independent of the load because there is a constant relationship between the volume strained (plastic strain) and the area of the indentation. If the size of the indentation is increased, the material is not strained more, but rather more material is strained.

The ultimate tensile strength (UTS) is related to the diamond pyramid hardness, the relation depending on the Meyer index.[9] The ratio is given by

$$\frac{\text{UTS}}{\text{DPH}} = \frac{1-x}{2.9}\left(\frac{12.5x}{1-x}\right)^x \qquad (7)$$

where $x = n - 2$ and n is the Meyer index derived experimentally from the relation:

$$L = kD^n \qquad (8)$$

where L is the load applied to a ball indenter of fixed diameter, D is the chordal diameter of the indentation, and k and n are constants for the material under examination.[9]

For a variety of materials the ratio UTS/DPH varied from 0.35 to 0.5 (UTS in kg/mm^2).[9]

V. MICROINDENTATION HARDNESS TESTERS

Bückle[2] has distinguished three types of microindentation hardness testers on the basis of the method used to apply the load: (*a*) spring-loaded, (*b*) beam-loaded, and (*c*) direct-loaded instruments. A wide variety of testers has been reviewed by Mott[1] and by Brown and Ineson.[3] Equipment suitable for high temperature testing has been described by Westbrook,[16-17] O'Neill,[5] and Domagala.[18] (See also Refs. 19 through 24.)

Spring loading has been used in many testers, including the Zeiss Haneman,[1,25-26] Safety in Mines Research Establishment (SMRE),[3,27] Girschig,[1,28] Reichert,[1,29-31] Eberbach,[1,32-33] and Miniload (Durimet)[1,34-36] testers. The Eberbach tester was later adapted to the Bausch and Lomb Balphot microscope.[37] Beam-loading instruments include the Kentron,[38] Bergsman,[1,39-40] Cook, Troughton, and Simms,[1,41] Stewart and Lloyds,[3] Guest, Keen and Nettlefold (GKN),[3] and Tukon[42-44] testers. The Perryman tester[45-46] is an example of the direct-loaded kind; Mott[1] gives references to additional instruments.

We shall briefly describe a typical representative of each class.

A. Zeiss Microhardness Tester

The Zeiss MHP microhardness tester[47] is designed as an attachment for use with the Zeiss Universal photomicroscope or Ultraphot II microscope (Figure 4). The diamond pyramid or other indenter is attached to the face of an objective lens mount, forming an integral unit which is supported in special leaf springs so designed that the load is linearly proportional to the displacement. The latter is measured with a differential transformer and galvanometer. After selecting the area to be tested, a push button activates an electrical circuit that moves the normal objective away and brings the indentor into exact position within ± 0.3 μ. The load is applied by manually moving the specimen up against the indenter, using the fine focus control, until the desired load is indicated on the galvanometer scale. After a desired time has elapsed, the load is removed in the same way, and the indentation is measured after moving the objective electrically into position.

Load calibration is carried out by attaching a beam scale and a set of weights; the galvanometer sensitivity can be changed for different load scales, 0 to 50 g, 0 to 100 g, and 0 to 200 g. Good measurement accuracy can be obtained as a result of the high quality optical system. The visibility of the indentation is greatly aided by the use of differential-interference Nomarski illumination which gives color contrast. Bright field or lateral bright field illumination can also be used. A less desirable feature is the fact that the load cycle is manually controlled, so that the loading rate is variable and exact timing is difficult. If creep occurs, the load will tend to vary, unless corrected by the operator.

B. Kentron Microhardness Tester

The Kentron microhardness tester[38] is a beam-type machine provided with a loading pan (inside the housing in Figure 5) and weights. The range covered is from 1 to 10,000 g. Relatively friction-free motion of the beam is provided by a flexure plate suspension which limits indenter motion to a predetermined arc. The indenter is attached to the lever arm. An area is first selected with the stage in position under the measuring microscope. The stage is then manually moved along a slide against a stop. The depression of a button initiates the loading cycle by freeing the beam, causing the indenter to descend on the sample at a rate controlled by a preset hydraulic dashpot. The lever at the side of the machine is used to remove the load after the required dwell time has

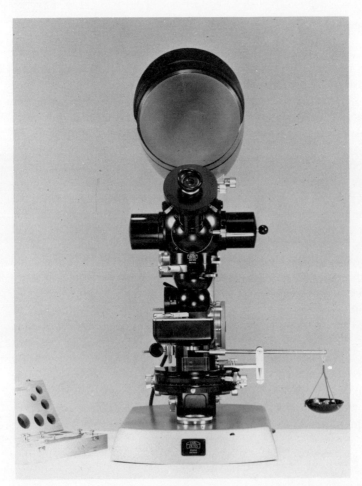

Figure 4 Zeiss microhardness tester MHP attachment mounted on microscope. (Courtesy of Carl Zeiss, New York.)

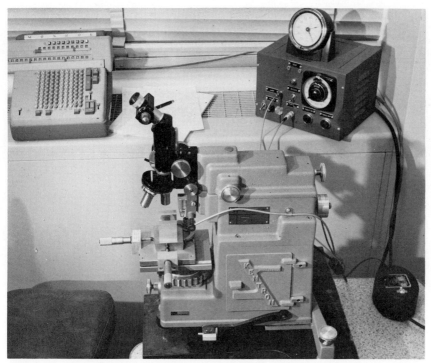

Figure 5 Ametek[48] Riehle Kentron microhardness tester with photoelectrically actuated timing device after Russell and Holik.[49] (Courtesy of General Electric Company.)

elapsed, and the stage is slid back against a stop into the measurement position. Reproducibility is improved by the addition of a timing mechanism,[49] as shown in Figure 5.

A filar eyepiece is provided to facilitate measurements of the indentation. Very small indentations may be advantageously measured by other methods to obtain improved accuracy. The instrument is reliable and simple to use. The design appears to be optimized for the middle of the range covered, rather than for the high or low load ends.

C. BNFMRA Microhardness Tester

The tester developed by Perryman[1,45-46] at the British Non-Ferrous Metals Research Association (BNFMRA) is of the dead-loading type and is used as an attachment on an upright microscope. It consists (Figure 6) of a brass body carrying a ground and lapped hard steel plunger, to which is attached a diamond pyramid indenter head. When the sample is brought into contact with the indenter, lifting its weight (about 17 g), an electrical circuit is broken which extinguishes a signal light, thus providing a means of timing. Somewhat higher loads can be obtained

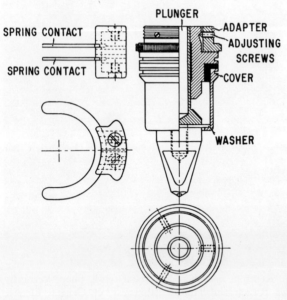

Figure 6   BNFMRA microhardness tester. After Perryman[1,45-46].

by placing weights inside the plunger. The adjusting screws permit the indenter to be centered roughly. A rotating lens turret is used, in order to select an area for indentation and then to move the indenter into position. A trial indentation is made, and the x-y stage translation necessary to move this into alignment with the eyepiece cross hair is noted. The necessary translation is factored into subsequent impressions. The impressions can be measured on the same microscope or a metallograph.

The design is based on that of Lloyd and Geoffrey,[27] who employed spring loading which has the advantage of simplicity and cheapness with little to go wrong. However, it is very limited in load range and requires considerable patience to use. This tester is not commercially available.

D. GE High Temperature Microhardness Tester

There are at least two commercially available high temperature micro-hardness testers.[50-51] Two testers developed at the General Electric Research Laboratory were described by Westbrook.[16,17,52] In the improved version,[17,52] as shown in Figure 7, the specimen was raised by a motor drive against a deadweight-loaded indenter mounted on a counterbalanced beam. Diamond, sapphire, and titanium diboride were used as indenter materials. Loads of 5 g to 1000 g could be applied. The specimen was inserted into, and removed from, the vacuum chamber through vacuum locks without cooling down the equipment. A temperature range from −190°C to 1500°C was covered. The rate of loading was adjustable from 0.2 to 1.2 mm per minute and the dwell time from 2 to 60 sec.

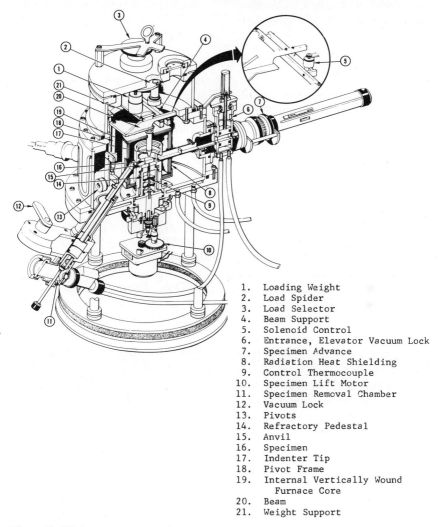

1. Loading Weight
2. Load Spider
3. Load Selector
4. Beam Support
5. Solenoid Control
6. Entrance, Elevator Vacuum Lock
7. Specimen Advance
8. Radiation Heat Shielding
9. Control Thermocouple
10. Specimen Lift Motor
11. Specimen Removal Chamber
12. Vacuum Lock
13. Pivots
14. Refractory Pedestal
15. Anvil
16. Specimen
17. Indenter Tip
18. Pivot Frame
19. Internal Vertically Wound
    Furnace Core
20. Beam
21. Weight Support

Figure 7 High temperature microhardness tester. After Westbrook.[52]

In this sophisticated design the entire load cycle was automated with electronic control. The sample was advanced automatically a small distance between indentations. The indentations were measured at room temperature after removal of the sample.

VI. FIELDS OF APPLICATION

We shall not attempt to review the literature concerned with applications of hardness and microindentation hardness measurements which is covered in standard textbooks.[1,4,5,53] Our purpose is simply to indicate the principal kinds of application.

A. Bulk Property Determination

Average (bulk) properties can be studied if the indentation is large enough to provide a true sample. Thus microindentation hardness may be used as a nondestructive control test in the processing and manufacture of (small) objects. The nondestructive and localized nature of the test means that, for example, a complete aging curve can be obtained by repeated heat treatment of a single sample if so desired. Similarly, the state of deformation can be repeatedly checked, as deformation proceeds in stages.

1. Inclusions

Microindentation hardness provides a unique, cheap, and efficient technique for studying the mechanical properties of small particles such as oxides and intermetallics in metallographic samples. The mechanical properties might not otherwise be accessible. Quite brittle materials such as glasses[54-55] can be studied. Hardness can be determined as a function of temperature (for the relationship between hardness and test temperature, see Refs. 56 and 57). Hardness has proved particularly valuable as a guide to the high temperature strength of oxides[52] and intermetallic phases.[16,17,58,59]

Hardness is a useful guide to the identification of inclusions and other second phase particles and in the identification of microstructural features such as martensite, and can be used in conjunction with other information such as the etching behavior. Hardness is normally listed along with other properties in handbooks and standard data collections.

2. Kinetic Studies

The kinetics of recovery, recrystallization, order hardening, phase transformations, precipitation hardening, and other phenomena involving hardening or softening as a function of time and temperature lend themselves to study using hardness testing.[1,5,60] Depending on the shape and form of the samples used, microindentation hardness may be appropriate.

B. Microproperty Determination

By the suitable choice of load, the indentation size can be made as small as a few microns in most materials and yet is still large enough to measure with reasonably accuracy. Microindentation hardness, therefore, is a very powerful tool for looking at variations in mechanical properties on a microscale. Some of the sources of these variations will now be considered.

1. Hardness Gradients

Hardness gradients whether produced by nonuniform deformation, surface hardening processes such as carburizing, nitriding or shot peening, diffusion gradients, partial phase transformations such as the tempering of steels, segregation or other causes may be determined by microindentation hardness testing, using microsections when appropriate.[1,5] A particularly interesting research application is the study of grain boundary segregation (see, e.g. Refs. 61 through 63) and segregation near surfaces.[64]

2. Thickness of Thin Layers

The approximate thickness of a thin, hard or soft surface layer on a substrate can be readily determined from a hardness traverse at constant load on a section made normal to the surface. If the layer is fairly thin, a simpler approach is simply to prepare a plot of hardness versus load from a series of indentations made on the surface. A break in the curve indicates when penetration into the substrate occurs. The corresponding depth is calculated from the indenter geometry.

3. Grain Size Dependence of Strength

Information relating to the grain size dependence of hardness, and hence strength, can be obtained by studying the effect of indentation size on the hardness, varying the indentation size from much smaller than the grain size to much larger.

C. Crystallographic Studies

Slip markings are often observed around an indentation. Mechanical twinning may be observed in certain materials. These markings lend themselves to formal crystallographic analysis for the identification of the slip and twinning systems, using an appropriate technique such as microbeam x-ray back reflection for orientation determination. Cracking effects related to the orientation may likewise be studied.

Anisotropy of deformation in, for example, hexagonal crystals may be studied from shape measurements on microindentations. The Knoop indenter offers some advantage, since the long axis may be set parallel to selected directions.[65]

D. Creep

Lead and low melting point alloys typically show an indentation hardness dependent on the time of load application, due to creep that occurs under the indenter.[66-68] Attempts have been made[69-74] to correlate this (accelerated) indentation creep with normal (slow) creep in metals; in some cases good correlations have been obtained.

E. Anomalous Creep and Effects Due to Adsorbed Water

Anomalous indentation creep has been observed[75-77] in certain minerals and in materials such as lithium fluoride, potassium bromide, alumina, magnesium oxide, titanium carbide, and germanium, which would not be expected to creep at room temperature. This was shown by Westbrook and Jorgensen[77] to be a Rebinder-type effect, associated with water adsorbed on the surface, and disappeared when the water was absent.

Adsorbed water also tended[77] to reduce the microindentation hardness of materials such as those above, having covalent, partially TiC type, or completely ionic bonding. The same effect was found in a variety of minerals including oxides, silicates, sulfides, fluorides, carbides, and carbonates, and was shown to vary with the crystal face exposed.[78]

The difference in microindentation hardness between the polar surfaces of noncentrosymmetric crystals such as the AB compounds of the zinc blende or wurtzite structure has been shown[79] to be caused by the difference in the absorption of water between the two faces which have either all A atoms or all B exposed, respectively. The hardness difference disappeared if absorbed water was absent.[79]

Anomalous creep due to absorbed water could be removed by heating to temperatures of as little as 60°C in dried argon to desorb water; however, the anomalously low hardness did not disappear until higher desorption temperatures of 200 to 300°C were used.[79] After desorption the samples were quenched into, and tested under, fresh anhydrous toluene, although several other techniques were also used.[79] Thus the anomalous creep was attributed to thick layers of loosely bound absorbate, and the anomalous hardness to thin layers of tightly bound absorbate.[79] There is evidence[79] that the water-softening effects are confined to a layer at the surface about 3 μ thick and so are only observable with small indentations.

F. Determination of Dislocation Mobilities

Vaughan and Davisson[80] developed a method for determining dislocation mobilities in a dislocation-free region of a crystal using a microhardness indentation. After indenting, the dislocation rosettes running out from the indentation were revealed by means of a dislocation etch-pitting reagent. The propagation distance is then a measure of the material strength, since the leading dislocation runs out until the stress exerted on it by the succeeding dislocations drops to the stress required to move it through the crystal.

Westbrook[79] demonstrated that the mobility in a lithium fluoride crystal was enhanced by the presence of adsorbed water. He attributed this to a force field effect; that is, the absorption of water relaxes the local change in elastic modulus, lattice parameter, or surface double layer, at the surface facilitating dislocation motion near the surface. The effects are accentuated in covalently and ionically bonded solids as a result of the presence of unsatisfied bonds.

G. Photomechanical and Electromechanical Effects

The microindentation hardness of a semiconductor has been shown (See Refs 81–100) to be reduced 20 to 30% or more in a surface layer of 1 to 3 μ when the test surface is illuminated, a photomechanical effect. A similar but electromechanical effect is found[87,97,99,101–109] when a very small potential of the order of 0.1 V is impressed between the indenter and the sample surface. The two effects are reversible. Westbrook confirmed these effects (see Ref. 79) and showed that both were again caused by adsorbed water. In this case the amount of water adsorbed is enhanced by the action of the light or electric field. The effects have been observed in semiconductors such as Ge, Si, Bi, Sb, InSb, CdS, and SiC, but not in ionic crystals.

VII. REFERENCES

1. B. W. Mott, "Micro-Indentation Hardness Testing," Butterworth & Co., London, 1956.

2. H. Bückle (a) "Progress in Micro-Indentation Hardness Testing," *Met. Rev.* 4, 49–100 (1959); (b) "*L'Essai DeMicrodureté et Ses Applications*," Publications Scientifiques et Techniques de l'Air, Service de Documentation et D'Information Technique de L'Aéronautique, Paris, March 1960.

3. A. R. G. Brown and E. Ineson, "Experimental Survey of Low-Load Hardness Testing Instruments," *J. Iron Steel Inst.* (*London*), 169, 376–388 (1951).

4. V. M. Glazov and V. N. Vigdorovich, *Microhardness of Metals and Semiconductors*, Metallurgiya Press, Moscow, 1969 G. D. Archard translator, Consultants Bureau, New York, 1971.

5. H. O'Neill, *Hardness Measurement of Metals and Alloys*, 2nd ed., Chapman and Hall, London, 1967.

6. *Metals Handbook*, American Society for Metals, Philadelphia, 1948, 93–104.

7. "ASTM Standard Method of Test for Vickers Hardness of Metallic Materials," E92-67, ASTM Standards, Part 31, 382 (1967).

8. "Method for Vickers Hardness Test," B.S. 427, Part 1: 1961, Testing of Metals; Part 2: 1962, Verification of Vickers Hardness Testing Machines, British Standards Institution, London.

9. D. Tabor, *The Hardness of Metals*, Clarendon Press, Oxford, 1951.

10. "Specification for Knoop Indenters," Natl. Bur. Standards Circular Letter LC819, April 1946.

11. R. J. Ellis, *J. Sci. Instrum*, 38, 105 (1961).

12. Instron Engineering Corp., Canton, Mass.

13. S. Tolansky, *Surface Microtopography*, Interscience Publishers, New York, 1960, p. 195.

14. *Metals Handbook*, 8th ed., Vol. 1, Taylor Lyman, Ed., American Society for Metals, Philadelphia, 1961, p. 1234.

15. "Hardness Conversion Tables for Metals (Relationship Between Brinell Hardness, Vickers Hardness, Rockwell Hardness, Rockwell Superficial Hardness and Knoop Hardness)," ASTM Standards, Part 31, 564 (1967).

16. J. H. Westbrook, *Proc. ASTM*, 57, 873 (1957).

17. J. H. Westbrook, *ASTM Bull.*, No. 246, 53 (1960).

18. R. F. Domagala, in *High-Temperature Technology*, I. E. Campbell, Ed., John Wiley & Sons, New York, 1956, p. 402.

19. F. P. Bens, *Trans. ASM*, 38, 505–545 (1947).

20. A. Brenner, *Plating*, 38, 363 (1951).

21. M. Semchyshen and C. S. Torgerson, *Trans. ASM*, 50, 830 (1958).

22. K. L. Maurer, *Berg- und Hüttenmännische Monatshefte*, 109, (3), 99 (1963).

23. T. N. Loladze, G. V. Bokuchava, and G. E. Davidova, *Zarodski Lab.*, 33, 1005 (1967).

24. J. L. Kamphouse, J. C. Blake, and J. Moteff, *Rev. Sci. Instrum.* 40, 321 (1969).

25. H. Haneman and E. O. Barnhardt, *Z. Metallk.*, 32, 35 (1940).

26. H. Haneman, *Metallurgia*, 32, 62 (1945).

27. H. Lloyd and R. Geoffrey, *J. Sci. Instrum*, 24, 186 (1947).

28. R. Girschig, *Rev. Met. Mémoires*, 43, 95 (1946).

29. E. M. Onitsch, *Mikroscopie*, 2, 131 (1947).

30. P. Ramsthaler, *Mikroscopie*, 2, 345 (1947).

31. "Microhardness, Its Theory and Practice with the Reichert Microhardness Tester," Wien, 1950.

32. D. P. Holbrook and C. O. Snudberg, *Metal Progr.*, 64 (5) 108 (1953).

33. M. Hill and J. C. McGee, *Trans. ASM*, 33, 140 (1944).

34. H. Benninghoff, *Ind. Diamond Rev.*, **10**, 303 (1950).
35. H. Broschke, *Microtecnic*, **6**, 15 (1952).
36. K. Walz, *Draht*, **4**, 176 (1953).
37. R. Steinitz, *Metals and Alloys*, **17**, 1183 (1943).
38. F. Robertson and W. J. VanMeter, *Econ. Geol.*, **46**, 541 (1951).
39. E. B. Bergsman, *Metal Industry*, **69**, 109 (1946).
40. E. B. Bergsman, *Metal Progr.*, **54**, 183 (1948).
41. E. W. Taylor, *J. Inst. Metals*, **74**, 493 (1948).
42. D. R. Tate, *Trans. ASM*, **35**, 374 (1945.)
43. V. E. Lysaght, *Materials and Methods*, **22**, 1079 (1945).
44. V. E. Lysaght, *ASTM Bull.*, No. 138, 39 (1946).
45. E. C. W. Perryman, *Metal Industry*, **76**, 23 (1950).
46. E. C. W. Perryman, *Metal Industry*, **79**, 131 (1951).
47. J. Gahm, Zeiss Information, No. 62, 121 (1966).
48. Ametek/Testing Equipment Systems, Lansdale, Pa.
49. R. R. Russell and A. S. Holik, *Rev. Sci. Instrum.*, **37**, 330 (1966).
50. Akashi Products, Tokyo, Japan.
51. Marshall Products Co., Columbus, Ohio.
52. J. H. Westbrook, *Rev. Hautes Temp. Réfract.*, **3**, 47 (1966).
53. V. E. Lysaght, *Indentation Hardness Testing*, Van Nostrand Reinhold Co., New York, 1949.
54. D. M. Marsh, *Proc. Roy. Soc., Ser. A.*, **279**, 420 (1964).
55. J. H. Westbrook, *Phys. Chem. Glasses*, **1** (1), 32 (1960).
56. J. H. Westbrook, *Trans. ASM*, **45**, 221 (1953).
57. E. R. Petty, *Metallurgia*, **65**, (387), 25 (1962); **66**, (398), 267 (1962).
58. J. H. Westbrook and E. R. Stover, in *High-Temperature Materials and Technology*, I. E. Campbell and E. M. Sherwood, Eds., John Wiley & Sons, New York, 1967.
59. E. K. Storms, *The Refractory Carbides*, Academic Press, New York, 1967.
60. W. Chubb, *J. Metals*, **7**, 189 (1955).
61. J. H. Westbrook and D. L. Wood, *J. Inst. Metals*, **91**, 174 (1962–1963).
62. A. U. Seybolt and J. H. Westbrook, *Acta Met.*, **12**, 449 (1964).
63. S. Floreen and J. H. Westbrook, *Acta Met.*, **17**, 1175 (1969).
64. K. T. Aust, A. J. Peat, and J. H. Westbrook, *Acta Met.*, **14**, 1469 (1966).
65. H. Winchell, *Amer. Mineral.*, **30**, 583 (1945).
66. F. Hargreaves, *J. Inst. Metals.*, **39**, 301 (1928).
67. V. P. Shishokin, *Zh. Tekh. Fiz.*, **8**, 1613 (1938).
68. J. Pomez, A. Royez, and J. P. Georges, *Rev. Met.*, **56**, 215 (1959).
69. V. P. Shishokin and Y. V. Shestopalova, *Zh. Tekh. Fiz.*, **8**, 1613 (1938).
70. C. Rubenstein, *Proc. Phys. Soc. Ser. B*, **67**, 563 (1954).
71. J. W. Goffard and R. G. Wheeler, *Trans. Met. Soc. AIME*, **215**, 902 (1959).
72. J. Pomerz, A. Royez, and J. P. Georges, *Rev. Met.*, **56**, 215 (1959).
73. E. E. Underwood, *J. Inst. Metals.*, **88**, 266 (1960).
74. T. O. Mulhearn and D. Tabor, *J. Inst. Metals*, **89**, 7 (1960-1).
75. W. W. Walker and L. J. Demer, *Trans. Met. Soc. AIME*, **230**, 613 (1964).
76. R. Mitsche and E. M. Onitsch, *Mikroskopie*, **3**, 257 (1948).
77. J. H. Westbrook and P. J. Jorgensen, *Trans. Met. Soc. AIME*, **233**, 425 (1965).
78. J. H. Westbrook and P. J. Jorgensen, *Amer. Mineral.*, **53**, 1899 (1968).
79. J. H. Westbrook, in *Environment Sensitive Mechanical Behavior*, Gordon and Breach, New York, 1967, pp. 247–268.

80. W. H. Vaughan and J. W. Davisson, *Acta Met.*, **6**, 554 (1958).
81. G. C. Kuczynski and R. F. Hochman, *Phys. Rev.*, **108**, 946 (1957).
82. G. C. Kuczynski and R. F. Hochman, *J. Appl. Phys.*, **30**, 267 (1959).
83. M. Kikuchi and M. Saito, *J. Phys. Soc. Japan*, **14**, 1642 (1959).
84. T. Figielski, *Phys. Status Solidi*, **1**, 306 (1961).
85. G. C. Kuczynski, R. F. Hochman, and C. Allen, in *Mechanical Properties of Engineering Ceramics*, W. W. Kriegel and H. Palmour, III, Eds., Interscience Publishers, New York, 1961, p. 495.
86. N. Y. Garid'ko, P. P. Kuz'menko and N. N. Novikov, *Sov. Phys.—Solid State*, **3**, 2652 (1962).
87. J. H. Westbrook and J. J. Gilman, *J. Appl. Phys.*, **33**, 2360 (1962).
88. P. P. Kuz'menko, N. N. Novikov, and N. Y. Gorid'ko, *Sov. Phys.—Solid State*, **4**, 1950 (1963).
89. P. P. Kuz'menko, N: N. Novikov, and N. Y. Gorid'ko, *Ukr. Fiz. Zh.*, **8**, 116 (1963).
90. V. M. Beilin and U. K. Vekilov, *Sov. Phys.—Solid State*, **5**, 1727 (1964).
91. V. M. Beilin, U. K. Vekilov, A. Kadyshevich, N. V. Pigusov, and R. Rattke, *Sov. Phys—Solid State*, **5**, 1726 (1964).
92. N. N. Novikov, N. Y. Gorid'ko, and A. G. Rudenko, *Izv. Vyssh. Ucheb. Zaved., Fiz.*, **4**, 22 (1964).
93. V. N. Lange and I. I. Lange, *Sov. Phys.—Solid State*, **6**, 942 (1965).
94. P. P. Kuz'menko, N. N. Novikov, and N. Y. Gorid'ko, *Sov. Phys.—Solid State*, **6**, 2056 (1965).
95. P. P. Kuz'menko, N. N. Novikov, N. Y. Forid'ko, and L. J. Fedorenko, *Sov. Phys.—Solid State*, **8**, 1381 (1966).
96. N. Ya Fonid'ko, P. P. Kuz'menko, and N. N. Novikov, *Ukr. Fiz. Zh.*, **11**, 1016 (1966).
97. D. B. Holt, in *Environment Sensitive Mechanical Behavior*, A. R. C. Westwood and N. S. Stoloff, Eds., Gordon and Breach, New York, 1966, p. 269.
98. V. A. Drozdov and L. A. Mozgovaya, *Phys. Status Solidi*, **22**, K109 (1967).
99. R. E. Hanneman and P. J. Jorgensen, *J. Appl. Phys.*, **38**, 4099 (1967).
100. N. Ya. Fonid'ko, P. P. Kus'menko, and N. N. Novikov, *Ukr. Fiz. Zh.*, **12**, 484 (1967).
101. Y. V. Mil'man, V. I. Trefilov, and G. E. Khanenko, *Sb. Nauch. Tr. Inst. Metallovfiz. Akad. Nauk Ukr. SSR*, No. 19, 51–53 (1964).
102. N. N. Novikov, N. Y. Gorid'ko, and A. G. Rudenko, *Izv. Vyessh. Ucheb. Zaved. Fiz.*, **4**, 22 (1964).
103. T. A. Kontorova, *Sov. Phys.—Solid State*, **6**, 1760 (1965).
104. M. S. Ablova, *Sov. Phys.—Solid State*, **6**, 2520 (1965).
105. M. S. Ablova, *Fiz. Tverd. Tela*, **7** (9), 2740 (1965).
106. T. A. Kontorova, *Sov. Phys.—Solid State*, **6**, 1760 (1965).
107. G. P. Upit, S. A. Varchenya, and J. P. Szalvin, *Phys. Status Solidi*, **15**, 617 (1966).
108. G. P. Upit, S. A. Varchenya, and I. P. Spalvin, *Phys. Status Solidi*, **15**, 617 (1966).
109. T. A. Kontorova, *Fiz. Tverd. Tela*, **9**, 1235 (1967).

3

OPTICAL MICROSCOPY*

I. General Background 54
  A. Introduction 54
  B. Light Sources 55
  C. Angular Aperture 57
  D. Laws of Refraction and Reflection 57
  E. Numerical Aperture 58
  F. Resolution 58
  G. Depth of Focus 59
  H. Aberrations of Lenses 59
    1. Chromatic Aberration 59
    2. Spherical Aberration 60
    3. Coma 61
    4. Astigmatism and Curvature of the Field 61
    5. Distortion 62

II. Microscope Components 63
  A. Objectives 63
  B. Eyepieces 65
  C. Vertical Illuminators 66
  D. Practical Light Sources 69

III. Optical Methods of Enhancing Contrast 70
  A. Dark Field Illumination 71
  B. Polarized Light 72
    1. Sources of Polarized Light 72
      a. Reflection 72
      b. Double Refraction 72
    2. The Sensitive Tint Plate 76
    3. Identification of State of Polarization 76
    4. Applications of the Polarizing Reflection Microscope 77

* Adapted from V. A. Phillips.[1]

C. Phase Contrast 78
   1. Phase Detail 78
   2. Nature of Normal (Absorption) Contrast 79
   3. Nature of Phase Contrast 80
   4. Sensitivity 83
   5. Quantitative Phase Contrast Measurements 83
D. Interference 83
   1. Criteria for Interference 83
   2. Interference at a Wedge 85
   3. Measurements Possible with Interferometers 85
   4. Two-Beam Interferometry 86
     a. Reflection Microscopes Using Unpolarized Beams 86
       (1) Zeiss Reflection Interference Microscope 86
     b. Reflection Microscopes Using Polarized Beams 88
       (1) Nomarski Interference Microscope 88
     c. Transmission Microscopes Using Unpolarized Beams 90
       (1) Leitz Interference Microscope 92
     d. Transmission Microscopes Using Polarized Beams 92
       (1) Baker Transmission Interference Microscope 93
       (2) Zeiss Transmission Interference Microscope 94
   5. Multiple Beam Interferometry 96
     a. Optimizing Multiple Beam Fringes 100
     b. Thickness Measurement Using Multiple Beam Fringes 100
   6. Light-cut and Light-Profile Methods. 101
E. Filters 102
F. Examples of Results 104

IV. References 109

I. GENERAL BACKGROUND

A. Introduction

Light can be considered as having certain properties of a stream of particles and others of a wave motion. The former fact is the basis of geometrical optics theory which will be the approach used in discussing lens aberrations, while the latter leads to physical optics which will be used in discussing polarizing, phase contrast, and interference microscopy. Geometrical optics is based on the following fundamental laws: (*a*) light travels in a straight line; (*b*) parts of light beams can be treated as separate entities or rays; (*c*) the law of reflection; and (*d*) the law of refraction. These laws emerge as approximations from the electromagnetic wave theory which is the basis of physical optics. The reader who is interested in the details of the theory should consult the standard textbooks such as Refs. 2 through 6.

B. Light Sources

For the present purposes, light may be considered as a wave motion having amplitude, wavelength, and frequency. A *monochromatic light source* will emit waves of a single wavelength (ideal source) or narrow band of wavelengths (actual source) in all directions with a certain amplitude or intensity. Examples of such a source are a sodium flame, a mercury or thallium light source with a suitable filter, or a laser. A *white light source,* for example, a tungsten filament lamp or carbon arc, will emit light of many wavelengths (continuous emission). With the use of a suitable prism, or diffraction grating, white light can be split into its component wavelengths. In light microscopy, we shall be concerned with the spectrum visible to the eye; however, a photographic plate is sensitive to a much wider range of wavelength λ than just the visible spectrum (Table 1).

When two or more light waves are superimposed, no simple description of the observed phenomena is in general possible, as pointed out by Françon.[7] When two waves are emitted from a single monochromatic point source, the intensity of the light varies in the region of superposition and *interference* is observed. In the general case the light vibration is elliptical and can be considered to be the resultant of two vibrations directed along mutually perpendicular axes x and y, with the third perpendicular axis z representing the distance of propagation. Then we find that

$$x = x_1 \cos (\omega t + \theta) \tag{1}$$
$$y = y_1 \cos (\omega t + \psi) \tag{2}$$

where x_1 and y_1 are the amplitudes referred to the x and y axes, θ and ψ the corresponding phases, ω the angular frequency, and t the time.

Table 1 Spectral Range Used in Microscopy

Type of Radiation	Wavelength λ (angstrom = 10^{-8} cm)
Electrons (electron microscope)	$\sim$0.05
X-rays	0.10 to 150
Ultraviolet	150 to 4000
Shortest ultraviolet used in photomicrography	$\sim$2000
Visible spectrum (blue-green-yellow-red)	$\sim$4000 to 7000
Infrared used in photomicrography	7000 to 8600

From Phillips.[1]

The vibration from the atom is emitted during approximately 10^{-9} sec, corresponding to a wave packet of 10 to 100 waves, and then disappears. Other atoms of the source emit vibrations but these have no relation to previous vibrations that have stopped. Thus x_1, y_1, θ, and ψ alter in a completely irregular manner as a function of time. After a certain time the relation between x and y given by Equations 1 and 2 no longer exists, so that for interference experiments in which the times are much greater than 10^{-9} sec, there are now two incoherent vibrations in the x and y directions. Thus, in order to characterize the observed effects, it is necessary to consider the averages in time. The two vibration intensities will be proportional to the averages of x_1^2 and y_1^2 and these averages are equal for natural light.

Generally speaking, interference and diffraction phenomena can be considered without regard to the vectorial character of light. Instead, the light waves can be represented by a scalar function of the form:

$$x = b \cos (\omega t + \phi) \tag{3}$$

where the phase ϕ is related to the optical path Δ by the relation

$$\phi = \frac{2\pi\Delta}{\lambda} \tag{4}$$

The distinction between *coherent* and *incoherent* light sources is a difficult one to grasp, and involves quantum concepts and the uncertainty principle, so that it goes beyond the boundaries of either geometrical or physical optics. A coherent source of white light is one in which there is a constant phase difference maintained between the different waves emitted. A practical point source of white light may be regarded as a coherent source, although it is probably more correct to regard it as emitting both coherent and incoherent light, that is, as a semicoherent source. Light from such a source, if divided into two beams, can be made to interfere. Light from two separate sources (point or extended) is always incoherent and cannot be made to interfere. Under certain circumstances—which will be discussed with interference—light from an extended source can be regarded as coherent. The most important practical coherent source is the *laser*, which has so far been little employed in metallography. The basic laser types are solid state, semiconductor and gas. Examples of the solid state type are ruby crystals and yttrium aluminum garnet (YAG) crystals, which operate at 6943 Å and 10,600 Å, respectively. Gallium arsenide, which generates 8400 Å radiation in the lasing mode, is an example of the semiconductor type. Helium-neon is an example of the gas type and emits laser light at 6328 Å.

C. Angular Aperture

The angular aperture α of a lens or lens system is the angle of the bundle of rays from a point in the object that can enter the front lens when the lens is focused on the specimen. The apertures of objectives for optical microscopes vary from about 10 to 140°. The resolution of detail is directly proportional to the angular aperture; as the latter increases, it is clear that the lens takes in more light rays from a given object point, but more intense illumination is nevertheless needed because of the smaller field of view. The focal length and working distance, although not directly related, both tend to decrease with an increase in angular aperture.

D. Laws of Refraction and Reflection

Snell's *law of refraction* states that the refracted ray lies in the same plane as the incident ray and the normal to the surface (the plane of

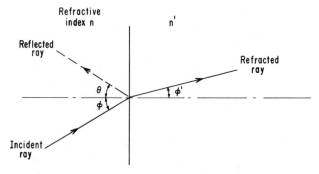

Figure 1 Illustration of the laws of refraction and reflection. From Phillips.[1]

incidence), as in Figure 1, and that the following relation holds:

$$n \sin \phi = n' \sin \phi' \tag{5}$$

where ϕ is the angle of incidence in the medium of the refractive index n and ϕ' is the angle of refraction in the medium of the index n'; $n = 1$ for vacuum, $n \approx 1$ for air.

There will also be a reflected ray which lies in the plane of incidence, the angle of reflection θ (Figure 1) being equal to the angle of incidence ϕ. These two facts constitute the *law of reflection*.

E. Numerical Aperture

Since the angular aperture of a lens can be increased by increasing the refractive index n of the medium between the lens and object, it is more convenient to define a numerical aperture that includes this factor:

$$NA = n \sin\left(\frac{\alpha}{2}\right) \tag{6}$$

where α is the angular aperture and n the refractive index of the medium between the lens and the object (1 for vacuum, ~ 1 for air, 1.33 for water, 1.52 for oil used with oil immersion objectives, 1.66 for naphthalene monobromide, and 1.74 for methylene iodide).

The practical limit for the numerical aperture for objectives is approximately 0.95 for air, 1.25 for water, 1.40 for oil immersion, and 1.60 for naphthalene monobromide immersion. These values are less than the above-mentioned theoretical limits because it is necessary to have a working distance greater than zero.

Although immersion objectives are available for use with methylene iodide which has the highest refractive index listed above, there is little further gain in numerical aperture over napthalene monobromide. The principal advantage gained is one of contrast for the immersion of an object of refractive index differing greatly from that of the immersion medium.

F. Resolution

The resolution of a microscope is defined as the minimum distance between two points in the object that are just distinguishable. The image of a light point formed by an objective is not a point but a diffraction disk. The greater the resolving power of a lens, the smaller will be the diffraction disks and the greater the ability of the lens to separate the disks corresponding to two points on the object. Provided that there is a high degree of object contrast, the linear resolution d is

$$d \approx \frac{0.61\lambda}{NA} \tag{7}$$

where λ is the wavelength in vacuo.

In practice, the fine object detail rarely has sufficient inherent contrast to enable this resolution to be reached. Equation 7 shows that the resolving power can be increased, that is, d decreased, by decreasing λ or by increasing the numerical aperture of the objective lens (but see the effect of spherical and other lens aberrations below). By using white light illumination with a green filter giving λ as approximately 5600 Å and with a lens of NA 1.4, the limit of resolution is approximately 2400 Å. Although the resolving power can be halved using the ultraviolet light microscope, lenses suitable for reflection microscopy have not apparently been developed, and one would normally resort to the electron microscope with which several orders of magnitude of improvement can be obtained.

The increase in resolution with the increase in numerical aperture can be understood after Abbé (see Ref. 6) if the object is regarded as a diffraction grating, viewed in transmission by coherent Köhler illumination. Light that is scattered forms first, second, third, . . . order diffraction spectra in the back focal plane of the objective. The direct beam only contains information on the shape of the object. The direct and one or more diffracted beams must enter the objective, in order to resolve detail in the object. The more diffracted beams enter the objective, the closer will the image represent the object. If first, second, and third order spectra enter, the representation is complete.

G. Depth of Focus

The depth of focus corresponds to the height range of object detail that is in focus at a given lens setting. The depth of focus is roughly inversely proportional to the square of the numerical aperture of the lens. For a 1.40 NA objective, the depth of focus is only about λ/2, that is, about 2500 Å.

H. Aberrations of Lenses

Aberrations fall into two main types: (*a*) those caused by the variation of the refractive index of the lens with wavelength, *chromatic aberrations*, and (*b*) those that arise even if the light is monochromatic, *monochromatic aberrations*. The aberrations are not caused by lens defects but are simply a consequence of the laws of reflection and refraction at spherical surfaces.

1. Chromatic Aberration

Since the focal length of a lens is a function of the refractive index of the material of which it is made, and the refractive index varies with the wavelength or color of the light, it follows that the focal length is different for different colors. An image formed by white light consequently consists not of a single image but of a separate image for each wavelength present, focused at different distances from the lens, giving rise to *longitudinal chromatic aberration* (Figure 2*a*). Since the magnification depends upon the focal length, there is a variation of the image

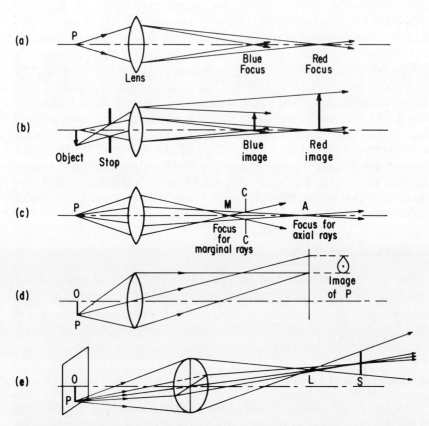

Figure 2   Lens aberrations: (*a*) longitudinal chromatic aberration: (*b*) lateral chromatic aberration; (*c*) spherical aberration; (*d*) coma; and (*e*) astigmatism. From Phillips.[1]

size referred to as *lateral chromatic aberration* or *chromatic magnification difference* (Figure 2*b*).

2. Spherical Aberration

Spherical aberration arises when rays from an object point *P* (Figure 2*c*) on the axis of a lens are more strongly refracted by either the center or margin of the lens, so that there is no single focus but rather a succession of focal positions. The object point P then appears in a given image plane as a circle of finite area, not as a point. Position C (Figure 2c) at which the bundle of refracted rays has its minimum cross section is known as the *circle of least confusion;* the best image is secured by placing a screen here. Spherical aberration may be greatly reduced by introducing an aperture, so that only the central portion of the lens is used, and by the appropriate design of a compound lens.

It may be desirable to employ cover glasses with samples suitable for transmission illumination. The spherical aberration is sensitive to the cover glass thickness, particularly with dry objectives of high numerical aperture, and the correct value for which the lens is designed (commonly 0.18 mm) should be employed. Immersion objectives are much less sensitive to the cover glass thickness, since the immersion oil employed has a refractive index similar to that of the cover glass, so that there is essentially a continuous optically homogeneous medium between the specimen and the objective.

3. Coma

A lens may be corrected for spherical aberration but still show coma. This aberration only affects rays from object points off the axis of the lens, but, as with spherical aberration, it arises from differences in the refraction of rays from an object point P (Figure 2d) passing through the inner and outer zones of the lens. The point P is imaged as a comet-shaped figure; hence the name coma. Coma may be reduced by introducing a suitable lens aperture. An optical system free of both coma and spherical aberration is said to be *aplanatic.*

The most troublesome aberrations in objectives of large aperture, apart from the chromatic one, are spherical aberration and coma. These are suppressed when the following condition is fulfilled:

$$d\, n \sin u = d'n' \sin u' \tag{8}$$

where d is the distance from the axis, n the refractive index, and u the angle between the ray and the axis. Unprimed quantities refer to the object space, primed quantities to the image space.

4. Astigmatism and Curvature of the Field

These two aberrations have the same origin. If the ray paths from the off-axis tip P of a vertical line object OP are considered (Figure 2e), it will be seen that rays passing through the horizontal plane of the lens converge to a horizontal line L, the primary image, while rays passing through the vertical plane of the lens converge to a vertical line, the secondary image S. The cross section of the refracted beam is elliptical, degenerating to a line at L and S, and forming a circle of least confusion somewhere between them. The distance between the two focii is a measure of the astigmatism. Lens combinations corrected for astigmatism are known as *anastigmats*.

Since the astigmatism is zero for a point O on the lens axis (Figure 2e), that is, the focii coincide, and increases for an object point as it moves away from the axis, it follows that, for a planar object, lines L and S become curved focal planes which intersect on the lens axis. Since the surface of best focus is not planar, but curved, this aberration is known as the *curvature of the field*. The correction for astigmatism brings the two images together, but the image plane is still curved. Curvature of the field can be eliminated if the two image planes have equal and opposite curvature, but astigmatism remains. Special *flat field objectives* (see Section II.A) are sometimes employed for high magnification photomicrography. Some compensation of the curvature is possible by the suitable choice of an eyepiece.

5. Distortion

Whereas spherical aberration, coma, and astigmatism involve the failure of the lens to form a point image of a point in the object, distortion involves, not a lack of sharpness of the image but a variation in magnification with the distance of a point in the object from the axis of the lens. The image of any straight line in the object plane through the axis will be imaged as a straight line, but others are distorted. *Barrel distortion* results if the magnification decreases with increasing axial distance, *pincushion distortion* if the magnification increases away from the axis (Figure 3).

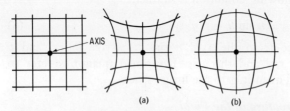

(a) (b)

Figure 3 (a) Pincushion distortion; (b) barrel distortion. From Phillips.[1]

Distortion can be avoided if

$$\frac{\tan u}{\tan u'} = \text{const} \tag{9}$$

where u is the angle between the ray and the object space and the axis, and u' the angle between the ray in the image space and the axis. To avoid distortion as well as spherical aberration and coma, it is necessary to use only beams slightly inclined to the axis in order to satisfy this condition, and also the sine condition.

II. MICROSCOPE COMPONENTS

Reference may be made to Chamot and Mason[8] for a more detailed discussion of this subject.

A. Objectives

Objectives are compound lenses, consisting of several lenses mounted coaxially, and are designed to correct for various aberrations. They may be grouped in the order of increasing degree of optical correction, complexity, and cost into achromats, semiapochromats, and apochromats.

Achromats have longitudinal chromatic aberration corrected for two colors (generally red and green) and, in addition, have spherical correction for one color (generally yellow green). Their performance is optimized by employing yellow green light and orthochromatic plates for photography. Achromats have fewer lenses than apochromats and, consequently, have smaller reflection losses. This is an advantage when monochromatic illumination is necessary, as in interference microscopy. Their use is desirable for fluorescent microscopy to avoid autofluorescence from fluorite components. Achromats have a larger working distance than other types and can more easily be made strain-free which is highly desirable for polarizing work, in order to avoid unwanted polarizing effects. Achromats are adequate for much straightforward metallographic work in black and white, in conjunction with a green filter.

Apochromats have longitudinal chromatic aberration corrected for three colors (red, green, and violet) and also have spherical aberration corrected for two colors (green and violet). As a result of their high spherical correction, they produce lateral chromatic aberrations, evident as color bands on the side of objects near the edge of the field, and should be used with appropriately designed *compensating eyepieces* to correct this. These objectives are particularly suitable for high magnifications and photomicrography. Their numerical apertures are larger than achromats but their depth of field is less, so that achromats and semiapochromats may be preferable for low and medium magnification work. An important advance of recent years has been the development of *apochromatic planoobjectives*, with which, in combination with appropriate matched eyepieces, both the curvature of the field and astigmatism have been eliminated, increasing the usable field area by a factor of 4 or, in some instances, 10. In some cases it is possible to use high power eyepieces up to $25\times$ advantageously.

Semiapochromats, often referred to as *fluorite objectives* when they contain one or more fluorite lenses, are intermediate in degree of correction between achromats and apochromats. They are usually corrected chromatically and spherically for two colors. *Neofluors* employ synthetic, rather than natural, fluorite and can be obtained absolutely strain-free.

It is convenient if, when changing objectives, the specimen remains approximately in focus. Sets of objectives designed for this purpose are known as *parfocal*.

When an oil immersion objective is used, oil is placed between the specimen and the objective which matches the refractive index of the lowest lens of the objective. Achromatic water tank objectives and fluorite water immersion objectives are available for studies of corrosion and electrode processes in aqueous solutions. Immersion gives increased aperture and also increased brilliance, since it eliminates an interface. In addition to these important advantages, immersion may be used to remove undesired topographical features in reflection or transmission.

Since modern objectives may have a large number of component lens elements, multiple partial reflection of the light, as it passes repeatedly between glass and air, may tend to cause glare and reduce image contrast. This had led to the development of *bloomed objectives*, in which the lenses are coated with material of such thickness and refractive index that the reflectivity approaches zero. The theory of coating has been discussed by Françon[7] and by Born and Wolf.[6] Taylor[9] and Benford[10] have given examples of the improvement obtainable in metallography.

The *working distance*, that is, the separation between the object and the front surface of the lens is related to the numerical aperture and the power. The greater the power, the smaller the working distance. The working distance has no direct relation to the focal length of an objective and is a matter of design. Usually, the working distance is much less than the focal length. Some typical working distances are 39 mm for a dry achromat of $2.5\times$ and 0.08 NA, 7 mm for a $10\times$ achromat of 0.25 NA, 0.08 mm for a dry achromat of $63\times$ and 0.85 NA, and 0.08 mm for an oil immersion achromat of $100\times$ and 1.25 NA.

For a given aperture, considerably greater working distances may be obtained by the use of *reflecting objectives*, facilitating the design of hot stage microscopes for use when it is necessary to avoid the heating of the lens, and having more working space for manipulation in fractography. For example, a coated mirror objective of 40× magnification and 0.52 NA may be obtained with a working distance of 8.1 mm. The working distance can be further increased, if necessary, by scaling up in size. Reflecting systems have the additional advantage of being achromatic, permitting a free choice of illumination wavelength.

Burch[11,12] was the first to see the advantages of using only reflecting components in a microscope. He showed that at least one of the mirrors (Figure 4) must be aspherical if the numerical aperture is to exceed

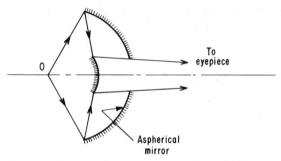

Figure 4 Burch reflection objective. From Phillips.[1]

0.5. Numerical apertures as high as those of the best oil immersion refracting objectives cannot be achieved. The working distance may be as great as the focal length of a mirror objective. Aspherical mirrors are difficult to manufacture and align, unlike concentric spherical pairs. A disadvantage of mirror objectives is the obstruction due to the secondary mirror which decreases the numerical aperature by amounts up to 45%. Data on a number of alternate systems have been tabulated by Winterbottom and McLean.[13] Spherical mirror objectives with additional zero or low power lenses to correct the aberrations have been constructed, also catadioptic systems in which both reflection and refraction contribute appreciably to the convergence of light. Finally, attachments have been developed for use with ordinary objectives that increase the working distance by producing a unit or low magnification real image which is then examined with the ordinary microscope.[14]

B. Eyepieces

The compound microscope achieves its final magnification in two steps; the primary enlarged image produced by the objective is observed with a magnifier of relatively simple design known as the eyepiece or ocular. This is necessary even if the objective is ideally corrected, since the finest detail resolved by the objective must be magnified to at least a size that can be recognized and resolved by the human eye which unaided has a resolution of only about 0.1 mm. The eyepiece cannot reveal detail that is not present in the primary image. However, the higher the objective magnification, the more difficult it becomes to approach ideal correction. The final image quality can be greatly improved if the eyepiece is designed with built-in aberrations which exactly compensate those remaining in the objective, that is, if it is chromatically overcorrected. The lateral chromatic aberration shown by apochromats can thus be corrected by suitable *compensating eyepieces*.

Compensating eyepieces should be employed with all objectives having numerical apertures over 0.65, with apochromats, and with planoobjectives. Special *negative-compensating eyepiece systems* designed to flatten the field of a particular objective are made specially for photomicrography but cannot be used for visual work, since they give a virtual image. The simpler noncompensating *Huygens-type* negative eyepiece is commonly used for visual work with achromats of numerical aperture lower than 0.65. This in its simplest form consists of two separated planoconvex lenses, with a field diaphragm located between the lenses, and, since it is not optically corrected, is unsuitable for use with objectives of high numerical aperture. *Hyperplane* and *periplane* negative eyepieces are partially corrected and are suitable for use with high aperture achromats and semiapochromats, giving a slightly flatter field than compensating eyepieces.

In general, in visual work, there is little point in using an eyepiece of below 10X magnification. High power eyepieces can be used advantageously in combination with low power objectives, when the surface under study is rough and good depth of field is needed. Eyepiece magnifications of 15X, 20X, and even 25X can usefully be used with well-corrected objectives that have a relatively high ratio of numerical aperture to magnification.

C. Vertical Illuminators

Vertical illuminators are used for the examination of opaque samples by incident light. The reflected light of the vertical illuminator is axial, so that a plane surface appears bright, whereas inclined surfaces such as scratches, pits, and grain boundaries tend to appear dark. Two main types of illuminator are employed. For convenience, these have been drawn (Figure 5) so that the objective lens also serves as a condenser.

The illuminator in Figure 5a has a partially reflecting 45° glass slip and thus does not reduce the numerical aperture. It is desirable to employ filtered light and coated objective lens surfaces, in order to reduce glare. Glass slip illuminators normally reflect only about 7% of the incident light onto the specimen. This may be increased several times by coat-

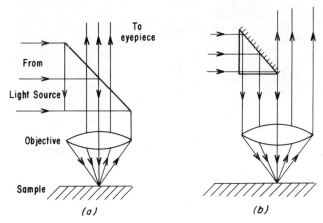

Figure 5 Types of vertical illuminator: (*a*) 45° glass slip; (*b*) 45° prism (silvered). From Phillips.[1]

ing.[6,7] Since a 45° glass reflector works near the Brewster angle for glass, it tends to introduce appreciable polarization. This is greatly reduced in modern reflectors which are coated for high reflection on one face and bloomed for antireflection on the other. The illuminator in Figure 5b employs a 45° prism. It provides more light with less glare than an (uncoated) plane glass illuminator but reduces the numerical aperture by a half; therefore it is only suitable for low power use, for example, with 10× and 20× objectives. Prism reflectors are less commonly used than the plane-glass type.

Oblique or dark field illumination may be obtained in a number of ways, for example, by tilting the vertical illuminator, so that no direct light enters the objective lens. This is discussed in more detail in Section III.A.

Vertical illuminators such as the glass-slip or prism types do not permit any control of the field or aperture. For optimum objective resolution, the light source must be imaged in the plane of the object and the cone of incident light should just fill the aperture of the objective, giving *critical illumination*. The illuminated field is reduced to that of the area visible in the microscope. One of several systems suitable for this purpose, based on that proposed by Köhler,[15] is illustrated in Figure 6. Similar systems may be used for the substage illumination of transparent objects.[8,16] The image of the light source is formed by the collector lens

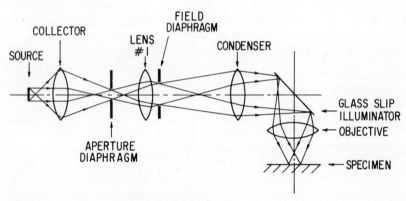

Figure 6   Köhler-type illuminating system.

on the aperture diaphragm which thus acts as a secondary light source and is, in turn, imaged in the back focal plane of the objective by two more lenses via the glass slip reflector. For maximum resolution, the aperture diaphragm is opened until its rim is not visible in the field of view, thus filling the full aperture of the objective with light. A field diaphragm is imaged by the condenser and objective in the plane of the specimen and is adjusted so that only the field of view is illuminated. Its function is to improve contrast by cutting out scattered rays. In practice, the condenser and objective lenses are compound lenses.

A high condenser aperture is not as advantageous as indicated above if the object lacks high inherent contrast because it is then usually necessary to close the aperture stop partially to obtain sufficient image contrast.[17] The degree of condenser correction affects the image contrast and thus the obtainable resolution. An ideally corrected system satisfies the sine condition and in addition is corrected spherically, chromatically, and for chromatic difference of the spherical aberration. A sharp reduced image of the field diaphragm is then projected onto the object plane, and each point of the aperture diaphragm projects a parallel pencil of light onto the object up to high apertures. Condensers with only spherical correction may be used for black and white photomicrography, provided that a strictly monochromatic green filter is employed. Systems that fulfill the sine condition, so that all condenser zones give the same reduction ratio, and in which the chromatic difference is eliminated, are called *aplanatic*. Since chromatic aberrations lead to several colored images at different focal distances from the condenser, chromatic correction is necessary for a condenser used in color photomicrography. The best systems are *achromatic-aplanatic* condensers that have spherical aberrations corrected for a wide range of wavelengths, while fulfilling the sine condition. A high degree of condenser correction is necessary, in order to make advantageous use of high condenser apertures.

D. Practical Light Sources

A large variety of light sources is employed by the microscopist. The low voltage, high current tungsten filament lamp is still commonly used for direct observation but is being replaced by the quartz iodine lamp, in which the tungsten filament is contained in an atmosphere of iodine. This type of lamp has also replaced the Pointolite tungsten arc lamp, once popular for direct viewing. It is very good for color and has a color temperature of about 2000°K.

Direct or alternating current, carbon arc lamps, once extensively used for photomicrography, since they have high intensity and good color temperature of about 4000°K, have tended to be replaced by the zirconium arc lamp. This lamp is relatively expensive and typically has a life of only 200 hr but does not suffer from the instability of the carbon arc lamp and the need for replacement of the carbons. The zirconium arc lamp is being supplanted by the xenon d-c arc lamp which has a longer life and lower cost. This lamp provides white light low in the mercury green wavelengths but quite good for color (color temperature about 6000°K). Its intensity is equivalent to, or better than, the carbon arc lamp and is adequate for display on the large screens used in some of the newer metallographs. When fed from a regulated constant current supply, the light output from a xenon lamp can be maintained constant within ±1%.

High pressure metal halide arc lamps are now available that have a larger arc than xenon lamps and are free from the inhomogeneities common to xenon lamps. These are mercury arc lamps with added metal halides that increase the continuous spectrum, while suppressing the normal mercury ultraviolet radiation and decreasing the blue emission. Almost white light is produced which is excellent for projection, reflected light work, polarization microscopy, and color photomicrography. (color temperature about 3800°K). Since this source is operated by alternating current, it cannot be used for cinematography.

High intensity mercury capillary arc lamps find employment as sources for fluorescent and ultraviolet microscopy, using a filter to screen out the visible range. The light should be fully enclosed to avoid skin burns and the eyes protected using special goggles (e.g., Corning "Noviol" glass), although this is unnecessary for dark field fluorescence observations. Sodium lamps may be used for monochromatic illumination but have a very low luminance.

Since the heat from a light source may result in damage, a heat-absorbing glass filter should be used. In the case of the more powerful sources employed in metallographs, the filter may be immersed in water in a small cell.

III. OPTICAL METHODS OF ENHANCING CONTRAST

In order to achieve high resolution, there must be sufficient contrast in the image for different amounts of light to reach the eye from the regions to be resolved. In metallurgy, image contrast is frequently improved by etching to obtain different local coefficients of reflection and scattering. Etching is not always desirable or efficacious in revealing the required structural detail. However, a number of purely optical methods of enhancing contrast applicable to any kind of surface, etched or otherwise, are available. These may be classified as:

1. Dark field illumination
2. Polarized light
3. Phase contrast
4. Interference
5. Filters

Two types of features may be distinguished on metal surfaces, namely, amplitude features and (optical) phase features. *Amplitude features* can exist on reasonably flat surfaces as a result of different reflectivities, for example, of two alloy phases. *Optical phase features* can exist on surfaces of the same reflectivity, but different levels, since there is a path difference and hence a phase difference between light reflected from different points. Alternatively, a phase feature can exist on a flat surface, even with no difference in reflectivity, because of a different magnitude of phase change on reflection from different points. Since the eye or photographic plate can only directly discern amplitude differences, phase differences must be converted into (visible) amplitude differences. This is the basis of the phase contrast (see page 78) and interference techniques (see page 83).

A. Dark Field Illumination

A vertical illuminator can be tipped to secure nonvertical, that is, *oblique illumination* or *dark field illumination*. In dark field illumination, the light beam striking the object is of such obliquity that no direct light, that is, specularly reflected light, enters the objective. Object detail that reflects or scatters the light then appears light on a dark background. This is the opposite situation to bright field illumination. Oblique pencil illumination is obtained by tilting the illumination system, so that illumination is from one side only. Much better dark field images are obtained by using an annular cone of rays which is brought to a focus in the object plane, instead of tilting the illumination. A stop (circular disk) is used to remove the central beam. With suitable surfaces, for example, etched, the gain in contrast over the bright field is often striking. The falsification of fine detail by diffraction effects is sometimes a problem. An azimuthal diaphragm may be used to improve the contrast. All but a small sector of the annular illumination is blocked, so that unidirectional dark field illumination in an azimuth selected to bring out particular object features, is employed, or two sectors 180° apart may be used.

B. Polarized Light

In the wave theory of light, a ray is considered to be vibrating perpendicular to the ray in all planes containing the ray (Figure 7). The sinus-

Figure 7 Light considered as a wave motion. From Phillips.[1]

oidal wave can be regarded as rotated through 2π about the ray to generate a three-dimensional figure. It should be noted that the phase is the same, regardless of the plane selected. In the more general case the cross section is elliptical, rather than circular.

If light is vibrating in only one plane, for instance, the plane of this paper, it is said to be *plane-polarized*. If two such waves with the same phase A and B, vibrating in planes at right angles, are combined, they are equivalent to a third wave C which also has the same phase but is in a different plane. It follows that any plane-polarized wave C can be resolved into two component waves of the same phase, lying in two arbitrary planes at right angles. Now, if two waves A and B, which are out of phase by $\lambda/4$ or $3\lambda/4$, are combined, a *circularly polarized* wave C is obtained. This can be represented by a right-handed or left-handed helix, with its axis along the ray direction. Thus the plane of polarization of wave C rotates into a different position, as one moves from point to point along the ray. The point of maximum amplitude of C also traces out a helical line, as one moves along the ray. *Elliptically polarized* light can be obtained if the phase difference is, for example, $\lambda/8$ or $3\lambda/8$. Unless otherwise stated, we shall be concerned with plane-polarized light.

1. Sources of Polarized Light

a. REFLECTION

The reflection from a glass plate at the polarizing angle is the simplest means of obtaining polarized light. Such light is polarized in a plane at right angles to the plane of incidence. Brewster's law states that:

$$\tan \phi_p = \frac{n'}{n} \tag{10}$$

where ϕ_p is the polarizing angle and n, n' are the refractive indices. If $n = 1$ (in vacuo or approximately for air) and $n' = 1.5$ (typical glass), then $\phi_p = 56°$. It will be seen from Equation 10 that the polarizing angle is that for which the reflected and refracted rays are at right angles to one another.

Since only about 4% of the incident light is reflected, it would be necessary to use a stack of parallel-sided glass plates, in order to secure a reasonable intensity of the reflected polarized light. Even so, the reflected light would be only partially polarized, so that this is not a very useful method of producing plane-polarized light. However, the refractive index of an opaque sample such as a metal may be determined from Brewster's relation by measuring the polarizing angle.

b. DOUBLE REFRACTION

Certain crystals, for example, calcite (Iceland Spar), are naturally birefringent. A natural cleavage rhomb of calcite will split a ray into an ordinary ray O and an extraordinary ray E, vibrating in two planes at right angles to one another and widely separated. Thus, if some method of separating the two waves can be found, a doubly refracting crystal will serve as a polarizer. This can be accomplished in a number of ways. William Nicol in 1828 devised the *Nicol prism* by suitably grinding and cementing together two Iceland spar crystals (Figure 8). The emergent extraordinary ray *E* is plane-polarized. The ordinary ray *O* is totally reflected at the interface, since the angle of incidence exceeds the critical angle and hence the angle of refraction exceeds 90°. Nicol prisms are now obsolete.

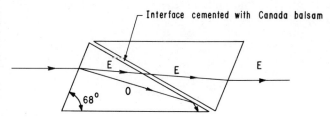

Figure 8   Principle of the Nicol prism. From Phillips.[1]

Certain doubly refracting crystals such as tourmaline exhibit *dichroism*, that is, one of the polarized components is absorbed preferentially; hence, by choosing a proper thickness, one component can be made predominant (Figure 9). Dichroism is the basis of the *polaroid* sheet, developed by Land in 1934. In one form microscopic crystals (needles) of dichroic

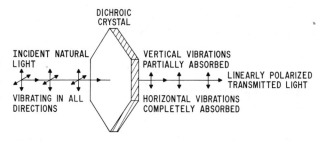

Figure 9   Illustration of the linear polarization of light by transmission through a dichroic crystal.

iodoguinine sulfate were oriented parallel in sheets in a plastic film. More recently, the molecules in thin sheets of polyvinyl alcohol were strain-oriented and rendered dichroic by iodine staining. The modern materials now used are known as *polars*.

In practice, two polars are used (Figure 10): a *polarizer* which polarizes the light incident on the sample, and an *analyzer* which is used to examine the light from the sample and is commonly fitted inside the eyepiece.

The analyzer is usually rotatable from 0 through 90° and is furnished with an angle scale. If the polars are "crossed," that is, set at 90°, the light is completely extinguished, unless optically active sample detail rotates the polarized light. If set parallel, the maximum amount of polarized light is transmitted. At settings in between, partial extinction occurs and sample detail appears gray, as the ends of the visible spectrum are not completely polarized.

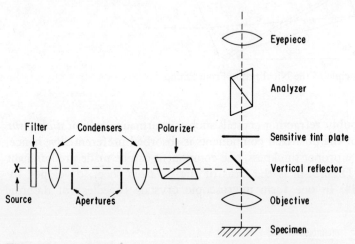

Figure 10 Schematic arrangement of polarizing reflecting microscope. From Phillips.[1]

If plane-polarized light is allowed to pass through a crystal plate at an angle to the optic axis that lies in the plane of the plate, the emergent ordinary and extraordinary polarized beams may or may not be in phase, depending on the thickness of the plate and the wavelength λ of the incident light. Hence one can make a *quarter-wave plate*. Cellophane film can be used as a cheap, readily available material and several thicknesses combined by trial and error to give the desired relative retardation. Cleaved mica, gypsum, selenite, or quartz plates can also be used. The insertion of a wedge or stepped wedge gives a means of progressively changing the phase retardation.

If the refractive indices of the extraordinary and ordinary rays are n_0' and n_0, respectively, then the thickness t of a quarter-wave plate is

$$t = \frac{\lambda}{4(n_0' - n_0)} \tag{11}$$

The difference $(n_0' - n_0)$ is called the *birefringence*. Transmission of mono-chromatic plane-polarized light through a quarter-wave plate gives elliptically polarized light, provided that the optic axis of the plate is not parallel to the plane of the incident light, in which case no change occurs. Quartz has the advantage that $n_0' - n_0$ does not vary greatly with the color of the light used.

A quarter-wave plate may be employed to analyze elliptically polarized light. Since any elliptical motion may be built up from two plane vibrations with a phase difference of 90°, plane-polarized light emerges if the plate is positioned with its optic axis parallel to one of the axes of the ellipse.

The phase difference may be continuously changed by a device called the *Babinet compensator*. This consists of two wedge-shaped sections of crystalline quartz, mounted so that one can be slid over the other (Figure 11). The optic

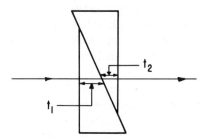

Figure 11   The Babinet compensator. From Phillips.[1]

axis is lengthwise in one section and transverse in the other. Thus a wave that travels through the first crystal as an ordinary wave, travels through the second as an extraordinary wave, and vice versa. The resultant phase difference ϕ_d is

$$\phi_d = \left(\frac{2\pi}{\lambda}\right)(t_1 - t_2)(n_E - n_0) \tag{12}$$

where n_E is the refractive index for an extraordinary wave, and n_0 the refractive index for an ordinary wave.

A variety of other compensators may be employed to measure optic path differences such as the calibrated quartz wedge, Berek compensator, quarter-wave compensator, and rotary mica compensator.[18] These bire-fringent devices are all used in a similar manner, that is, they are inserted below the analyzer and adjusted so as to neutralize exactly the specimen influence, so that light leaves the compensator as if it had never interacted with the specimen; it is again linearly polarized and is completely extin-guished by the analyzer. From previous calibration of the compensator, the nature of the specimen action is determined.

2. The Sensitive Tint Plate

The sensitive tint plate, as pointed out by Conn and Bradshaw,[19] is so named because it provides a very sensitive means of detecting double refraction or birefringence. It is identical with a quarter-wave plate, except that it is four times the thickness, so that the path difference between the ordinary and extraordinary rays is equal to one wavelength λ. The wavelength chosen is near the middle of the visible spectrum in a critical position such as magenta, where a slight change in birefringence will result in completely different colors. The plate is usually made of gypsum or quartz; however, a reasonably satisfactory plate may be made of several layers of cellophane, with the rolling direction inclined at 45° to the polarizer and analyzer axes. Additional layers are added until a purple tint is seen.

If an optically isotropic sample is viewed in white light between crossed polars, no light is seen. If a sensitive tint plate is placed between the polars, no light of wavelength λ is seen because the compounded vibration is in a plane normal to the principal plane of the analyzer. The light of wavelengths differing from λ will be seen, since there is a resultant component parallel to the principal plane of the analyzer. The field will thus show the complementary color. If a birefringent detail is present in a transparent sample, it will change the path difference between the ordinary and extraordinary rays for the given λ and therefore alter the wavelength eliminated by the sensitive tint plate, changing the field color locally. Any ellipticity introduced by a specimen viewed under reflected light will also alter the color, so that the grains or phases will assume different tints.

Birefringent uniaxial crystals may be classified as positive or negative crystals by using a sensitive tint plate. They are positive if the ordinary ray has a slower speed than the extraordinary ray, that is, if $n_0' < n_0$. The rotation direction for positive crystals is usually shown on the mounting of the tint plate.

3. Identification of State of Polarization

As pointed out by Winterbottom and McLean,[13] *plane-polarized* light can be extinguished completely with two positions of a polar. If the light consists of a *mixture* of polarized and unpolarized light, or is *elliptically polarized*, variations of intensity occur on rotating the analyzer. If the light is elliptically polarized, then interposition and rotation of a quarter-wave plate will give complete extinction in some positions. If the light is *unpolarized, circularly polarized*, or a mixture, no variation in intensity will occur on rotating the analyzer. If a quarter-wave plate gives extinction in some positions, circularly polarized light is indicated. If only variations in intensity are observed, a mixture is present; if no variations occur, the light is unpolarized. These tests are sensitive to degrees of polarization above about 20%.

4. Applications of the Polarizing Reflection Microscope

This technique deserves more application in metallographic work than it has received. The quantitative use of polarized light has been reserved mainly for transmission work in the petrographic and biological fields, although in principle a variety of quantitative measurements are possible in reflection on metals. Their description falls outside the scope of the present work, but the reader may refer to other sources.[13,19-22] Since the nature of the metal surface and the presence of surface films, etch pits, and so forth, affect the nature of the quantitative results obtained, surfaces are best examined in the as-polished (unetched) condition, preferably using mechanical rather than electrolytic polishing. The use of a coated glass-slip type of vertical illuminator is generally preferable, with the vibration direction of the polarizer horizontal, since the reflecting power is greater for the component vibrating parallel with the surface than for that in the plane of incidence. The plane in which a light beam vibrates is, in fact, easily found by examining light reflected from glass at the polarizing angle, which vibrates parallel to the glass surface. Since reflection tends to produce polarization, a certain degree of polarization is almost unavoidable in optical systems. Reference has already been made to the need for strain-free glass reflectors and lenses in polarizing systems.

Cubic metals are optically isotropic, when clean, and reflect polarized light unchanged, when it is at nearly normal incidence. All grains appear equally dark between crossed polars, regardless of orientation. However, they can often be made active under polarized light either by etching to produce pits or grooves,[21,23,24] so that the (vertical) illumination is elliptically polarized because of oblique reflection at the etch facets or grooves, or by coating with an anisotropic film, for example, by anodizing in the case of aluminum.[25] Grain contrast can be observed if the film orientation or thickness is a function of the orientation of the underlying grain. The effect can be enhanced by the use of a sensitive tint plate.[26]

Noncubic metals are anisotropic, even in the clean condition, so that grain contrast can be obtained under polarized light. Inclusions, even if isotropic, may be active, since light is reflected obliquely at the metal-inclusion interface. A fuller discussion will be found in Refs. 19 and 21.

Since orientation-dependent contrast can often be obtained, semiquantitative studies may be made of preferred orientation.[13,19,21] In principle the optical constants may be measured at normal incidence on clean surfaces, orientations may be determined,[23,27,28] and phases may be identified.[13,19,21]

C. Phase Contrast

Phase contrast has found its principal application in transmission and has proved particularly valuable for obtaining contrast in biological samples and revealing fine detail; see the work of Bennett et al.[29] Application to opaque samples has proved limited; however, examples of its application in metallography have been given by Winterbottom and McLean,[13] Cuckow,[30] McLean,[31,32] and Perryman.[33] Other references are given by Bennett et al.[29] The technique is useful in producing contrast when there are surface level differences but identical color and reflectivity, for example, at slip steps or surface fatigue striations. It has been employed occasionally to distinguish between surface elevations and depressions at grain boundaries and particles in etched samples of, for example, temper embrittled steel.[13] Replica electron microscopy is more commonly employed and capable of higher resolution. Replicas may also be usefully examined in the phase contrast microscope.[29]

Since phase contrast is a basic method of obtaining contrast, the principles will be discussed. It is also the basis of an interference technique useful in measuring small optical path differences, that is, differences in thickness or refractive index, of transparent materials. Phase contrast principles are applicable in the electron microscope.

1. Phase Detail

Transparent or opaque samples may show no structure in normal bright field because the samples do not show differences in light absorption. There may, however, be differences in refractive index and in optical path that result in changes of phase of the light waves. The change of phase is a consequence of the fact that the velocity of light increases with an increase in refractive index. Such "phase detail" can be made visible by the *phase contrast method* due to Zernike,[6] in which an intensity distribution is produced which is directly proportional to the phase changes introduced by the object.

Consider a transparent slice of material of refractive index n' containing a dephasing object, such as an inclusion, of thickness t and refractive index n (Figure 12a). The difference in optical paths Δ of rays A and B is given by

$$\Delta = (n - n')t \tag{13}$$

and the difference of phase by

$$\phi = 2\pi \frac{\Delta}{\lambda} \tag{14}$$

Now consider a reflecting dephasing object such as an etched metal surface (Figure 12b). The optical path difference is $2t$, with respect

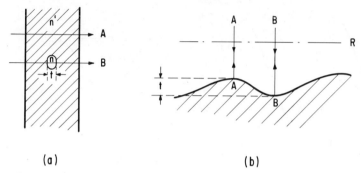

(a)   (b)

Figure 12   Phase contrast due to (a) transparent dephasing object; (b) reflecting dephasing object. From Phillips.[1]

to a reference plane R, where t is the difference in level between two points A and B on the metal surface. If it is assumed that the phase change on reflection at points A and B is identical, then the phase difference ϕ is

$$\phi = 4\pi \frac{t}{\lambda} \tag{15}$$

In order to render phase detail visible, it is necessary to convert the phase changes into differences of light intensity, that is, amplitude differences. This can be done in qualitative fashion by phase contrast, or quantitatively by interference contrast, in which case the image intensity is a quantitative measure of the phaseshift. The difference between normal, that is, absorption contrast, and phase contrast is most simply understood in terms of wave diagrams. The treatment used follows that in Ref. 34. A fuller account is given by Bennett et al.[29]

2. Nature of Normal (Absorption) Contrast

In the case of normal contrast, the intensity, that is, amplitude, is reduced by passage through an absorbing specimen, but the wavelength λ

is unchanged (Figure 13*a*). The effect of the specimen upon the amplitude of wave *L* is equivalent to superimposing on it a wave *M* which is shifted in its phase by $\lambda/2$ or $180°$ relative to *L*, so that destructive interference occurs (Figure 13*b*). The amplitude of *M* is such that $L + M = N$ (Figure 13*b*).

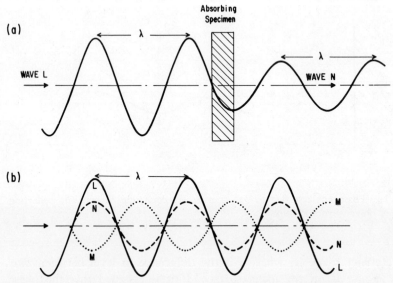

Figure 13 Wave diagram illustrating normal, that is, absorption contrast. From Phillips.[1]

3. *Nature of Phase Contrast*

Phase detail, in contrast to amplitude detail, does not change the amplitude. The wave *N* (Figure 14*a*) has the same amplitude and wavelength λ as wave *L* but is shifted in phase by an amount $\delta\phi$. The specimen effect is again equivalent to superimposing a suitable wave *M* over *L*, so that *N* results (Figure 14*b*); *M* can be constructed graphically as the vertical distance between *L* and *N*. Wave *M* turns out to be a wave of smaller amplitude, with about $\lambda/4 = 90°$ phase shift relative to *L*. The amplitude of *M* is such that amplitudes $L + M = N$.

Now, if the phase difference between *L* and *M* could be increased to about $\lambda/2 = 180°$, then the phase detail would have the same effect as if it were absorption detail. If the amplitude of *L* were also reduced to that of *M*, complete destructive interference would result, throwing the phase detail, for instance, an inclusion embedded in a transparent crystal, into high contrast, so that it would appear similar to a light absorbing structure (Figure 14c). One method of doing this is to place

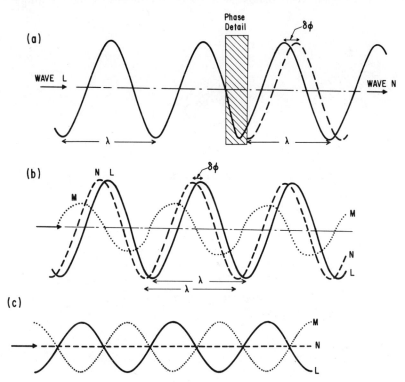

Figure 14   Wave diagram, illustrating phase contrast. From Phillips.[1]

an annular ring diaphragm (Figure 15a) in the condenser, which, with monochromatic light, will act as a coherent source of light, and also to put a phase-retarding absorption ring (Figure 15b) in the back focal plane of the objective (Figure 16). The annular ring blocks out all light, except that passing through the clear annulus. The inner annulus of the phase-retarding absorption ring reduces the intensity of beam L relative to beam M by means of an evaporated metal coating, so that the amplitudes become comparable. The outer annulus retards beam M by $\lambda/2$ relative to beam L.

Phase retardation plates can be made of very thin magnesium fluoride, the phase adjustment being made by thickness change, as discussed earlier.

Clearly, many variations of technique are possible, since one can adjust the relative intensities of the two beams and the relative phases to produce various contrast effects. Such terms as positive, negative, minus, bright, and dark phase contrast are used and lead to much confusion, since the contrast effect obtained depends not only on the details of the technique

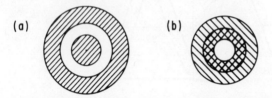

Figure 15   (*a*) Annular ring condenser diaphragm; (*b*) phase-retarding absorption ring. From Phillips.[1]

but also on the specimen, that is, on whether the object detail has a greater or lesser refractive index than the surrounding material.

The intensity variation observed in the image is only qualitatively proportional to the optical path variations in the object; one reason is that a portion of wave *B* penetrates the portion of the phase-retarding ring supposedly reserved for wave *A*.

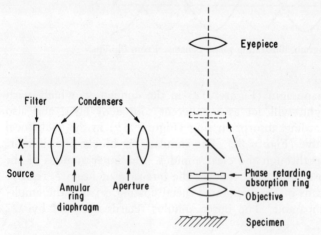

Figure 16   Schematic arrangement of phase contrast reflecting microscope. From Phillips.[1]

4. Sensitivity

The optimum contrast is obtained if the direct and diffracted beams are of equal strength. If the phase detail is weak, the diffracted beam is weak. Contrast can then be heightened by increasing the absorption of the phase plate to reduce the strength of the direct beam, but very little light then gets through. In principle, path differences of only 1 Å could be distinguished with a first class ideal optical system. In practice, the sensitivity is much poorer than this and one often gets diffracted halos, as well as stray light.

5. Quantitative Phase Contrast Measurements

For quantitative measurements it is necessary to use an interference method, in which a measuring beam and a reference beam interfere. A controlled phase shift is introduced into the reference beam by means of a nearly parallel-sided wedge which is moved by a screw equipped with a scale and vernier calibrated directly in $m\mu$ optic path length. In Fig. 17a M and R represent the measuring and reference beams. The phase of R is adjusted until it is $\lambda/2$ different from M and the field appears dark (Figure 17b). A phase detail in the specimen gives an additional shift represented by MS in Figure 17b, so that the detail appears light against the dark background. The wedge is now moved until R and MS interfere destructively (Figure 17c). The wedge movement required to extinguish this detail gives the optical path length through the detail. This leads us to a discussion of interference techniques.

D. Interference

1. Criteria for Interference

In discussing light sources, it was pointed out that interference could, in general, he obtained if two beams from a single point source were superimposed, but not if the beams were from two separate sources, since the fluctuations were then usually uncorrelated. In the case of division of an extended source, partial coherence tends to exist and the visibility of interference phenomena decreases as coherence of the two beams decreases. An *interferometer* divides an incident wave from a point source into two or more waves, which after traversing different paths are superimposed, resulting in interference phenomena.

It is at first glance puzzling why interference cannot occur between beams from two separate sources, whereas it can if the two beams come from one source, even, under some conditions, if this is a large single source. A useful practical criterion is that interference will only occur if it cannot be decided which of the two paths the light followed. If conditions are such that the photon could have gone either way, interference will occur, that is, we can regard the light as interfering with itself. Clearly, interference cannot occur between light from two separate sources. With a large source, if some kind of screen is introduced in

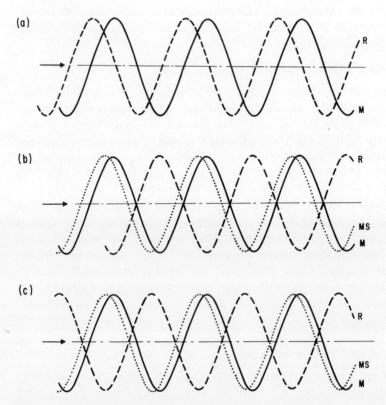

Figure 17 Wave diagram illustrating interference technique used for quantitative phase contrast measurements. From Phillips.[1]

order to be able to tell which half of filament the photon came from, that is, which of the two paths it follows, interference will no longer occur. The only exception is the instance of the laser; here interference can be obtained between the beams from two lasers suitably coupled, for example, by a third laser, so that the phases are coordinated.

Two general methods exist for dividing a single beam. The first, division of the wave front by apertures placed side by side, is useful only with sufficiently small sources. The second is division of the amplitude by one or more partially reflecting surfaces at which part of the light is reflected and part transmitted, so that an extended source can be used and intensity gained.

Before considering interferometers, we shall discuss interference phenomena at a simple wedge.

2. Interference at a Wedge

If n is the refractive index of a wedge that is illuminated normally from above with monochromatic light of wavelength λ_0 in air ($n_0 \sim 1$), then the wavelength of light within the wedge is λ_0/n. Light incident on the wedge is reflected from the back surface and interferes with the incident light. Black bands, known as fringes of equal thickness, are seen on viewing by reflection, corresponding to regions in which the thickness is 0, $\lambda_0/2n$, λ_0/n, $3\lambda_0/2n$, $\cdots$ (Figure 18). Light traverses the wedge twice; hence the path length is

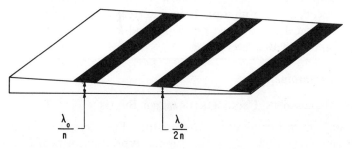

Figure 18 Interference at a wedge, viewed by reflected light. From Phillips.[1]

0, λ_0/n, $2\lambda_0/n$, $3\lambda_0/n$ $\cdots$. In reflection the fringes are black rather than white because, when a wave is reflected from a surface of a medium of higher refractive index, it undergoes a phase change of π (i.e., $\lambda/2$). The visibility of the fringes is good in reflection, since the interfering beams are of comparable amplitude; this would not be the case in transmission.

3. Measurements Possible with Interferometers

All interferometers divide a beam of light into two or more parts, which follow different optical paths and then recombine to form interference fringes. Thus interferometers measure differences in the optical path. Since this is the product of the geometric path and the refractive index, a difference in the geometric path can be measured when the two beams travel in the same medium, or a difference in the refractive index can be measured when geometric paths. are equal but the beams travel in different media. Since the unit in which the path difference is measured is the wavelength of light, this can also be measured. An interferometer disperses a beam of light, as does a prism. The interferometer can measure small differences better than absolute values. For transparent objects, both the refractive index and the thickness may be measured simultaneously from two measurements, using either two different embedding media, or two wavelengths, or two temperatures. In the image there is no halo effect such as appears in phase contrast microscopy.

Interference methods are very important to the metallographer for quantitative studies of the microtopography of reflecting specimens and for the measurement of the thickness of thin films. If an interferometer is adjusted so that one fringe is spread over the entire field, detail present which would normally be barely visible can be thrown into sharp interference contrast. Black and white contrast is obtained if monochromatic illumination is employed, and differences in color if white light is used.

Various types of interferometer will now be discussed, comprising two-beam instruments and the multiple beam interferometer. The former have a resolution typically about $\lambda/10$ in depth and λ laterally, the latter about $\lambda/200$ in depth and 10 λ laterally.

4. Two-Beam Interferometry

a. REFLECTION MICROSCOPES USING UNPOLARIZED BEAMS

(1) *Zeiss Reflection Interference Microscope.* The Zeiss two-beam microscope is an interferometer of the Linnik type. Schematically, it consists of a light source and lens C (Figure 19a), providing parallel illumination that falls on the 45° partially reflecting surface of the beam splitting prism P. Two mirrors M_1 and M_2, equidistant from the prism, each reflect half of the light. The combined beams are viewed in the direction indicated and can interfere with each other. If M_1 and M_2 are accurately perpendicular and one is slightly displaced, the entire field will appear light or dark, depending on the phase difference in the usual way. If one mirror is slightly tilted with reference to the other, a set of parallel fringes spaced $\lambda/2$ is observed, whose spacing can be adjusted.

In practice, one mirror consists of the opaque specimen surface (Figure 19b), which is often silvered to increase the reflectivity. It is important to match the specimen reflectivity with that of the reference mirror surface, in order to obtain beams of comparable amplitude. This is aided by providing several comparison mirrors of different reflectivities, from which one is selected. Identical objectives O_1 and O_2 are inserted, with the comparison mirror and specimen surface at their respective focal planes; the two interfering beams are viewed through an eyepiece. It is not convenient to adjust the interference conditions by moving the specimen or comparison mirror, since this would cause them to move out of focus. Two identical parallel-sided glass plates G_1 and G_2 are therefore inserted. The spacing of the fringes is adjusted by tilting one of these; the fringe direction is set by rotating the plate.

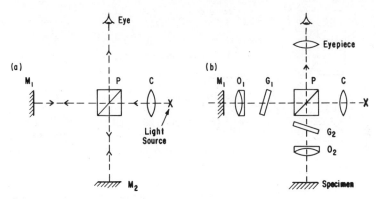

Figure 19 Principle of Zeiss two-beam reflection interference microscope. From Phillips.[1]

For monochromatic thallium light, the fringe spacing $\lambda/2 = 0.27$ μ, and black and white fringes are produced. The pattern is photographed for measurement.

A step higher than $\lambda/2$ on the specimen surface results in a displacement of more than one fringe. Since all the fringes appear identical, only a fractional displacement of a fringe can be distinguished. The fringe pattern is therefore observed a second time in white light of similar wavelength ($\lambda/2 \approx 0.3$ μ) which gives fringes, each of which is dispersed into spectral colors. Since each fringe now has a somewhat different color appearance, the number of integral fringe displacements can be counted and added to the fractional displacement which is more accurately measured with the sharper monochromatic fringes.

The *refractive index* of a transparent specimen, for example, a parallel-sided glass plate of known thickness ($\sim$0.025 in.), can be measured in the simple cell shown in Figure 20, using an oil of known refractive index. Alternatively, the refractive index of the oil can be measured using a standard glass of known refractive index. Interference occurs between the beam reflected from the bottom of the cell which is aluminized to produce a mirror surface. The fringe displacement δ is given by the relation:

$$\delta = \left(\frac{2d}{\lambda}\right)(n_1 - n_2) \qquad (16)$$

where d is the sample thickness and n_1, n_2 are refractive indices of the oil and specimen, respectively. The fringe pattern provides a map of

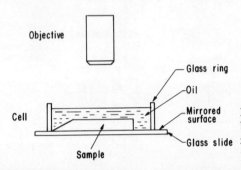

Figure 20   Cell used with Zeiss two-beam reflection microscope for refractive index measurement of transparent sample. From Phillips.[1]

the lateral variation in the refractive index, averaged through the sample thickness. Since a multiple fringe displacement may occur, it is convenient to taper one side of the sample, so that the fringes can be followed from the oil across the sample edge. Alternately, a small heating coil can be wound round the glass ring and a thermocouple immersed in the oil. Using a standard oil of known refractive index and temperature coefficient, the temperature can be adjusted until the refractive index of the oil and specimen match; this means that the fringe pattern undergoes zero displacement at the edge of the sample.

The limit of absolute accuracy of this technique lies in the fourth decimal place. Small differences in refractive index in the fifth decimal place and possibly in the sixth place are detectable within a single field, provided that the refractive index through the thickness can be regarded as constant and the temperature is uniform within 0.1°C. For fused SiO_2 the refractive index n is about 1.46 for $\lambda = 540$ mμ, and the temperature variation of n is about 1×10^{-5} per °C.

b. Reflection Microscope Using Polarized Beams

(1) Nomarski Interference Microscope. This relatively inexpensive device, marketed by Reichert and based on a design of Nomarski,[13,35-37] converts an ordinary microscope into an interference contrast system

and employs a single objective as in the previous device. It is inserted between the normal incident light achromatic or fluorite objective and illuminator in a polarizing reflection microscope (Figure 21). The polarizer and rotatable filter analyzer are set in a "crossed" position and the

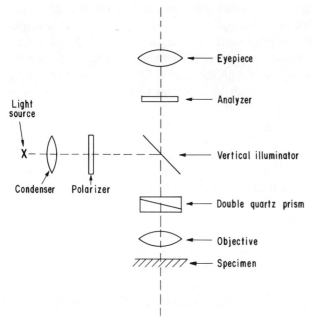

Light source

Condenser   Polarizer

— Eyepiece

— Analyzer

— Vertical illuminator

— Double quartz prism

— Objective

— Specimen

Figure 21   The Nomarski reflection interference microscope. From Phillips.[1]

objective focused. A high intensity light source such as a zirconium arc lamp is desirable.

The principal feature of the device is the beam-splitting, double-quartz prism (Figure 21), consisting, as in the Wollaston prism, of two wedges with mutually perpendicular optic axes, cemented together. The prism splits the beam into two separate inclined beams that are linearly polarized perpendicular to each other. Unlike the Wollaston prism, the wedges are cut in such a manner that the convergence point of the two beams lies outside the prism in the region of the focal plane of the objective lens. Each ray passes through the prism twice; an incident ray which passes on one side of the axis, after reflection from the specimen, passes through an equal distance on the other side of the axis, thus following a total path equal to traveling twice through the center, permitting the two beams to interfere coherently in the image plane. Two images of the object are thus produced, slightly displaced laterally, and differing in phase by half a wavelength, analogous to the situation described in Figure 17. One disadvantage, in comparison with the Zeiss two-beam microscope, is that every point of the surface is compared with its surroundings and not with a well-defined plane. Thus it is impossible to know whether an interference band displacement is a result only of the roughness of the point in question, or also of the roughness of its neighborhood which just coincides with it because of the image-doubling effect. Thus, if the surface is generally irregular, only an approximate estimate can be made of the roughness depth. On the other hand, very accurate height measurements can be made on simple surface configurations, for example, from the fringe displacement across a slip step or an edge of a film.

When using monochromatic light, all the bands show the same light-dark contrast, and their spacing and width may be adjusted using a lever that moves the prism along the microscope axis. If white light is used, the central achromatic bands are accompanied on both sides by colored bands whose intensity decreases with distance from the center. If the displacement of the bands at any point is $1/n$ of a band spacing, this corresponds to a height variation of $\lambda/2n$. Under favorable conditions a displacement of $\frac{1}{20}$ of a spacing may be measured corresponding to a height difference of about 150 Å for sodium illumination ($\lambda/ = 5890$ Å). Corrections are necessary if light departs substantially from normal incidence on the specimen surface, for example, if a microhardness indentation is viewed.

If the coincidence point of the two beams is made to coincide with the focal plane of the objective, then one interference band covers the entire field uniformly, and any level differences become visible as variations in brightness of color, that is, *interference contrast* is obtained. Since the device design is such that the image separation is comparable in magnitude with the microscope resolution, all changes in specimen level become visible as variations in brightness proportional to the tangential slope, appearing brighter or darker according to the sign of the slope. This is often referred to as *differential interference contrast*. With the use of a second lever, the prism may be moved laterally, so that the interference image passes through the range of the Newton colors. Parts that are similar in character than assume a similar color, which is easier to recognize than a similar degree of brightness in the light-dark contrast position.

c. Transmission Microscopes Using Unpolarizing Beams

The principle of transmission interference microscopes using unpolarized beams has already been indicated in the discussion of quantitative phase contrast measurements using a wave diagram. A more succinct representation can be made using a schematic wave surface diagram, following Françon,[7] as in Figure 22. Here the broken lines represent the relative positions of the wave fronts. W_0 represents the plane incident monochromatic wave which is split in the microscope into two beams that are in phase, represented by W_1 and W_2. W_2 is retarded locally as it passes through a dephasing object, giving a deformed wave front W_2'. The means is provided in the interferometer for regulating the path difference Δ between the recombined waves (Figure 22).

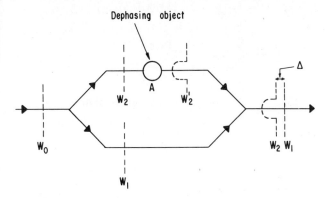

Figure 22   Schematic wave surface diagram for interference microscope with transmitted unpolarized light. From Phillips.[1]

If this is zero, the two waves are in phase, except in the region deformed by the object, and the light intensity is I_0. If the path difference introduced by object A is δ, the resulting image intensity I is

$$I = I_0 \cos^2 \pi \frac{\delta}{\lambda} \qquad (17)$$

where λ is the wavelength. The image intensity I is different from the background intensity I_0; hence the transparent object shows a visible contrast. The object will appear black on a white background if $\delta = \lambda/2$. For the general case, when Δ is not zero, the intensity of the object is

$$I = I_0 \cos^2 \pi \frac{\Delta \pm \delta}{\lambda} \qquad (18)$$

and the background intensity is given by setting $\delta = 0$. If white light is used, and Δ and δ are small, Newton interference colors are apparent, so that the phase detail shows a different color from the field.

The Leitz interference microscope is an example of this kind of instrument and is based on the Mach-Zehnder interferometer.[6,7]

(1) Leitz Interference Microscope. The optical system of the Leitz interference microscope contains a beam-splitting prism and a similar prism that combines the two beams, so that the two images can be observed with the lens *L* and the eyepiece (Figure 23). One beam passes

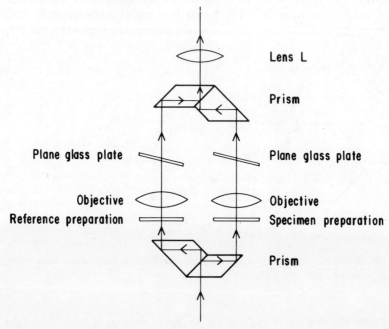

Figure 23 Schematic of Leitz transmission interference microscope. From Phillips.[1]

through the specimen in the focal plane of an objective lens; the other passes through a reference preparation, consisting of identical glass supports but no specimen, which is located in the focal plane of a matched second objective. A tiltable, rotatable, plane glass plate inserted in each path permits the adjustment of the phase difference between the two beams. No ghost images are formed. Since the two optical paths can be made identical, white light may be employed instead of monochromatic light, when it is desired to obtain colored images.

d. Transmission Microscopes Using Polarized Beams

With two polarized beams it is possible to operate in white light and conveniently observe phase variations of transparent isotropic objects. Reference may be made to Françon[7] for a detailed discussion in terms of wave surface diagrams. The Baker[13] and Zeiss interference microscopes will now be discussed to illustrate the principles involved.

(1) *Baker Transmission Interference Microscope.* A birefringent plate in the substage illuminating train splits the plane-polarized beam into two components, polarized in mutually perpendicular directions (Figure 24). A second birefringent plate is mounted on the front of the objective

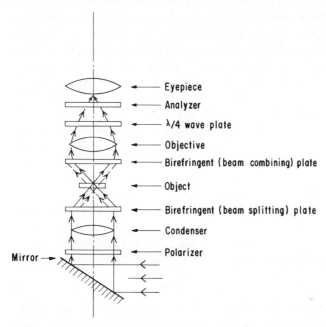

Figure 24   Baker transmission interference microscope system. From Phillips.[1]

and recombines the two beams into the same optical path. Since the refractive index of this plate differs with the polarization direction of the incident beams, the two beams are not focused on the same point. Two arrangements are possible. In one called the "shearing system" the two beams are focused in the same plane but are laterally displaced; thus one is composed of light passing through the object and the other of light passing around the object. The shearing system is said to be more suitable for the examination of isolated features surrounded by featureless matrix. In the other arrangement, illustrated in Figure 24, called the "double focus" system, one beam is focused on the object, the other on a different plane below the object.

In the double-focus system, both beams contribute to each point in the final image. The image formed by the beam focused on the object has its phase determined by the surface level of the object, whereas the image formed by the other defocused beam has a phase that is the average of many points in the object and thus serves as a reference plane, as far as phase is concerned. Alternatively, if the object is small, the reference beam passes around it. A quarter-wave plate above the objective changes the two oppositely polarized beams into left- and right-handed, circularly polarized light, whose resultant is plane-polarized light, and whose plane depends on the phase difference between the two circularly polarized beams. Thus the direction of polarization for each point in the final image depends upon the optical thickness of the corresponding point in the object. When viewed through an analyzer and eyepiece, the brightness with polarized light, or color with white light, is directly related to the optical thickness of each object point.

Quantitative measurement of height differences is possible with monochromatic light, since the analyzer rotation required to extinguish two successive object points is equal to half the phase difference between them. One problem is that the total optical path is included, that is, the slide, mounting medium, specimen, and cover glass. Artifacts and errors occur because of variation in the optical path through the slide, due to imperfections and uneven surfaces. The use of a known optical path standard such as a magnesium fluoride film of less than one wavelength optical path thickness, is desirable and permits the evaluation of errors.

(2) *Zeiss Transmission Interference Microscope.* As in the Baker microscope, a birefringent plate in the substage illuminating system splits the plane-polarized beam into two mutually perpendicular components (Figure 25). However, these are separated laterally, so that only one passes through the object, corresponding to the "shearing" system available in the Baker microscope. The two beams are then recombined into the same optical path by a second birefringent crystal plate and pass through the objective. The distance between the observation and comparison paths in object space depends on the objective chosen and varies between 0.05 and 0.55 mm. These values correspond to the diameter of the observation field free from overlapping.

Difference up to 1λ in optical path of object points may be measured, in a manner similar to that described for the Baker microscope, by inserting a quater-wave plate above the objective and viewing the resultant plane-polarized beam through an analyzer and eyepiece. For smaller phase difference measurements up to $\lambda/10$, $\lambda/20$, or $\lambda/30$, a

Brace-Köhler type of compensator, which consists of a calibrated, rotatable birefringent lamina, may be substituted for the quarter-wave plate used in the Senarmont method. For larger phase difference measurements up to 122λ, the analyzer may be replaced with a calibrated Ehringhaus-type compensator which has combination calc-spar plates.

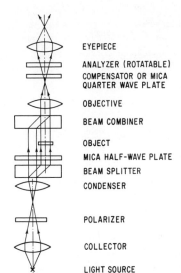

EYEPIECE

ANALYZER (ROTATABLE)
COMPENSATOR OR MICA
QUARTER WAVE PLATE

OBJECTIVE

BEAM COMBINER

OBJECT
MICA HALF-WAVE PLATE
BEAM SPLITTER
CONDENSER

POLARIZER

COLLECTOR

LIGHT SOURCE

Figure 25 Zeiss transmission interference microscope system.

It is necessary, of course, to use monochromatic light for exact measurements of phase differences; this may be obtained by the insertion of an interference filter, such as one to isolate the spectral line having $\lambda = 5461$ A, or by using a wedge interference filter with a film of graded thickness with which a suitable wavelength can be selected. A bright field image will appear if the analyzer or polarizer is removed. For the most absolute monochromasy, a high pressure mercury lamp may be employed with a filter (see page 70 for safety precautions). With an ideal object, phase differences as small as $\lambda/200$ may be measured.

Phase differences in a birefringent object are directional, necessitating the use of a crosshair eyepiece which is set parallel to the direction of the linearly polarized incident light and acts as a reference direction. The object is then rotated in turn into each of its principal vibration directions and interferometric measurements are made.

5. Multiple Beam Interferometry

The technique of multiple beam interferometry has been developed to a high state of perfection by Tolansky,[38,39] to whom reference may be made for a detailed discussion of both the theory and applications. While the technique can be used in transmission, the present discussion will be confined to the reflection technique. A suitable optical system is illustrated in Figure 26 and involves only minor inexpensive modifica-

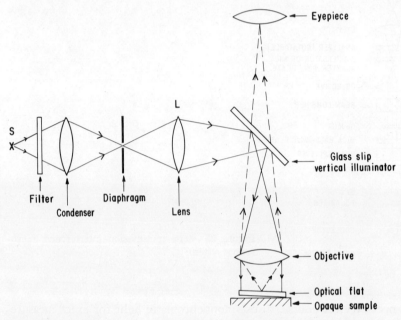

Figure 26 Optical system suitable for producing multiple beam fringes in reflection. From Phillips.[1]

tion of a typical metallurgical microscope system. For greater convenience, commercial multiple beam instruments may be purchased such as the Angstrometer, marketed by the Sloan Instruments Corporation, or the Varian Å-Scope, marketed by Varian. A multiple beam attachment is available from E. Leitz. Monochromatic light from a source S is used; the green line ($\lambda = 5461$ Å) from a filtered mercury vapor lamp is suitable. A condenser images the source on an iris diaphragm. The lens L produces a small image of the diaphragm aperture via the vertical illuminator in the backfocal plane of the objective lens. The latter is focused on the object, so that it is illuminated by parallel light at normal incidence which passes through a glass flat in contact with the sample. The image is viewed through an eyepiece. At higher magnifications, the size and position of the diaphragm aperture are critical.

The principle of the technique is illustrated in Figure 27. The silvered

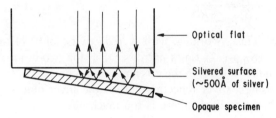

Figure 27   Schematic illustrating principle of multiple reflection. From Phillips.[1]

surface of the optical flat is placed in close contact with the specimen surface; a slip of tissue paper is used to raise one end slightly off the specimen. The evaporated silver layer (~500 Å thick) has a reflectivity of about 0.90, reflecting 90% of the incident light and transmitting the rest, except for a few percent absorbed. Multiple reflection occurs as many as 60 to 80 times between the specimen and the flat, a portion being transmitted each time through the silver and interfering with the original beam reflected from the silver.

The existence of many mutually interfering beams causes the fringe patterns to be much sharper than they would be with an unsilvered flat (Figure 28), in the same way that a diffraction grating gives sharper

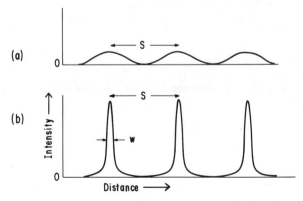

Figure 28   Intensity distribution due to multiple reflection: (a) two beams only, that is, low reflectivity R; (b) many beams, that is, high reflectivity. From Phillips.[1]

diffracted maxima than does a single pair of slits. The halfwidth W, that is, the width at half-maximum intensity, divided by the fringe spacing S (Figure 28), is a measure of the sharpness of a fringe and is given by:[38]

$$\frac{W}{S} = \frac{1 - R}{\pi R^{\frac{1}{2}}} \tag{19}$$

where R is the reflectivity. Thus, working with a reflectivity of 0.94, W/S is about $\frac{1}{50}$, and the fringes are 25 times narrower than for two-beam fringes for which W/S is $\frac{1}{2}$ and $R = 0.24$. In the green region a reflectivity of 0.94 can be readily secured, using a silver reflecting film. If the silver film is too thin, W/S decreases; if it is too thick, there is too much absorption. A good range is 500 to 700 Å.

It will be assumed that light is incident normally on the silvered surface of the flat and that the refractive index of air is equal to one. Light reflected, and that transmitted back through the silver from the specimen, will interfere and will "apparently reinforce" if the path difference $2t$ (Figure 29) equals $m\lambda$,

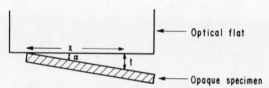

Figure 29 Geometry of multiple reflection. From Phillips.[1]

where m is an integer and λ is the wavelength of light. Actually, t is slightly different from that drawn, but the difference is in practice much smaller than λ and can be neglected here. Now

$$t = x \tan \alpha \tag{20}$$

where α is the wedge angle and x the distance from the apex of the wedge. Reinforcement occurs if

$$x \tan \alpha = \frac{m\lambda}{2} \tag{21}$$

as in the case of interference at a wedge, discussed previously. Thus a set of parallel fringes spaced $\Delta/2 \cot \alpha$ is formed. Since $\alpha = \tan^{-1}(m\lambda/2x)$ if λ is known, the wedge angle α can be calculated from the number of fringes per unit length in the appropriate direction.

Alternatively, $\alpha = \tan^{-1} (\lambda/2S)$, where S is the average spacing of the fringes which is independent of the magnification used. The smaller the wedge angle α, the larger the spacing, so that it is important to get a very small α to improve the sensitivity to small height differences on the sample. More generally,

$$t = \frac{m\lambda}{2n \cos \theta} \tag{22}$$

where n is the refractive index of medium and θ is the angle of incidence on the silvered surface.

If the beams reinforce, one would at first sight expect to see bright fringes, since they are viewed by reflection, but we have neglected the fact that there is a phase change of π (i.e., $\lambda/2$), when light is reflected from the specimen surface, so that dark fringes are seen. Each fringe is the locus of points of equal gap thickness, so that, if the glass surface is flat, they depict contours of equal level on the specimen surface. Each fringe represents a gap change of $\lambda/2$ which for the green mercury line is 2730 Å. The sense of the level change can be decided by moving the specimen slightly toward the optical flat, causing the fringes to move toward the valleys.

Since, under ideal conditions, the width of a black fringe may be about 1/100th of the fringe spacing, and a displacement of half a fringe width may be detected locally, a height difference of only 2730 ÷ 200, that is, approximately 10 Å, might be measured. Tolansky claims,[39] in fact, to have measured 2.3 Å in one instance. If the specimen surface is very uneven, it is impossible to make the surface closely approach the optical flat in all regions; the fringes become diffuse and considerable sensitivity is lost.

In interpreting the fringe patterns, it should be remembered that they represent the intercept with the specimen surface of a series of planes parallel to the optical flat spaced $\lambda/2$. Hence the fringe pattern changes when the inclination of the specimen surface to the optical flat is changed. Since a change in the specimen level parallel to the fringe patterns produces no visible displacement, two separate photographs of the patterns, produced by tipping the specimen successively about two different directions roughly at right angles, are needed in order fully to map the surface topography.

Newton's rings, which are a set of concentric circular fringes produced when a spherical surface is matched against a flat surface, are a special case of multiple beam interference, in that the fringe pattern is unaffected by tilting the optical flat.

a. Optimizing Multiple Beam Fringes

Silver is particularly favorable for producing fringes; it is frequently desirable not only to use a silvered flat but also to coat the specimen surface with about 1000 Å of evaporated silver to improve the reflectivity. Tolansky[38,39] has demonstrated that this can be done without appreciably altering the topography which it is desired to measure. A further advantage can be gained, if the optical flat is coated not with silver but with up to nine alternate layers of zinc sulfide and cryolite, each $\lambda/4$ thick, giving a reflectivity up to 0.97 with much less absorption than with silver.

The quality of the optical flat or glass slide used is important. It is possible to select a 1-in. disk of glass to be flat within $\lambda/50$ within a 1-mm field, or $\lambda/500$ within a 0.1-mm field. Fire-polished glass is remarkably smooth locally. Selected muscovite mica sheets can be smooth even to molecular dimensions over large areas.

b. Thickness Measurement Using Multiple Beam Fringes

The thickness of evaporated films can be measured down to about 50 Å with high precision (± 3 Å), provided that they are continuous (not islands). The technique used[38] is indicated in Figure 30.

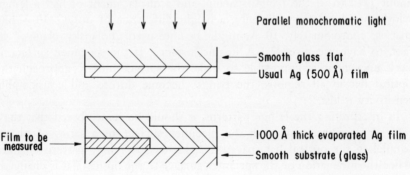

Figure 30 Arrangement for measuring film thickness, using multiple beam fringes. From Phillips.[1]

The film to be measured covers part of the substrate and ends with a sharp edge. The entire area is coated with an opaque reflecting film 1000 Å thick of evaporated silver which accurately contours the step height t. This is now measured in the usual way, setting the fringes normal to the step edge and viewing the fringes in reflection from above.

The advantage of this arrangement is that the phase shift on both sides of the step is the same, and is that of bulk silver, since the silver thickness is large. If the silver coating is not applied and the specimen film to be measured is thin, the magnitude of the phase shift would depend in an unknown way on both the thickness and material of the film, and would be different on opposite sides of the step.

6. Light-Cut and Light-Profile Methods

Although not really methods of enhancing contrast, the light-cut and light-profile methods of determining profile are appropriately considered to be alternatives to multiple beam interferometry. The light-cut and light-profile methods of Tolansky[40,41] represent a considerable advance on the original light-cut technique of Schmaltz.[42] In the light-cut method, a slit, placed close to the field iris in the illuminating system, is imaged as a light line on the surface under examination which is obliquely illuminated by a pencil of light and appears dark. An instrument of this kind is marketed by Carl Zeiss. In the light-profile and multiple profile methods, the slit is replaced by an opaque line or set of parallel lines on a bright field, giving a dark line or set of dark lines profiling the surface. The greatest advantage over the light-cut method is that the surface and profile may be viewed simultaneously. The technique has been used at a magnification of 2000, giving a vertical and lateral resolution better than 0.25 μ. Thus the vertical resolution is poorer, but the lateral resolution better, than the two-beam and multiple beam interferometric techniques which are difficult to apply at high magnifications above about 300. Monochromatic light is employed, and the profile magnification M' is given by

$$M' = \left(\frac{2M}{n}\right) \tan i \qquad (23)$$

where M is the microscope magnification, n is the refractive index of the immersion medium, and i is the effective angle of incidence, determined by the lens aperture and the position of the metal tongue sector reflector. In practice, it is convenient to determine M' using a specimen with a step of known height, rather than calculating it from Equation 23.

The light section microscope may be employed to measure surface profiles, the thickness of partial films, and the embossment depth. The roughness of internal surfaces may be measured on a suitable replica. The light-cut method provides a rapid method of measuring the thickness t of anodized layers. If S is the observed distance between the two light bands, obtained by 45° reflection from the coating and subsurface, respec-

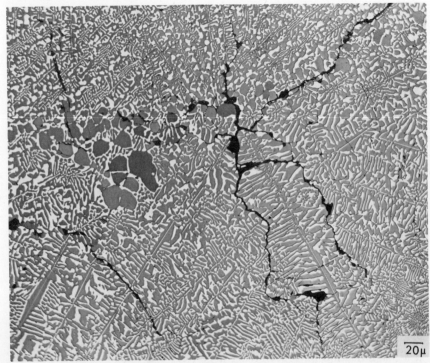

Figure 31 Bright field optical micrograph of cast columbium-oxygen sample, etched in hydrofluoric, nitric, and lactic acids etchant (2:30:50). Taken using vertical illumination and reflected light. Micrograph by R. R. Russell. (Courtesy of W. C. Hagel and General Electric Research and Development Center.)

tively, then, after Illig:[43]

$$t = S(2n^2 - 1)^{\frac{1}{2}} \qquad (24)$$

where n is the refractive index of the anodized layer (usually between 1.59 and 1.60 for aluminum). Similarly, the thickness of other transparent layers such as glaze coatings may be measured simultaneously with the distortion of the substrate.[44] The refractive index of the glaze may be measured using a critical angle refractometer (Rayner Co., London, England; available from Ward's Natural Science Establishment, Rochester, N.Y.)

E. Filters

Colored illumination is a useful method of contrast enhancement, particularly in the study of colored objects by reflected or transmitted light,

100μ

Figure 32 Dark field optical micrograph of as-polished sintered thoria, taken by reflected light. Micrograph by R. R. Russell. (Courtesy of P. Jorgenson and General Electric Research and Development Center.)

whether by bright field or dark field illumination. By the appropriate choice of illumination color, the contrast between differently colored constituents may be enhanced. An object will tend to appear dark and opaque if illuminated with light of a complementary color, and pale and transparent if illuminated by light of similar color to its own. Light of the color desired is conveniently obtained by placing a color filter in front of the light source. The common types of filter consist of colored glass or of colored gelatin mounted between glass plates. Alternatively, interference filters, consisting of single or multiple layer evaporated films on glass, may be obtained with minimum transmission near a desired wavelength.

If the features of interest are of similar color to their surroundings, tremendous contrast intensification in black and white may be obtained using light of appropriate wavelength. The most suitable method is to use a movable wedge interference filter, having a film of varying thickness

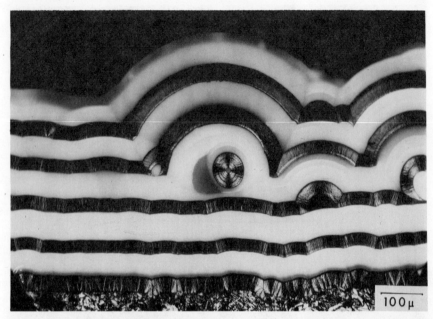

Figure 33 Optical micrograph of as-polished sample, consisting of alternating layers of graphite and boron nitride pyrolytically deposited, taken in polarized light using vertical illumination. Micrograph by Mrs. T. V. Brassard. (Courtesy of J. Diefendorf and General Electric Research and Development Center.)

which is apertured to select the region of the film desired to give mono-chromatic light of the wavelength required. This may be mounted below the condenser diaphragm.

F. Examples of Results

Examination under vertical bright field illumination after standard metallographic preparation and chemical etching is capable of revealing a wealth of structural detail in most metallurgical samples, as seen in Figure 31 of a cast columbium-62 at. % oxygen sample. The intergranular porosity seen in black contrast is best examined before etching which may cause some enlargement of the cavities. A dendritic two-phase struc-ture of CbO and CbO_2 is revealed by the etchant.

With normal, vertical bright field illumination, a sintered thoria sample would show little else but grain boundaries and inclusions or porosity intersecting the plane of sectioning, due to the high reflectivity. By the

Figure 34 Bright field optical micrograph of zinc crystal as-cleaved in liquid nitrogen, taken using reflected light and vertical illumination. Micrograph by A. S. Holik. (Courtesy of General Electric Research and Development Center.)

use of dark field illumination (Figure 32) a lot of additional detail lying below the surface of the translucent sample is clearly revealed.

With polarized light the layer detail in the optically anisotropic, hexagonal graphite phase (dark contrast) is seen clearly in an as-polished section in Figure 33, whereas the alternating layers of optically isotropic, cubic boron nitride (white contrast) show no internal detail. The latter

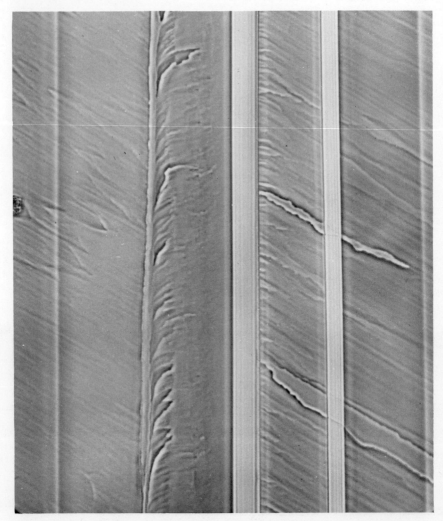

Figure 35 Same field as Figure 34, but optical micrograph taken using phase contrast. Micrograph by A. S. Holik. (Courtesy of General Electric Research and Development Center.)

could be brought out by chemical or other appropriate etching techniques, so that a combination of etching with polarized light examination would be beneficial.

The potentialities of several different optical techniques are illustrated in Figures 34 to 37, taken by A. S. Holik, which show the same area of a cleaved zinc crystal displaying a series of parallel twins, and "river

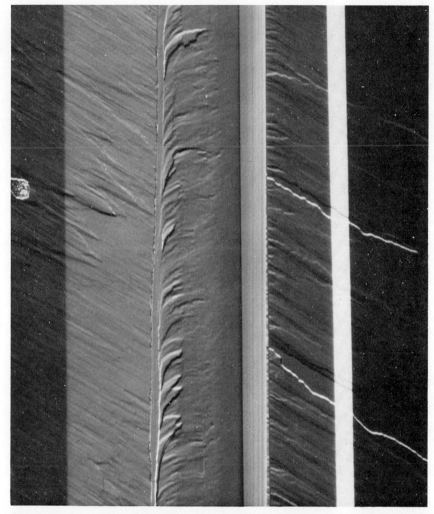

Figure 36 Same field as Figure 34, but optical micrograph taken using Nomarski interference contrast. Micrograph by A. S. Holik. (Courtesy of General Electric Research and Development Center.)

markings" indicating the brittle cleavage mode of fracture. Phase contrast (Figure 35), employing the phase contrast attachment on a Bausch and Lomb metallograph, gave improved contrast at both the twins and at the river markings over that obtained with bright field illumination (Figure 34), even though the latter was slightly obliqued to optimize the contrast. Interference contrast (Figure 36), obtained using a Reichert

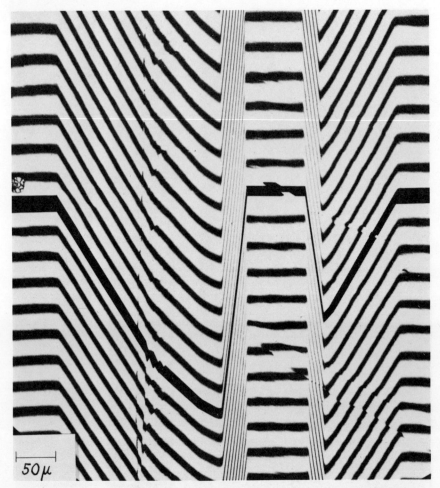

Figure 37 Same field as Figure 34, but two-beam interferogram taken using monochromatic thallium illumination ($\lambda \sim 5400$ Å). Micrograph by A. S. Holik. (Courtesy of General Electric Research and Development Center.)

Nomarski attachment adapted to the same metallograph, gave greatly enhanced contrast at the twins and river markings. Even more information is revealed in color micrographs taken with this form of illumination, where regions showing the same color have the same slope. The nature of the surface level differences giving rise to the contrast effects in Figures 34 to 36 is shown in a two-beam interferogram, taken with a Zeiss reflection interference microscope (Figure 37). The twinning resulted in surface tilts that tend to give little contrast in bright field illumination, unless very steep, as at the two narrowest twins. Only the largest steps at river markings, producing jogs in the interference fringes in Figure 37, are easily visible in bright field (Figure 34).

Step heights down to about 200 Å, corresponding to about one-tenth of a fringe displacement (fringe spacing $= \lambda/2$) are measurable (i.e., detectable) by the two-beam interference technique. A multitude of finer steps, not detectable in Figure 37, are revealed by phase contrast (Figure 35), and with greater sensitivity and improved contrast by the interference contrast technique (Figure 36). It should be pointed out that finer steps could be measured using the multiple beam interference technique, although the linear magnification and resolution would place a limit on the information obtainable because of the close spacing of the steps and recourse would be necessary to replication electron microscopy. It should particularly be noted that Nomarski interference contrast illumination gives a clear visual indication of the presence of surface tilts which is absent in both bright field and phase contrast illumination, where the observer is likely to infer erroneously that the surface is flat.

IV REFERENCES

1. V. A. Phillips, "Optical Techniques for Metallographic Examination," in *Techniques in Metals Research*, Vol. 1, R. F. Bunsah, Ed., John Wiley & Sons, New York, 1967, pp. 25–67.
2. W. T. Welford, *Geometrical Optics*, North-Holland Publishing Co., Amsterdam, 1962.
3. F. A. Jenkins and H. E. White, *Fundamentals of Optics*, 3rd ed. McGraw-Hill Book Co., New York, 1957.
4. F. W. Sears, *Optics*, 3rd ed., Addison-Wesley Publishing Co., Reading, Mass., 1958.
5. J. K. Robertson, *Introduction to Optics Geometrical and Physical*, 4th ed., Van Nostrand Reinhold Publishing Co., New York, 1959.
6. M. Born and E. Wolf, *Principles of Optics*, Pergamon Press, New York, 1959.
7. M. Françon, *Modern Applications of Physical Optics*, Interscience Publishers, New York, 1963.
8. E. M. Chamot and C. W. Mason, *Handbook of Chemical Microscopy*, Vol. 1, John Wiley & Sons, New York, 1966.
9. E. W. Taylor, *J. Roy. Microscop. Soc.*, **65**, 1–7 (1945).
10. J. R. Benford, *Amer. Soc. Metals*, **36**, 452–470 (1946).
11. C. R. Burch, *Proc. Phys. Soc.*, **59**, 41 (1947).
12. C. R. Burch, *Proc. Phys. Soc.*, **59**, 47 (1947).
13. A. B. Winterbottom and D. McLean, *The Physical Examination of Metals*, 2nd ed., B. Chalmers and A. G. Quarrell, Eds., Edward Arnold & Son, Leeds, England, 1960, pp. 1–80.

14. H. T. Heal, *Conference on the Examination of Metals by Optical Methods,* British Iron and Steel Institute, London, 1949.
15. A. Köhler, Z. F. Wiss, *Mikrosk.,* **10,** 433–440 (1893); **16,** 1 (1899).
16. D. Birchon, *Optical Microscope Technique,* George Newnes, London, 1961.
17. Technical Information Bulletin, E. Leitz Inc., New York, **6,** (4), 4 (1966).
18. A. F. Hallimond, *Manual of the Polarizing Microscope,* Cooke, Troughton and Simms, York, England, 1956.
19. G. K. T. Conn and F. J. Bradshaw, Eds., *Polarized Light in Metallography,* Academic Press, New York, 1952.
20. B. W. Mott, *Polarized Light in Metallography,* Butterworth & Co., London, 1952.
21. B. W. Mott and H. R. Haines, *Research,* **4,** pp. 24–33, 63–73 (1951).
22. B. W. Mott and H. R. Haines, *J. Inst. Metals,* **80,** pp. 629–636, (1952).
23. P. Dunsmuir, *Brit. J. Appl. Phys.,* 3, p. 264–267 (1952).
24. R. E. Reed-Hill, C. R. Smeal and Linda Lee, *Trans. Met. Soc. AIME,* **230,** 1019 (1964).
25. P. Lacombe and L. Beaujard, *Chim. Ind.,* **53,** 222 (1945).
26. A. Hone and E. C. Pearson, *Metal Progr.,* **53,** 363 (1948).
27. S. L. Couling and G. W. Pearsall, *Trans. AIME,* **209,** 939 (1957).
28. R. E. Reed-Hill and D. H. Baldwin, *Trans. Met. Soc. AIME,* **233,** 842 (1965).
29. A. H. Bennett, H. Jupnik, H. Osterberg, and O. W. Richards, *Phase Microscope—Principles and Applications,* John Wiley & Sons, New York, 1951.
30. F. W. Cuckow, *J. Iron Steel Inst. (London),* **161,** 1 (1949).
31. D. McLean, *Metal Treatment,* **18,** 51 (1951).
32. D. McLean, *J. Inst. Metals,* **81,** 133 (1952).
33. E. C. W. Perryman, *Metal Industry,* **79,** 51 (1951).
34. Technical Information Bulletin, E. Leitz, New York, **3,** (2), 5 (1963).
35. G. Nomarksi and A. R. Weill, *Bull. Soc. Fr. Mineral. Cristallogr.,* **77,** 840 (1954).
36. G. Nomarski and A. R. Weill, *Rev. Metallurgie,* **52,** 121 (1955).
37. F. Herzog, *Industrie-Anzeiger,* No. 60, 27 July (1962).
38. S. Tolansky, *Multiple-Beam Interferometry,* Clarendon Press, Oxford, 1948.
39. S. Tolansky, *Surface Microtopography,* Interscience Publishers, New York, 1960.
40. S. Tolansky, *Nature,* **169,** 445–446 (1952).
41. S. Tolansky, in *Properties of Metallic Surfaces,* a symposium held at the Royal Institution, London, November 1962, published by Institute of Metals, London, 1953, pp. 12–15.
42. G. Schmaltz, *Technische Oberflächenkunde,* Springer, Berlin, 1936 ed., Edwards Brothers, Ann Arbor, Mich., 1944, pp. 73–81.
43. W. Illig, *Metalloberflache,* **13,** (2), 32 (1959).
44. R. Brown, *Ceramic Bull.,* **45** (2), 206 (1966).

4

LOW ENERGY ELECTRON DIFFRACTION
AND AUGER ELECTRON ANALYSIS

I. Low Energy Electron Diffraction 112
  A. Introduction 112
  B. Equipment 113
  C. Sample Preparation 117
  D. Theory 118
    1. Diffraction Pattern Analysis 118
    2. Intensity Analysis 126
  E. Applications 128
    1. Arrangement of Atoms on Clean Surfaces 128
    2. Faceting 128
    3. Structures Formed by Vapor Deposition 128
    4. Oxidation and Absorption of Gases on Metals 129
    5. Catalysis 130
    6. Surface Diffusion 130
    7. Inner Potential 130
    8. Root Mean Square Displacements of Surface Atoms . . . 132
    9. Miscellaneous 132
II. Auger Electron Analysis 132
  A. Introduction 132
  B. Principles 133
  C. Equipment 133
  D. Applications 137
III. References 143

I. LOW ENERGY ELECTRON DIFFRACTION

A. Introduction

The recent availability of commercial equipment for low energy electron diffraction (LEED) or combination LEED/Auger electron analysis will undoubtedly give considerable impetus to studies in this field. The subject of low energy electron diffraction has been reviewed by May,[1] Lander,[2] Germer,[3] MacRae,[4] Morabito and Somorjai,[5] and Bauer.[6] Extensive bibliographies are given in Refs. 1, 2, 5, and 6.

The wavelike character of electrons was proposed in 1924 by deBroglie. Thomson demonstrated the correctness of this idea by showing that diffraction effects occurred when high energy electrons passed through metal foils. Davisson and Germer[7-9] were the first to show that slow electrons are mainly reflected by the first few atomic layers. Farnsworth[10] was one of the first to explore the applicability of the technique to surface studies of metal crystals. Farnsworth and co-workers have continued these studies up to recent years (see May[1] for additional references), still employing the original method of Davisson and Germer[7-9] in which each spatial diffraction beam is detected as an electric current by means of a Faraday cage.

The field of low energy electron diffraction developed very slowly in comparison with high energy electron diffraction. This was partly caused by the experimental difficulties of the Davisson and Germer Faraday cage technique and partly by serious difficulties encountered in the interpretation of the results. The field received new impetus from the development by Ehrenberg[11] in 1934 of the "postacceleration" technique, in which many of the diffracted beams were simultaneously displayed on a fluorescent screen.

The Ehrenberg technique was improved by Germer and co-workers[12-13] and by Lander, Morrison, and Underwald[14] (see Ref. 1 for additional references). The other development of importance was the incorporation of high vacuum techniques. In this type of equipment the fluorescent screen was mounted in the back reflection position, concentric with the electron gun. This led to the difficulty that, at normal incidence on the crystal, the zero order beam was reflected into the gun and lost. Sproull[15] overcame this by bending the incident beam through 90°, using a magnetic field. A movable Faraday cage was used to detect the diffracted beams after they were again bent by the field. Tucker[16] incorporated this idea in a postacceleration imaging-type system.

The recent renewal of interest in low energy electron diffraction can be ascribed to the availability of high vacuum techniques and pure single crystals, and to the development of reliable means for reproducibly preparing and controlling the surface conditions of many materials.

B. Equipment

The principles of the commonly used type of equipment are illustrated in Figure 1. Electrons from a heated filament cathode are collimated

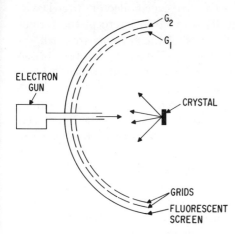

Figure 1 Schematic illustrating the commonest type of low energy electron diffraction arrangement.

and focused onto the crystal sample, usually with electrostatic lenses. The beam size at the crystal is typically about 0.5 mm or somewhat less in diameter, with a beam current of the order of 1 μA. Typical electron energies of the monochromatic beam range from 10 to 500 eV, with corresponding wavelengths λ (given by $\lambda = \sqrt{150/V}$, where V is the beam voltage in volts) of about 0.5 to 5 Å. Electrons backscattered from the sample are accelerated to the fluorescent screen which is maintained at a positive potential relative to the sample of about 5 kV. The pattern produced on the opaque hemispherical screen is observed or photographed through a window behind the crystal, since the whole is located in a chamber evacuated to a pressure of 10^{-9} to 10^{-10} torr in order to avoid the presence of adsorption on the specimen surface. Stainless steel chambers and ion pumps are commonly employed.

The two (or more) grids of 85% transparency, located between the crystal and the screen in Figure 1, have an important function. The grid G_1 nearest the sample is maintained at the same (ground) potential as the sample providing a field-free region, so that the paths of the incident and diffracted beams are not distorted. The second grid G_2 is biased slightly negative with respect to the cathode, so as to prevent passage of the unwanted inelastically scattered electrons, that is, electrons that have lost energy by interaction with the sample. Only the elastically scattered electrons have sufficient energy to penetrate this second grid and are then accelerated to the screen, producing a diffraction pattern. Special holders permit heating or cooling the crystal while it is under observation. Using a spot photometer, one can read the intensities of the diffraction spots from the photographic negative. Absolute current measurements necessitate the use of a Faraday cage collector, after Davisson and Germer,[7] which is mechanically scanned to intercept the various diffracted beams. This is time-consuming, since the currents are small.

At beam energies < 50 eV, the grid G_1 (Figure 1) is often biased positively with respect to the crystal in order, first, to improve the beam focus and, second, to deflect the diffracted electrons, some of which would otherwise miss the edges of the screen.[4] The resulting pattern is smaller but distorted—this does not unduly complicate the interpretation. The crystal must be tilted slightly if it is desired to obtain the zero order reflection on the screen.

Figure 2 shows an example of a modern low energy electron diffraction apparatus after MacRae.[4] This equipment is all-metal, high vacuum, and bakable. The diffraction pattern can be observed on the fluorescent screen which is visible through the large front window. The eight flanged side-ports, which are blanked off when not in use, permit the insertion of devices such as evaporators, crystal holders, an ion gauge, mass spectrometer, an ion bombardment gun, work function measuring equipment, and windows for optical measurements. The tube can be evacuated with either an ion pump and sublimation pump or a mercury diffusion pump. Pressures of 2×10^{-10} torr can be obtained after baking for several hours at $250°C$.

With certain types of poorly conducting samples such as thin epitaxial deposits on cleaved rock salt and mica, satisfactory electron diffraction patterns may not be obtained because of the accumulation of charge on the crystal which changes the incident energy and the focusing conditions. Johnson and MacRae[4] employed relays to alternate the beam energy at 60 cps from a low value suitable for diffraction to a high value (typically 400 eV) which caused sufficient secondary emission to discharge the sample. The grid nearest the fluorescent screen was biased during the discharge cycle to block off the electron signal. Provided that the phosphor persistence time was $> \frac{1}{60}$ sec, a continuous low energy diffraction pattern display could be obtained from an insulating sample for energies as low as 15 eV.

An alternative type of equipment built by Tucker[16] is shown in Figure

Figure 2 An all-metal, high vacuum, low energy electron diffraction apparatus. From MacRae.[4]

3. The fluorescent screen, consisting of a flat glass plate coated with Willemite and with conducting tin oxide, formed one end of a cylindrical glass chamber containing the crystal holder, with a commercial (glass) electron gun mounted on the opposite end. The chamber could be baked out, using a demountable furnace, so that 10^{-10}-torr pressure could be achieved with a standard vacuum system. Provision was made for introducing gases and for heating the sample crystal by electron bombardment or radiant heating, when desired. The electron beam emerging from the magnetically shielded gun was bent to a 7.5-cm radius onto the crystal by a magnetic field from a set of Helmholtz coils of approximately 1 meter diameter. Two other similar sets of coils mounted orthogonally

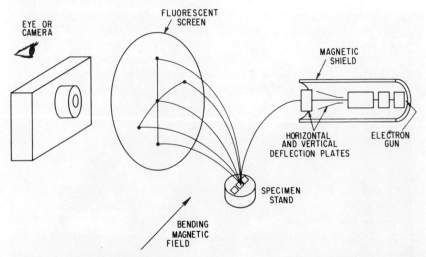

Figure 3   Schematic view of LEED apparatus, using postacceleration and magnetic deflection. From Tucker.[16]

to each other and to the first set were used to neutralize the earth's field.

The diffracted beams from the specimen crystal were deflected onto the fluorescent screen by the bending field, passing through a single copper screen (85% transmission) in front of the fluorescent screen maintained at 3 kV. This copper screen was held at the same potential as the crystal, the final gun electrodes, and the interior of the glass chamber. The diffraction pattern was visible on the rear of the fluorescent screen in a darkened room and could be photographed, although not accessible for measurements with a Faraday cage. Since the deflection depended on the orientation of a diffracted beam with respect to the magnetic field, the pattern was distorted, but in a calculable manner. A transparent polar mask was constructed to enable the scattering angles to be determined with 1 to 2% accuracy from a photograph.

The Tucker equipment has these advantages: (*a*) inelastically scattered electrons are deflected clear of the screen by the bending field; (*b*) the specimen is clear of the gun axis, removing a possible source of contamination; (*c*) zero order and high order reflections are simultaneously observable with the beam at normal incidence; (*d*) rocking curves and normal incidence are obtained without mechanical motions; and (*e*) several crystals may be studied simultaneously under virtually identical conditions. The principal disadvantage is the distortion of the pattern.

C. Sample Preparation

Proper surface preparation of the single crystal sample is crucial to successful examination by low energy electron diffraction, and proper techniques must be established at the start of any investigation. The diffraction pattern quality will depend on the size and number of domains with well-defined periodicity on the surface and on the cleanliness of the crystal. Cleavage inside the high vacuum LEED equipment is a reliable method of obtaining a clean surface on the few materials to which it can be applied such as graphite, mica, alkali halides, silicon, germanium, and III to V compound semiconductors. This technique is only applicable for preparing surfaces of specific orientations. More generally, the crystal must be sliced by some appropriate technique; deformed material removed from the surface; and the surface cleaned before examination, in order to remove absorbed material. It is normally desirable to have the crystal face oriented within 1° of the chosen (ideal) orientation to reduce the number of surface steps which tend to give a high background in the diffraction pattern. Ideal patterns will not be obtained unless the starting crystal is also highly perfect.

Polishing and or etching of the cut surface of a crystal are often used to remove the deformed layer. However, surface contamination by, for example, the atmosphere must still be removed after introduction into the high vacuum. The absorption of gases on many materials at pressures in the low 10^{-6} torr range will result in a monolayer forming in 1 sec. Heating to high temperature in vacuo has been successfully employed for many metals but may lead to the surface segregation of impurities initially present in the crystal. These will not be removed unless they are volatile. Ion bombardment, using argon or xenon ions of 100 to 400 eV at pressures of about 10^{-5} torr, in order to sputter away the surface atoms, is a useful cleaning technique for many materials.[17] The ion damage[18-19] produced can often be removed by a subsequent low temperature anneal. Alternate heating of the crystal in oxygen and hydrogen to oxidize and then reduce impurities to volatile form has been successful in some cases.

The cooling rate after annealing is a further variable, since the surface structure observed may be a function of temperature and a high temperature arrangement may be "frozen in" by rapid cooling.

The sensitivity of the technique even to fractions of a monolayer means that special precautions are necessary to avoid unwanted additional diffraction effects. Since Auger electron analysis can detect and identify very small concentrations of impurities on a surface, the use of combined LEED/Auger electron equipment is a recent development of much interest. The nature and amount of residual gases in the high vacuum may be monitored by a mass spectrometer. Carbon monoxide, water vapor, and hydrogen are usually dominant at 10^{-9} torr. The tungsten filament in the gun and degassing of the vacuum ion pump are two sources of carbon monoxide. Repeated baking of the vacuum chamber is usually necessary to reduce the amount of water vapor present, and a helium-cooled cold finger may be used to remove gaseous contaminants. In considering the results obtained, it is necessary to consider the reactivity of the sample with traces of contaminants remaining, and their solubility in the sample.

A "clean" sample is considered to be one that does not show extra diffraction effects caused by surface contaminants. This usually means that there are $< 10^{12}$ impurity atoms per square centimeter of surface. This may be difficult to establish in a particular case without experimentation to determine the reproducibility, the effect of specific contaminants, the effects of temperature, beam voltage, and so forth. Clearly, the addition of means for mass spectroscopy or secondary electron (Auger) spectroscopy is advantageous.

D. Theory

1. Diffraction Pattern Analysis

The fraction of incident electrons scattered elastically increases with decreasing electron energy, i.e., with increasing electron wavelength λ (angstroms); $\lambda = (150/V)^{1/2}$ where V volts is the beam voltage. It also depends on the scattering material. At 20 eV, 20 to 50% of the electrons may be scattered elastically; whereas, at 100 to 500 eV, this fraction will be only 1 to 5% of the total. The wavelength is larger than normal interatomic distances $< \sim20$ eV, but diffraction effects are observed, for example, on silicon, at much lower voltages, partly because periodicities larger than those of the substrate are present at the surface.

At low beam energies (< 50 eV) most of the electrons are back-scattered by the surface plane. However, some of them penetrate further, so that more than one layer participates in the diffraction process. About three layers contribute when the beam energy is 100 eV. This is still insufficient to define a regular substrate periodicity in a direction normal to the surface; therefore, true three-dimensional diffraction effects are not obtained. The second and third layers can thus participate in a two-dimensional fashion, so that the overall diffraction pattern may be quite complex.

Considering the surface layer of atoms, the diffraction patterns are potentially as sharp as those from a perfect two-dimensional grating. The spots follow precisely a form of Bragg relation valid for rows of atoms, rather than planes:[2]

$$n\lambda = d_{hk} \sin \theta_{hk} \tag{1}$$

which applies for normal incidence. Here θ is the angle between the normal to the surface and the diffracted beam hk; d_{hk} the interatomic row distance, and n the diffraction order. Also, $\lambda = (150/V)^{1/2}$, or more accurately $\lambda = [150/(V_i + V_0)]$, where V_i is the incident potential and V_0 is the inner potential (see pages 130–132).

The derivation of this type of relation is indicated for the simple case of a single row of atoms and normal incidence in Figure 4. Each atom

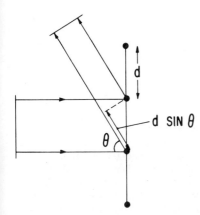

Figure 4   Diffraction from a single ordered row of atoms, which are d apart, under normal incidence.

is the source of scattered waves, spreading spherically, which reinforce in specific directions to produce zero-, first-, second-, and so forth, order diffracted beams. The condition for reinforcement is that the path difference, for rays scattered from adjacent atoms in the row, be an integral number of wavelengths so that the waves are in phase.

The diffraction patterns to a first approximation represent images of the reciprocal lattice, when a hemispherical screen is employed. We first consider a single row of equally spaced atoms and rewrite Equation 1 in the form:

$$\sin \theta_n = \frac{n/d}{1/\lambda} \tag{2}$$

This equation will be satisfied by all the generators of a cone, which has the semivertical angle $90 - \theta$, where θ is defined as in Figure 4 and is concentric with the line of atoms. The cones intersect the Ewald sphere of reflection, which has a radius proportional to $1/\lambda$ (actually c/λ, where c is an arbitrary constant), shown in Figure 5. The intersec-

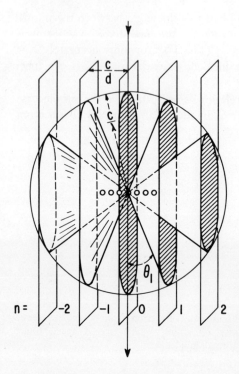

Figure 5 Schematic illustrating diffraction from a single-ordered row of atoms. The zero-, first-, and second order cones of diffracted radiation intersect the Ewald sphere of reflection. The reciprocal lattice elements are planes defined by the intersection of the cones with the sphere. From Lander.[2] (Copyright Pergamon Press, New York, 1965).

tions define a set of planes, called the reciprocal elements of the row, of spacing c/d. The row of atoms is normal to the planes.

The expected diffraction pattern for a single row of atoms or for multiple parallel rows with no regular periodicity normal to the row direction consists of streaks corresponding to the intersection of the cones with the reciprocal elements of the row (Figure 5). Lander[2] gives a number of examples of observations of this kind.

The Ewald construction can be extended to the purely two-dimensional diffraction case by adding, to the rows discussed above, other rows of ordered atoms in a direction appropriate to the second dimension of the two-dimensional array. A second set of cones is obtained which intersect the first set along lines corresponding to reinforcement in the diffracted beams. The reciprocal space of a two-dimensional grating is a set of parallel lines corresponding to the intersection of the planes in Figure 5 with the second set. The reciprocal lattice is illustrated in Figure 6 after MacRae.[4] The radii of the Ewald spheres shown are $1/\lambda_1$ and $1/\lambda_2$, where $\lambda_1 < \lambda_2$. Spot positions correspond to the intersection points of the reciprocal lattice rods with the Ewald sphere, so that it will be seen that they move nearer to the center of the screen with increasing energy, that is, decreasing wavelength.

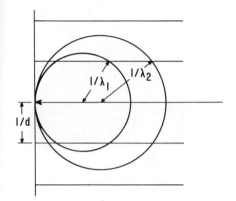

Figure 6 Reciprocal lattice of purely two-dimensional grating. The rods spaced $1/d$ intersect Ewald spheres, corresponding to two wavelengths λ_1, and λ_2. From MacRae.[4]

At high electron energies corresponding to thousands of volts, electrons can penetrate to a depth of many atomic planes, and we have three-dimensional diffraction conditions for which Bragg's law applies:

$$n\lambda = 2d \sin \phi \qquad (3)$$

The reciprocal lattice now consists of points, so that the diffraction conditions are very stringent. A diffracted beam only arises when the wavelength is such that a reciprocal lattice point intersects the Ewald sphere.

Most low energy diffraction cases are intermediate between two- and three-dimensional diffraction, so that some electrons penetrate below the surface layer. This is analogous to the case of high energy electron diffraction from a very thin crystal, that is, the reciprocal lattice spots are elongated normal to the crystal surface (thin direction) and the diffraction conditions are relaxed over the three-dimensional case.

It is apparent from the reciprocal lattice construction (Figure 6) that, if the sample were polycrystalline with a grain size much smaller than the beam size, so many diffraction beams would result that no pattern or no interpretable pattern would be obtained. The technique is thus limited to single crystals or highly textured polycrystalline samples.

Reference may be made to standard sources (see Refs. 20 through 23) for descriptions of plane nets and related symmetry operations. The crystallography of two-dimensional structures, including related symmetry operations and special conventions, have been discussed by Wood[24] and by Lander.[2] An arrangement of surface atoms, which is identical to that in the bulk unit cell as in Figure 7, is called the "substrate" structure and is designated by (1 × 1),

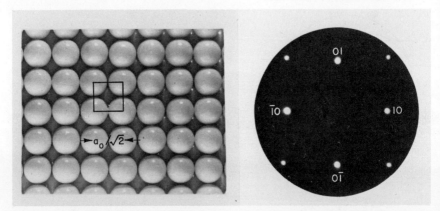

Figure 7 Hard sphere model showing (001) surface of face centered cubic nickel crystal and corresponding low energy (82 eV) diffraction pattern from a clean surface. From MacRae.[4] The 00 beam at the center of the pattern is obscured by the crystal.

since the interrow periodicities in the x and y directions are maintained up to the surface. The corresponding diffraction beams are called "normal." The horizontal rows of atoms in the solid sphere model of Figure 7 produce the spots indexed 01 and 0$\bar{1}$, while the vertical rows produce spots 10 and $\bar{1}$0. The surface unit mesh of side $a_0/\sqrt{2}$ is outlined.

If only normal features are observed in the diffraction pattern, it follows that the primitive translational periodicities of the surface structure are identical with those of a parallel substrate plane, so that identical "unit meshes" can be defined. The composition of the surface is a separate problem, since it may or may not be "clean." the (100) "substrate" structure of nickel is designated Ni(100)-(1 × 1).

If the arrangement of surface atoms differs from that in the bulk, it is called the "surface net" or "surface structure." The designation

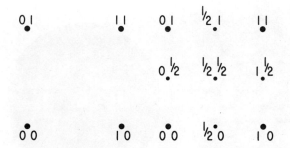

a. SUBSTRATE STRUCTURE   b. (2 x 2) SURFACE STRUCTURE

Figure 8 Diffraction pattern (schematic) from (001) face-centered cubic crystal. Large spots represent normal, and small spots extra, features.

(2×2) indicates that the unit cell of the surface structure is twice as long as the substrate structure along both the x and y crystallographic directions (Figure 8). The extra features are usually indexed with respect to the normal features and therefore have fractional values, for example, $\frac{1}{2}$, as in Figure 8b. Substrate and (2×1) surface structures are shown in Figure 9 for a (110) face-centered cubic (fcc) crystal. Figures 10 and 11 after MacRae[4] illustrate actual diffraction patterns, obtained on a (110) nickel surface, with the corresponding solid sphere models showing the arrangement of atoms. Exposure of a clean (110) nickel surface to oxygen at room temperature resulted in development of a (2×1) surface structure made up of nickel (white) and oxygen atoms (black),

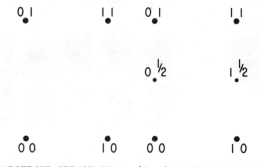

a. SUBSTRATE STRUCTURE   b. (2 x 1) SURFACE STRUCTURE

Figure 9 Diffraction pattern (schematic) from (110) face-centered cubic crystal. Large spots represent normal, and small spots extra, features.

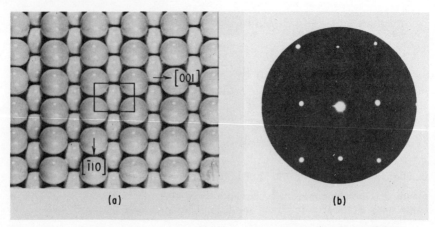

(a)   (b)

Figure 10 (*a*) Diffraction pattern obtained from a clean (110) nickel surface, employing 76 eV electrons. From MacRae.[4] (*b*) Hard sphere model of (110) nickel crystal.

as in Figure 11*b*. Since half of the surface layer sites are occupied by oxygen, this corresponds to a half-monolayer of oxygen atoms.

Much more complicated arrangements are frequently observed. Even the clean surfaces of silicon and germanium show extra features in patterns, indicating that the surface atoms depart from the ideal substrate arrangement. Extra features, ranging from $n/2$ to $n/7$ to $n/24$ order,

(a)   (b)

Figure 11 (*a*) Hard sphere model of (2 × 1) surface structure on a (110) nickel surface, where black spheres represent oxygen. (*b*) Diffraction pattern, obtained using 70 eV electrons on (110) nickel surface exposed to oxygen at room temperature, indicating (2 × 1) surface structure. From MacRae.[4]

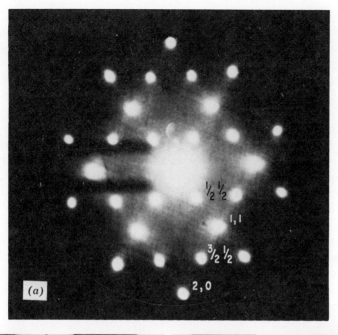

Figure 12 (a) Diffraction pattern, obtained using 122 eV electrons on a (110) tungsten surface on which half a monolayer of oxygen is absorbed. (b) and (c) Hard sphere models of two possible surface domain structures. From J. W. May.[1,25] (Copyright 1965, American Chemical Society.)

125

have been observed on silicon surfaces.[2] If a crystal surface possesses rotational symmetry, domain structures may nucleate and grow in any of the possible orientations. Since the beam size is usually large relative to the domain size, the resulting pattern may contain extra features from all possible domain orientations. Figure 12a shows an example of a pattern obtained by May[1,25] on a (110) tungsten surface, on which half a monolayer of oxygen was absorbed. The pattern was accounted for by a superposition of the extra features, due to the two surface domain configurations illustrated in hard sphere models (Figure 12b and c). In reference to Figures 8 and 9, it may be pointed out that identical patterns would be expected from (2 × 1) and (2 × 2) surface structures on the (001) surface of an fcc crystal if the domain size is smaller than the beam size. Discrimination may be possible if the surface coverage of the absorbate is known.

Although unique solutions for even complicated structures may be arrived at in x-ray diffraction analysis by consideration of the diffraction patterns, intensities, and composition, this is often not possible at present in low energy electron diffraction analysis, where ambiguity may persist. Bonding considerations, that is, whether bonding is principally covalent or ionic, influence the absorbate atom positions. Figure 13, after Lander,[2] illustrates eight of the simpler possible potential structures for a (2 × 2) surface array on a (111) silicon substrate. All give rather similar diffraction patterns if various possible domain orientations are included.

2. Intensity Analysis

Although the symmetry and dimensions of the surface structures may be determined from the positions of the diffraction spots in a low energy diffraction pattern, an analysis of the intensities is necessary to determine the actual locations of the atoms in the substrate. As in x-ray diffraction, the analysis becomes more complicated as the number of atoms per unit cell becomes larger. Unit surface meshes with dimensions of 40 Å on a side are not uncommon and may contain over 100 atoms.[4] At 100 eV, electrons typically penetrate a crystal to a depth of about three layers. Maxima in the intensity-energy curves result when the beams scattered from the surface layer and bulk atoms are in phase. Since the penetration depth increases with the incident electron energy, the spot intensities may vary with the latter. A full analysis could, in principle, determine the location of both the surface and interior atoms producing scattering.

The theory of intensity analysis has been considered in some detail by Lander,[2] to whom reference may be made for a bibliography, and

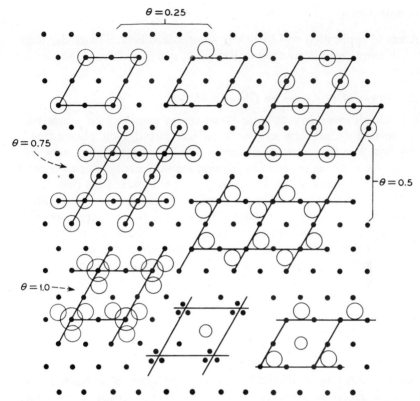

Figure 13 Examples of possible (2 × 2) surface arrays on a (111) silicon substrate. The θ values indicate fractional coverage. From Lander.[2] (Copyright Pergamon Press, New York, 1965.)

hence will only be discussed briefly here. Intensity analysis is clearly necessary for a rigorous understanding of the scattering mechanism and to determine the exact positions of the absorbate atoms. The theory necessary for structure analysis has not been fully developed yet, but some progress has been made, as in recent papers by Heine[26] and MacRae.[27] At the moment a completely unique analysis, even of simple diffraction patterns, is difficult. At low electron energies ($< \sim 100$ eV) multiple beam interactions may occur, since the intensity of the elastically scattered electron beam is high, compared to the incident intensity. Dynamical theory is then necessary. The simpler kinematical theory may be satisfactory at higher electron energies, where the amplitude of the backscattered electrons is relatively small.

E. Applications

Areas of application will be briefly summarized here. The reader interested in more detail is referred to Refs. 1, 2, and 5, and the bibliographies therein.

1. Arrangement of Atoms on Clean Surfaces

It is clearly necessary to understand the arrangement of atoms on a clean surface before one can hope to understand such phenomena as absorption. We have already defined a "clean" surface for present purposes as one showing no extra diffraction features attributable to ordered impurities. The simplest case encountered is that in which the two-dimensional structure that is stable is the same as the arrangement of the atoms in a parallel plane inside the body of the crystal. Surface structures with different translational symmetry than that of the bulk atoms have been observed within well-defined temperature ranges on the clean surfaces of such materials as gold, platinum, palladium, silver, antimony, bismuth, silicon, germanium, and aluminum oxide.[5] Surface structures identical with the bulk have been observed on clean surfaces of nickel, copper, tungsten, aluminum, antimony, bismuth, silver, chromium, niobium, tantalum, molybdenum, iron, and vanadium.[5]

2. Faceting

Extra features in a diffraction pattern can arise as a result of the presence of facets on the surface, parallel to specific crystal planes, which locally change the angle of beam incidence. Faceting is indicated by the presence of diffraction features that move outward, rather than inward, when the electron wavelength is decreased. Thermal etching by heating in vacuo is a well-known source of such effects. The importance of surface preparation methods is emphasized, since etching can produce crystallographic-type pits. Since the formation of facets of a different orientation to the general surface orientation is an attempt to minimize the surface energy, the composition of the surface is an important variable. MacRae[4] showed, in the case of a (111) nickel surface, that controlled oxidation to produce NiO on the surface resulted in the formation of pyramids of NiO with (100) faces which gave extra diffraction features.

3. Structures Formed by Vapor Deposition

The formation of epitaxial deposits, that is, of deposits having the same or a specific orientation relationship to the monocrystalline substrate, by evaporation of one metal onto the surface of another, is well known.

Such layers have a lattice constant and/or structure quite distinct from that of the substrate. However, this commonly occurs only after coverage exceeds that of a monolayer or after thermally activated surface migration has occurred to give thicker partial, that is, patch, coverage. During the initial stages of growth the absorbate atoms can assume a two-dimensional structure related to the substrate surface structure. Usually, the absorbate atoms are located on lattice sites of the substrate, so that the two-dimensional surface structures have dimensions that are integral multiples of the substrate site spacings. The relative size of the absorbate atoms, the lattice constant of the absorbate, and chemical bonding considerations all affect the value of the integer.

The deposition of a variety of metals onto cleaved {111} silicon surfaces has been studied.[2,5] Three ordered phases formed successively, when cesium was evaporated onto the {111} face of silicon.[28] Copper deposited on a clean {001} titanium surface formed oriented crystallites but was disordered if the surface was first covered with a chemisorbed layer of oxygen.[29] Other examples are given in Ref. 5.

4. Oxidation and Absorption of Gases by Metals

Lander[2] was one of the first to point out that surface structures had been relatively neglected in the chemisorption theory and discussed their importance in some detail. The subject has also been reviewed by May.[1] Prior to low energy electron diffraction studies, it was widely assumed that gas molecules or atoms filled up the available surface in a random way, and polycrystalline substrates were used without recognition of the effect of surface orientation and structure. Although, as already pointed out, it is difficult to determine the exact sites of absorbed atoms by low energy electron diffraction, it is often possible to determine whether the absorbed atoms are all located at a particular kind of site. The absorption of oxygen on metal surfaces of various orientations has been studied[1,2,5,30,31] on nickel, silicon, germanium, copper, tungsten, platinum, iron, molybdenum, tantalum, chromium, and rhodium. It was found that absorbed oxygen can often form ordered structures on the surface of the metal. Crystallographic orientation strongly influences the amount of gas absorbed, and its composition and arrangement on the surface. Reaction rates accordingly depend on the concentration of surface steps and facets.

The formation of extra diffraction features is apparently not spontaneous but appears to be associated with the ordering of a surface phase and has a formation rate influenced by temperature, time, gas pressure, and orientation. Simmons, Mitchell and Lawless[32] found initially extra diffraction features from (110) copper attributable to the random absorption of oxygen in the troughs of the (110) surface. On further absorption of oxygen, the streaks were replaced by spots attributable to the formation of an ordered structure which on further exposure changed to a second structure. One explanation of these results is that the surfaces undergo reconstruction during absorption, the metal surface atoms migrating to new sites and forming a new oxygen-metal structure. Clearly, such studies are valuable in understanding the early stages of oxidation. The absorption of other gases and gas mixtures such as nitrogen, hydrogen, carbon monoxide, and carbon dioxide, and of vapors such as iodine on metals, have been studied.[1,2,5]

5. Catalysis

The role of surfaces in bringing about or catalyzing chemical reactions is an important field. Although little used as yet, low energy electron diffraction is a potentially powerful tool, as discussed by May.[1] Clean surfaces are practically never obtained initially, except perhaps by cleavage in high vacuum. Low energy electron diffraction is necessary to define the state of the surface before catalysis begins.

6. Surface Diffusion

The technique has been little used for surface diffusion studies on clean surfaces but has considerable potentiality.[5] Surface self-diffusion rates are known to depend on the surface structure.[33] The surface of a crystal can be disordered by bombardment with high energy noble gas ions, resulting in a high concentration of metal ad-atoms, adsorbed gas atoms, and absorbed (soluble) gas atoms, and the disappearance of diffraction features previously evident at low voltage. Morabito and Somorjai[5] suggest that, by annealing at high temperature, the adsorbed gas could be desorbed and the diffusion rates of metal ad-atoms followed by monitoring the intensity of diffraction spot reappearance as a function of time and temperature. In this way the surface self-diffusion rate could be determined.

In a related type of kinetic study, Lyon and Somorjai[34] determined an activation energy for formation of the (5×1) surface structure on platinum.

7. Inner Potential

Incident electrons interact with the potential field of the atoms within the surface of a crystal. The atomic scattering factor f which is a function of both the scattering angle and the wavelength, describes this interaction. Unfortunately, accurate experimental values of f are lacking. As already described for the simple case of a single row of ordered atoms in Figure 4, the distribution of the atoms in a surface unit cell determines the amplitude of the resultant scattered wave. The amplitude and phase of this wave are described by the structure factor F. The intensity is proportional to F^2. The interpretation of the diffracted intensities must await the development of an adequate theory.

The incident energy of electrons entering the crystal is increased by the amount of the inner potential V_0 which typically ranges from 5 to 30 eV. Equation 1 thus should be rewritten as

$$n\lambda = d_{hk} \sin \theta_{hk} = n \sqrt{\frac{150}{V_i + V_0}} \qquad (4)$$

where V_i is the incident potential.

The inner potential V_0 may be measured by a number of methods. In a method due to Lander,[2] V_i and hence λ are varied; then the value of V_i is measured, at which a feature hk disappears at 90° to the surface normal (i.e., $\sin \theta_{hk} = 1$). The order of diffraction n is known. If d_{hk} is known from an independent measurement, V_0 may be determined. Alternatively, as pointed out by Morabito and Somorjai,[5] the intensity I_{00} of the specular reflection (00) may be determined as a function of V_i (Figure 14). I_{00} goes through a succession of maxima, some (indicated by arrows in Figure 14) associated with Bragg

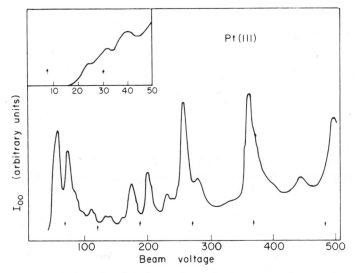

Figure 14 Intensity I_{00} of specular reflection versus beam voltage V_i for (111) surface of platinum. Arrows indicate calculated Bragg maxima. From Morabito and Somorjai.[5] (By permission of The Metallurgical Society of AIME.)

reflection. By measuring the actual positions that differ from the computed arrowed positions, V_0 may be determined.

8. Root Mean Square Displacement of Surface Atoms

Since surface atoms have fewer nearest neighbors, their mean square displacements are different from those of atoms in the interior. The intensities of the I_{00} beam decrease rapidly with increasing temperature. The root mean square (rms) displacements normal to the surface may be calculated from the measured temperature dependence (Debye-Waller factor). For face-centered cubic metals,[5] these are found to be greater than the displacements of interior atoms.

9. Miscellaneous

A good deal more information will probably be derivable from data such as Figure 4 when the theory is improved. Many peaks are present, in addition to the Bragg peaks. Some of these, usually referred to as "secondary Bragg peaks," are tentatively attributed to the asymmetry of the environment of surface atoms and may contain information on the electron density distribution. Multiple scattering may also contribute to these peaks.[5]

It may be pointed out, that the standard low energy electron diffraction equipment provided with double grid velocity selection, lends itself well to studies of secondary electron emission.

II. AUGER ELECTRON ANALYSIS

A. Introduction

Low energy electron diffraction and Auger electron analysis are complimentary techniques, the former giving structural information about the surface layers of a crystal, the latter permitting chemical analysis of the surface of a solid. The apparatus requirements are also similar and a number of commercial units are available, adaptable for either purpose. Although presently used as a spectrometry technique on macro-areas, approximately 3 mm in diameter, in principle, Auger analysis could be carried out in microprobe-type equipment and scanning images developed. The availability of a commercial scanning microscope with ion pumping would also permit the use of an Auger electron signal.

Auger discovered that materials, when bombarded with suitably energetic x-rays, emit certain electrons—the Auger electrons—which have energies characteristic of the originating atoms.[35-36] Later it was found that increased sensitivity could be obtained using electron excitation.[37]

The Auger electrons emitted under electron bombardment occur as small peaks on a large background in the secondary electron distribution. In 1953 Lander[37] suggested that measurement of the Auger electron energies could form a basis for an analytical technique analogous to x-ray fluorescence. However, experimental difficulties slowed development for many years until clean, high vacuum systems became available, and the method of electronic differentiation of the signal was developed[38] which enabled the Auger peaks to be distinguished from the background.

B. Principles

Under electron or photon bombardment of sufficient energy, electrons can be removed from an inner shell of an atom in a solid. The shell vacancy is filled by an electron from a higher shell, releasing a quantum of energy equivalent to the difference in levels which is either emitted as an x-ray photon or carried off by another electron from one of the outer shells. The energies of the latter Auger electrons are less well defined than the photon energies, since they tend to lose energy by scattering within the material and, in order to escape from the surface, give up an amount of energy at least equal to the work function. In view of this and the fact that the electron energy levels characteristic of an isolated atom are widened into bands in a solid, the observed energy spectrum should consist of comparatively broad peaks with a tail on the low energy side.[37]

X-ray emission decreases with a decrease in atomic number (see Chapter 9), so that the complementary Auger process of energy removal from the excited atoms increases.[39,40] Furthermore, the wavelength of the x-rays emitted decreases with a decrease in atomic number, making x-ray emission analysis more difficult and less sensitive because of adsorption problems. Auger analysis is thus most valuable for the light elements with atomic numbers below about 20. Since it is carried out in high vacuum, the comparatively low energy of the electron is advantageous in permitting velocity measurement with relatively compact electronic equipment operating at moderate voltage.

C. Equipment

The equipment used by Harris[38] for Auger analysis which resembled that previously described by others[37,41-43] is shown schematically in Figure 15. The apparatus shown was small enough to fit into a standard vacuum bell jar. The principle components are an electron gun, a sample wheel to permit sample charging inside the vacuum, a conventional 122°

electrostatic deflection energy analyzer,[44] and an electron multiplier. The gun employed a tungsten ribbon filament and provided an electron current of a few hundred microamperes, concentrated in a beam about 2 mm wide and striking the sample at about 15° from the grazing incidence. The analyzer had a mean radius of 2.5 in. and, with slit widths of 0.005 in., provided about 0.1% resolution. The secondary emission multiplier was shielded against the collection of stray electrons. Usually, an ion-pumped vacuum system was used, fitted with metal gaskets and pre-pumped with a molecular-sieve sorption pump. A clean vacuum of 10^{-8} torr or better is required.

A part of the secondary electron emission from the sample, excited by bombardment with electrons of a few keV energy, is received by the energy analyzer. By controlling the deflecting voltage applied to the curved plates of the analyzer, electrons within a selected narrow energy range are allowed to enter the electron multiplier for amplification. The electron-energy distribution could be determined by slowly sweeping the deflecting voltage through its range. In practice the derivative of the distribution curve was determined electronically by superimposing a small a-c voltage on the deflecting voltage and measuring the

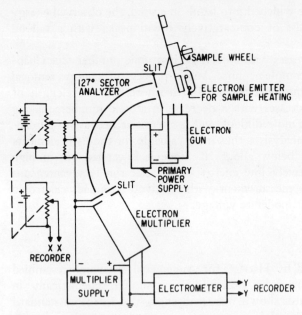

Figure 15 Simplified schematic, showing Auger analysis apparatus used by Harris. From Harris.[38]

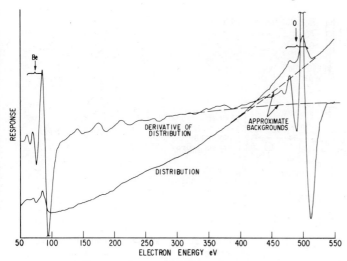

Figure 16 Effect of electronic differentiation on energy distribution curve, obtained on an oxidized beryllium sample. From Harris.[38]

a-c component of the corresponding frequency in the multiplier output current synchronously with a phase-sensitive detector. The amplitude of this signal is proportional to the slope of the actual energy distribution curve and is recorded after rectification. The noise background is reduced by using a very small band width in the detector. The signal gain is increased because there is little contribution from the slowly varying portions of the energy distribution curve. The enhancement of the Auger spectral features achieved by electronic differentiation[38] is illustrated in Figure 16. The structure of the Auger spectra, due to beryllium and oxygen, is much more clearly seen in the derivative curve.

Samples could be cleaned in position by sputtering with argon ions, after first turning off the ion pump and admitting a low pressure of argon. The argon was removed with the sorption pump and the ion pump restarted. Final cleaning could be carried out by heating the sample in the high vacuo to 700 to 800°C, by electron bombardment of the molybdenum backing plate in thermal contact with the sample.

Various other forms of electron spectrometer can be used. Palmberg and Rhodin[45] used an arrangement (Figure 17), devised by Weber and Peria,[46] which is an adaption of a LEED apparatus. Electrons are sorted according to energy by the spherically shaped grids, the potential of the second grid being depressed to provide a potential barrier. The current $I(E)$, due to electrons passing through this grid and reaching the

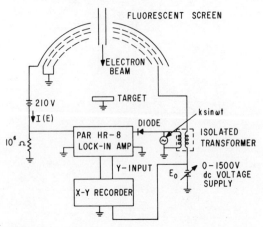

Figure 17 LEED apparatus, modified for Auger analysis. From Palmberg and Rhodin.[45]

collector, and measured as a function of energy by sweeping the grid voltage, is the integral of the energy distribution determined in a deflection-type analyzer. The second derivative of the transmitted current then gives the derivative of the energy distribution current and is obtained by an electronic perturbation technique, using a diode to generate a second harmonic reference signal. A phase-sensitive detector responds to the second harmonic of the perturbation frequency.

A number of commercial stainless steel equipments are available which may be used for either LEED or Auger analysis. The equipment shown in Figure 18 is typical and operates on a similar principle to that shown in Figure 16, except that the small a-c voltage used in signal differentiation is applied to the sample. The ultrahigh vacuum, ion-pumped system provides a vacuum of better than 10^{-10} torr.

Harris[47], using a modification of the apparatus illustrated in Figure 15, studied the effect of varying the angle of incidence of the primary beam on the sample, and the angle of signal pickup (observation angle). He verified that the observation depth was limited by the scattering of the low energy emitted electrons; by varying the angle of observation he was able to distinguish surface atoms from those distributed at various depths. Harris concluded that the primary beam should be normal to the surface and that the Auger electrons should be observed at angles far from normal. With the deflector type of analyzer (Figure 15), modified to permit variation of the observation angle, qualitative information can be obtained on the depth distribution of observed elements. On the

Figure 18 Commercial AES-200 Auger electron spectrometer with high precision manipulator. (Courtesy of Varian Vacuum Division.)

other hand, in the spherical grid, retarding field-type analyzer, the entire illuminated sample area is observed and the Auger electrons emitted in all directions are summated.

D. Applications

Auger electron spectroscopy has two distinctive features, compared with x-ray emission spectroscopy: (*a*) its greater sensitivity to the light elements, and (*b*) its extreme surface sensitivity, due to the lesser penetration combined with the short range of the Auger electrons inside the sample. Until more knowledge is gained and a sounder theoretical basis established, its principal use is for qualitative analysis in the parts per million range. The problem of obtaining suitable standards of known composition is even more difficult at these low concentration levels than in the case of the electron microprobe, where we are dealing with much higher levels. Palmberg and Rhodin[45] showed that the Auger electrons typically come from a layer only a few atoms thick. Weber and Peria[46] were able to detect a partial monolayer coverage of potassium or cesium atoms on a clean silicon surface, corresponding to about 10^{12} atoms in the area examined.

Auger spectroscopy has proved useful for the qualitative identification of surface contaminants such as residues from organic solvents and plating baths. The surface segregation of minor impurities such as sulfur has been detected in metals. While very valuable in certain cases, this also means that attempts to clean a metal surface by heating in the high vacuo inside the equipment can result in the migration and segregation of impurities at the surface, greatly enhancing the signals over those representing the bulk concentration (Figure 19). The sulfur line at about

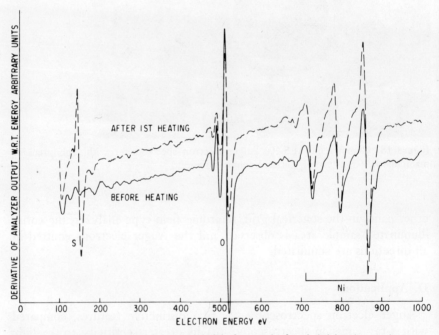

Figure 19 Auger spectra of pure nickel before and after brief heating to between 600 to 900°C, showing growth of sulfur line. The curves are arbitrarily separated vertically for clarity. From Harris.[48]

150 eV was identified with the $L_{II,III}M_I$ transitions in the sulfur atom.[48] Harris demonstrated[48] that the sulfur peak growth on heating resulted from the migration of sulfur from the interior rather than from the vacuum environment. He did this by showing that sulfur depletion occurred upon cyclical exposure to air, which removed most of the surface sulfur, and resegregation occurred on reheating.

Stein, Joshi, and Laforce[49] used Auger spectroscopy to look for impurity segregation at the fracture surfaces of temper embrittled and nonembrittled low alloy steels. The steel samples were cooled inside the Auger apparatus with liquid nitrogen and fractured in the high vacuo with a hammer device to avoid contamination. Their results on a high purity steel containing 0.39 wt % C, 3.5% Ni, 1.6% Cr, 0.003% Mn, 50 ppm P, 10 ppm Sn, < 6 ppm As, and 620 ppm Sb are shown in Figure 20. The spectrum from the fracture of the steel in the embrittled condition showed an antimony doublet at 460 V which is missing from the spectrum taken in the nonembrittled condition, providing evidence

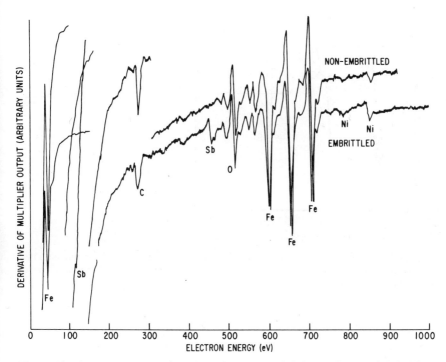

Figure 20 Auger spectra on fractures of temper-embrittled and nonembrittled low alloy steel with 625 ppm Sb addition. From Stein, Joshi, and Laforce.[49]

of the segregation of antimony to the grain boundary region in the former. The nickel peak was also more pronounced, suggesting similar segregation of the nickel.

Weber and Johnson[50] extended the work of Weber and Peria,[46] already cited, and showed that as little as 0.02 of a monolayer of cesium applied to a (100) silicon crystal surface inside the system by means of an alkali-ion gun could be easily detected (Figure 21). Their study included com-

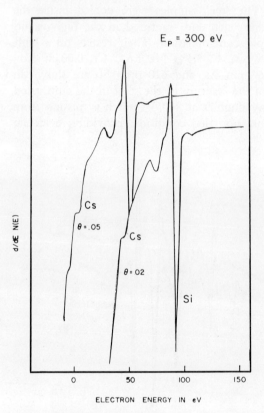

Figure 21   Auger spectra of 0.02- and 0.05 monolayer deposits of cesium on (100) silicon surface. The energy scale for the 0.05 monolayer curve is shifted for clarity. From Weber and Johnson.[50]

bined LEED and Auger analysis on germanium crystals, coated with potassium, and of the effect of heating on potassium desorption.

Palmberg and Rhodin[45] determined the Auger spectra believed characteristic of the clean surfaces of gold, silver, copper, palladium, and nickel, deposited at 10^{-10} torr onto an MgO substrate cleaved inside the vacuum system. By depositing one or more monolayers of one metal on another and observing the effect on the substrate spectra, they determined the mean escape depth for Auger electrons and concluded that, in silver, this varied between 4 and 8 Å for energies of 72 and 362 eV, respectively. The surface structures in some cases were determined by LEED.

A great deal of work will be necessary to develop a reference catalogue of spectra to facilitate interpretation. It is at present very difficult to obtain a surface representative of the bulk impurity analysis and contamination effects change the line amplitudes. Even at its present early stage of development, Auger spectroscopy is obviously a powerful tool for qualitative analysis with a sensitivity capable of detecting 10^{-12} gram of material. The Auger spectra may contain information regarding the chemical state of the atoms observed, particularly in the case of the light elements, in which the valence electrons are involved in the Auger transitions.[38] This will not be apparent without a more detailed and precise study of line shapes and positions.

Although the Auger spectra are determined primarily by the atomic nature of a material, and only to a small degree by the interaction between atoms, the characteristic losses in the energy of electrons reflected from or transmitted through a material are a result of collective oscillations of valence electrons or interband transitions and reflect the electronic nature of a solid.[51]

Ion-neutralization spectroscopy, developed by H. D. Hagstrum, also uses the Auger spectrum, but in a much lower energy range, making it even more sensitive to surface conditions and more difficult to interpret. Information is obtainable about the electronic state of a surface.[51]

Another technique related to Auger spectroscopy may be mentioned here, namely photoelectron spectroscopy, commonly referred to either as electron spectroscopy for chemical analysis (ESCA)[52,53] or as photoelectron spectroscopy (PES),[54] depending on the experimental approach used. In both, the energies of photoelectrons ejected by the interaction of a molecule with a monoenergetic beam is measured. The energy of radiation is sufficient to eject electrons from several molecular orbitals. The spectrum obtained reflects not only the elements present in the molecule but also the bonding configurations. In PES, the ionization is produced by a beam of photons from, for example, a helium discharge, and the electrons are principally ejected from the outer or valence orbitals of molecules; whereas, in ESCA, ionization is generally produced by a high energy x-ray beam and the photoelectrons are ejected mainly from inner orbitals. The samples in PES are generally introduced as vapor;[54] in ESCA, they are usually solid, gas, or liquid samples of low vapor pressure, stable under vacuo, although cryogenic probes may be used to reduce the vapor pressure.[53]

The above sample preferences reflect the current equipment designs used.[53,54] In principle, solid, liquid, or gas samples can be used for photoelectron spectroscopy (PES or ESCA); however, in the case of gas, or liquid or solid samples of appreciable vapor pressure, it would be necessary to pump the specimen chamber differentially, in order to maintain the necessary vacuum in the rest of the system. Irradiation energies of 10 to 12 eV are obtainable, using ultraviolet sources of short wavelength; this is more than sufficient to excite photoelectrons from the outer shell of most metals which typically requires 2 to 4 eV.

The principal components of a photoelectron spectrometer system are a source of ionizing radiation and an electron monochromator and detector to analyze the number and energy of the emitted electrons. Types of sources that have been used include x-rays, electron guns, and various kinds of ultraviolet photon sources. Gamma rays might also be used. Retarding field, magnetic, and electrostatic monochromators have been employed with a GM-tube, an electron multiplier, charge cups, or photographic detection. Degaussing is necessary to neutralize stray fields, in the case of magnetic monochromators. Shielding may be used with the electrostatic type: the field on the spectrometer may be increased continuously or stepwise as a function of time and the detector signal monitored by a rate meter, or a multichannel analyzer employed. The equipment is more elaborate and the spectra are more complex than in the case of Auger analysis which is a radiationless process. In the case of electron bombardment the spectra arise principally from Auger electrons.

Photoelectron spectroscopy is principally a surface technique, since the average penetration of a photoelectron is < 100 Å and typically 20 to 50 Å. Thus, unlike LEED, both the immediate surface and the underlying layer are examined. The technique has considerable potential for chemical structure determination, for the determination of ionization potentials in the case of PES, and for quantitative analysis, that is, for determining elemental ratios in a wide variety of materials including organic compounds. Samples as small as 10^{-8} g can be used and 1% precision should be realizable; trace quantities of one element in the 0.1% range should be detectable, even in the presence of many others if resolution can be further improved.[53] There are nevertheless considerable problems of calibration and standardization involved in quantitative analysis.

III. REFERENCES

1. J. W. May, *Ind. Eng. Chem.*, **57**, 19 (1965).
2. J. J. Lander, *Progress in Solid State Chemistry*, Vol. 2, H. Reiss, Ed., Pergamon Press, New York, 1965, pp. 26–116.
3. L. H. Germer, *Scientific American*, **212**, (3), 32 (1965).
4. A. U. MacRae, *The Use of Thin Films in Physical Investigations*, J. C. Andersen, Ed., Academic Press, New York, 1966, pp. 149–162.
5. J. M. Morabito, Jr., and G. A. Somorjai, *J. Metals*, **20**, 17 (1968).
6. E. Bauer, in *Techniques of Metals Research*, Vol. 2, Part 2, 1969, ed. R. F. Bunsah, Ed. Interscience Publishers, New York, pp. 559–639.
7. C. J. Davisson and L. H. Germer, *Phys. Rev.*, **30**, 705 (1927).
8. C. J. Davisson and L. H. Germer, *Proc. Nat. Acad. Sci. U.S.*, **14**, 317 (1928).
9. C. J. Davisson and L. H. Germer, *Proc. Nat. Acad. Sci. U.S.*, **14**, 619 (1928).
10. H. E. Farnsworth, *Phys. Rev.*, **34**, 679 (1929).
11. W. Ehrenberg, *Phil. Mag.*, **18**, 878 (1934).
12. L. H. Germer and C. D. Hartman, *Rev. Sci. Instrum.*, **31**, 784 (1960).
13. E. J. Scheibner, L. H. Germer, and C. D. Hartman, *Rev. Sci. Instrum.*, **31**, 112 (1960).
14. J. J. Lander, J. Morrison, and F. Underwald, *Rev. Sci. Instrum.*, **33**, 782 (1962).
15. W. T. Sproull, *Rev. Sci. Instrum.*, **4**, 193 (1933).
16. C. W. Tucker, Jr., *J. Appl. Phys.*, **35**, 1897 (1964).
17. H. E. Farnsworth, T. H. George, and R. M. Burger, *J. Appl. Phys.*, **26**, 252 (1955).
18. H. E. Farnsworth and K. Hayek, *Surface Sci.*, **8**, 35 (1967).
19. R. L. Jacobson and G. K. Wehner, *J. Appl. Phys.*, **36**, 2674 (1965).
20. W. H. Zachariasen, *Theory of X-Ray Diffraction by Crystals*, John Wiley & Sons, New York, 1945.
21. *International Tables for X-Ray Crystallography*, Kynoch Press, Birmingham, England, 1952.
22. M. J. Buerger, *Crystal Structure Analysis*, John Wiley & Sons, New York, 1960.
23. A. W. Loeb, *Acta Cryst.*, **17**, 179 (1964).
24. E. A. Wood, *J. Appl. Phys.*, **35**, 1306 (1964).
25. J. W. May, *Surface Sci.*, **4**, 452 (1966).
26. V. Heine, *Surface Sci.*, **8**, 426 (1967).
27. E. G. MacRae, *J. Chem. Phys.*, **45**, (9), 3258 (1966).
28. G. W. Gobeli, J. J. Lander, and J. Morrison, *J. Appl. Phys.*, **37**, 203 (1966).
29. R. E. Schlier and H. E. Farnsworth, *J. Phys. Chem. Solids*, **6**, 271 (1958).
30. C. W. Tucker, *Surface Sci.*, **2**, 516 (1964).
31. J. J. Lander and J. Morrison, *J. Chem. Phys.*, **37**, 729 (1962).
32. G. W. Simmons, D. F. Mitchell, and K. R. Lawless, *Surface Sci.*, **8**, 130 (1967).
33. N. A. Gjostein, *Metal Surfaces*, ASM, Metals Park, Ohio, 1962.
34. H. B. Lyon and G. A. Somorjai, *J. Chem. Phys.*, **46**, (7), 2539 (1967).
35. P. Auger, *J. Phys. Radium*, **6**, 205 (1925).
36. E. H. S. Burhop, *The Auger Effect and Other Radiationless Transitions*, University Press, Cambridge, England, 1952.
37. J. J. Lander, *Phys. Rev.*, **91**, 1382 (1953).
38. L. A. Harris, *J. Appl. Phys.*, **39**, 1419 (1968).

39. A. H. Compton and S. K. Allison, *X-Rays in Theory and Experiment*, 2nd ed., Van Nostrand Reinhold Co., New York, 1935 p. 477.
40. E. J. Calnan, *Phys. Rev.*, **124**, 793 (1961).
41. G. A. Harrower, *Phys. Rev.*, **102**, 340 (1956).
42. C. J. Powell, J. L. Robins, and J. B. Swan, *Phys. Rev.*, **110**, 657 (1958).
43. N. R. Whetten, *Appl. Phys. Lett.*, **8**, 135 (1966).
44. A. L. Hughes and V. Rojansky, *Phys. Rev.*, **34**, 284 (1929).
45. P. Palmberg and T. N. Rhodin, *J. Appl. Phys.*, **39**, 2425 (1968).
46. R. E. Weber and W. T. Peria, *J. Appl. Phys.*, **38**, 4355 (1967).
47. L. A. Harris, *Surface Sci.*, **15**, 77 (1969).
48. L. A. Harris, *J. Appl. Phys.*, **39**, 1428 (1968).
49. D. F. Stein, A. Joshi, and R. P. Laforce, *ASM Trans. Quart.*, **62**, 776 (1969).
50. R. E. Weber and A. L. Johnson, *J. Appl. Phys.*, **40**, 314 (1969).
51. L. A. Harris, *Industrial Research*, **10**, 53 (February 1968).
52. K. Siegbahn, et al, *ESCA—Atomic, Molecular and Solid State Structure Studied by Means of Electron Spectroscopy*, Ser. IV, Vol. 20, Nova Acta Regiae Societatis Scientiarum Upsalienssis, 1967.
53. D. M. Hercules, *Anal. Chem.*, **42**, 20A (1970).
54. D. Betteridge and A. D. Baker, *Anal. Chem.*, **42**, 43A (1970).

5

CONVENTIONAL ELECTRON MICROSCOPY

I. General Background 147
  A. Introduction 147
  B. Electron Wavelength 147
  C. Resolution 148
    1. Aberrations 149
      a. Spherical Aberration 149
      b. Astigmatism 150
      c. Chromatic Aberration 151
    2. Other Factors Affecting Resolution 152
    3. Resolving Power 153
  D. Depth of Field and Depth of Focus 153
  E. Fresnel Fringes 156
  F. Ray Paths 159
  G. Bright and Dark Field 160
  H. Selected Area Diffraction 162
  I. Calibration 164
    1. Magnification 164
    2. Resolution 165
    3. Electron Diffraction 166
    4. Rotation 168
  J. Stereoscopy 169

II. Instrumentation 170
  A. Electron Microscopes 170
  B. Accessories for Electron Microscopes 175
    1. Electron Microprobe Attachments 178
    2. Reflection Electron Diffraction Attachments 180
    3. Scanning Electron Microscopy Attachments 181

III. Specimen Preparation 182
  A. Replica Methods 182

       1. General Comments 182
       2. Plastic Replicas 183
       3. Carbon Replicas 184
       4. Oxide Replicas 187
       5. Fractography 188
       6. Shadow Casting 189
       7. Extraction Replicas 189
   B. Preparation of Thin Foils from Bulk Samples 190
       1. Electrothinning 190
       2. Ultramicrotomy 192
       3. Chemical Thinning 197
       4. Cleavage 200
       5. Ion Bombardment 200
       6. Miscellaneous 202
   C. Naturally Thin Samples 202
       1. Evaporated Films 202
       2. Miscellaneous 206
   D. Special Transmission Techniques 209
       1. One-Sided Thinning 209
       2. Etched Transmission Samples 209
       3. Surface Decoration Technique 211

IV. High Energy Electron Diffraction 212
   A. Transmission Electron Diffraction 213
       1. Reciprocal Lattice Concept 213
       2. Ewald Reflecting Sphere 214
       3. Structure Factor 218
       4. Double Diffraction 218
       5. Twinning 219
       6. Crystal Shape Effects 219
       7. Effects Due to Elastic Strain 220
       8. Kikuchi Lines 220
       9. Departure from Bragg Reflecting Condition 223
   B. Reflection Electron Diffraction 223

V. Contrast Theory 226
   A. Electron Scattering from Amorphous Solids 227
   B. Electron Scattering from Quasiamorphous Solids 229
   C. Electron Scattering from Crystalline Solids 230
       1. General Theory of Contrast without Lattice Resolution
          (Diffraction Contrast) 232
         a. Concept of the Wave Function 232
       2. Dynamical Two-Beam Theory for a Perfect Crystal 233
         a. Extinction Contours 238
          (1) Thickness Extinction Contours 239
          (2) Tilt Extinction Contours 239
       3. Kinematical Theory of Contrast Effects 241
         a. Perfect Crystal 241
         b. Imperfect Crystal 245
          (1) Stacking Fault 246
          (2) Dislocations 248

      (3) Screw Dislocations 249
      (4) Edge and Mixed Dislocations 251
      (5) Partial Dislocations 251
      (6) Precipitates 251
         (a) Coherent Precipitates 251
    4. Theory of Contrast with Lattice Resolution 256
      a. Dynamical Two-Beam Theory for a Perfect Crystal 258
      b. Imperfect Crystal 261
        (1) Dislocations 261
        (2) Grain Boundaries 265
        (3) Twins 265
        (4) Structure of Fibers 266
      c. Resolution Limit for Periodic Structures 267
      d. Resolution with Tilted Illumination 269
    5. Indirect Lattice Resolution—Multiply Periodic Structures 269

VI. Major Fields of Application 272
    A. Defects 272
    B. Solid State Transformations 274
    C. Structure and Properties of Thin Films 275
    D. Fracture 275
    E. Noncrystalline Materials 275

VII. References 276

I. GENERAL BACKGROUND

A. Introduction

Electrons, like light, can be considered as having certain properties of a stream of particles and others of a wave motion. Thus it is convenient to consider rays in discussing geometrical optics, aberrations of lenses, and so forth, whereas the wavelike characteristics are involved in contrast theory.

Unlike light, we are principally concerned with monochromatic beams. The usual electron source is a heated tungsten filament, and the wavelength of the electrons is determined by the accelerating potential applied, which is controlled within close limits in order to avoid chromatic aberration arising from the source.

Electron optics and the components of the electron microscope will not be discussed in detail here. Brief accounts are given in Refs. 1 and 2; more detail is to be found in Refs. 3 through 7.

B. Electron Wavelength

At high accelerating potentials the velocity of electrons may be an appreciable fraction of the velocity of light. The relativistic potential

V^* is related to the accelerating potential V in volts by the relation:

$$V^* = (1 + 0.9785 \times 10^{-6}V)V \tag{1}$$

The wavelength λ is given by

$$\lambda = \sqrt{150/V^*} \tag{2}$$

At 100 keV the relativistic correction amounts to only $+10\%$ and the wavelength λ is 0.037 Å (Table 1). At 1000 keV, the relativistic potential increases to 1978 keV and the wavelength is 0.0087 Å. At 3 million eV the relativistic voltage is increased by a factor of four and $\lambda = 0.0036$ Å.

Table 1.   Variation of the Relativistic Voltage V^* and Wavelength λ with the Acclerating Voltage V

V, keV	V^*, keV	λ, Å
50	52.4	0.054
100	109.8	0.037
250	311.2	0.022
500	744.6	0.014
750	1,300	0.011
1,000	1,978	0.0087
2,000	5,914	0.0050
3,000	11,806	0.0036

C. Resolution

Unlike the light microscope, the resolution of the electron microscope is not limited by the wavelength, but rather by the aberrations of the (objective) lens, fluctuations in the high voltage and lens supplies, vibration, and external fields. It should be pointed out that very often the sample is limiting, in that it may not contain detail approaching the resolution limit of a particular microscope; or there may not be a contrast mechanism available that is capable of showing such fine detail; or incoherent scattering of electrons, due either to excessive sample thickness (which varies with the material) or to the presence of "oxide" or contamination layers on the surfaces, may swamp the signal of interest. Of course, if a microscope is capable of resolving $< \sim2$ Å, then nearly all (crystalline) samples contain detail capable of resolution, namely, the lattice planes themselves, but this requires a special phase contrast technique which will be discussed later.

1. Aberrations

The electromagnetic lenses in the electron microscope suffer from the same principal aberrations as glass lenses in the light microscope which were discussed in detail in Chapter 3. From the point of view of resolution, the aberrations in the objective lens are the most important; the principal aberrations are spherical aberration, astigmatism, and chromatic aberration. We shall consider these in turn, assuming that the objective lens has only one aberration.

a. SPHERICAL ABERRATION

Spherical aberration is the most important defect in the objective lens, since at present there is no convenient way of correcting for it. The lens strength increases from the center to the outside of the lens. Electrons leaving a point P on the objective axis in the object plane are thus spread over a disk radius Δr_i in the image (Gauss) plane which for magnification M is given by $\Delta r_i = MC_s\alpha^3$ or referred back to the object:

$$\Delta r_i = C_s\alpha^3 \tag{3}$$

where C_s is the spherical aberration coefficient of the lens and α is the semi-apex angle, subtended by the lens at the object. If $C_s = 1.6$ mm, and $\alpha = 6.95 \times 10^{-3}$ radians, then $\Delta r_i = 5.35$Å.

The marginal rays are brought to a focus closer to the lens than the axial rays. The spherical aberration is slightly decreased if the image is slightly underfocused to coincide with the circle of least confusion which, as in the case of light, is slightly nearer the lens than the Gaussian plane. Spherical aberration affects all points on the object. Since electron lenses are always convergent, the power of the outer zones is always larger and one cannot correct by combining two lenses, as with glass lenses.

In practice the spherical aberration is reduced by stopping-down the objective lens as far as possible to reduce α in Equation 3, a limitation being imposed by diffraction effects at the edge of the objective aperture which cause the resolution to start to decrease again. The theoretically attainable resolution is therefore governed by two opposing factors. A minimum value P_0 of the distance between two points, just resolved in the image, occurs at a certain objective semiaperture α_0, where

$$P_0 = A(\lambda^3 C_s)^{1/4} \tag{4}$$

$$\alpha_0 = B\left(\frac{\lambda}{C_s}\right)^{1/4} \tag{5}$$

Here A and B are constants of the order of unity whose exact values depend on the illumination conditions; C_s is the spherical aberration coefficient, and λ is the wavelength. If $B = 1$, $\lambda = 0.037$Å (100 keV), and $C_s = 1.6$ mm, then $\alpha_0 = 6.95 \times 10^{-3}$ radians. For 100 keV electrons, the optimum objective aperture usually corresponds to an aperture diameter of about 20 μ.

b. Astigmatism

The nature of astigmatism is familiar from the discussion on glass lenses. Astigmatism in an electron lens is caused by the field strength's being slightly different in two perpendicular planes and results either from inaccuracies in manufacture or from inhomogeneities in the soft iron pole-piece. Thus a ray pencil, originating at an object point off the axis, is brought to two mutually perpendicular line foci. There is no sharp image point, but rather a circle of confusion at the image plane midway between the two line foci.

Some astigmatism is always present, and modern microscopes are provided with a stigmator that enables correction to be made by applying an equal but opposite astigmatism. In the case of the second condenser lens, correction is made while viewing the caustic image, the stigmator being adjusted to maximize the symmetry. Absolute cleanliness of the condenser system is necessary, in order to reduce this defect to a minimum for high resolution work. In a modern microscope provided with an anticontamination device and electromagnetic stigmator, the speed of correction is such that astigmation correction in the objective lens may be made as often as desired with the (self-cleaning) objective aperture and specimen in position, viewing near maximum magnification. The preferred procedure is to image the fine granularity, which has a scale factor of a few angströms, of the thin contamination layer or oxide layer present at a hole edge in a metal film near the area to be photographed. (The self-structure may be used in a carbon film.) The objective stigmator is then adjusted, so that minimum contrast is achieved at focus. In order to obtain optimum resolution, the correction is made at the instrument settings (kilovoltage, magnification, etc.) that one is going to use for photographing the sample.

Observation of the uniformity of width and the degree of asymmetry of the Fresnel fringes around a small hole in a carbon film in the slightly overfocused condition is also widely used as a means of checking astigmatism in the objective and as a guide in stigmating. This has been called the "edge-diffraction test."[8] A number of different procedures may be employed.[9] One is that the best stigmator setting is that which enables one to go from just having a complete white fringe to just having a complete black fringe with a minimum increase in focusing current.

Since a dirty objective aperture greatly increases the astigmatism, it is good practice to correct without the movable aperture in place, then to insert a clean aperture and assume no change. However, with the self-cleaning type of aperture this is unnecessary. With the new electromagnetic type of stigmator, correction can, in any case, be carried out so quickly that contamination of the movable aperture during stigmation is no longer a problem.

c. CHROMATIC ABERRATION

The focal length of an electron lens depends on the electron energy of the beam. Fast and slow electrons are effectively of shorter and longer wavelength, respectively. Electrons of lower energy are bent more by a given objective lens field and therefore give rise to a disk of confusion in the image plane of radius Δr_c referred to object space, given for a thin lens by

$$\Delta r_c = f \, \alpha \, \frac{\Delta E}{E} \qquad (6)$$

or for a thick lens by

$$\Delta r_c = C_c \, \alpha \, \frac{\Delta E}{E} \qquad (7)$$

where ΔE is the energy spread about electron energy E, f(cm) the focal length, α the semiangle of the pencil of rays from an axial point in the object plane, and C_c the chromatic aberration constant which is usually slightly less than f. If $C_c = 1.3$ mm, $\alpha = 6.95 \times 10^{-3}$ radians, $\Delta E = 0.05$ V, and $E = 100$ keV, then $\Delta r_c = 0.05$ Å.

It is important therefore to minimize ripple in the high voltage source. Alpha can be reduced by using a small objective aperture but this is necessary anyway to minimize spherical aberration. Chromatic aberration can arise in the sample itself, since electron energy is lost by inelastic scattering processes. Inelastic scattering increases with the sample thickness and the atomic number, and may be increased greatly by the presence of "oxide" or contaminant films on the sample surfaces. Fortunately, an appreciable fraction of the inelastically scattered electrons is intercepted by the objective aperture. Clearly, ripple in the objective lens supply produces a variation in focal length, giving an effect equivalent to chromatic aberration. The time period over which stability is required is that necessary to make a photographic exposure, typically a few seconds.

2. Other Factors Affecting Resolution

The basic design of the microscope is, of course, a matter of vital importance, and our simplified discussion assumes that one of the better commercial high resolution electron microscopes currently available is employed. The location of the microscope is important, in order to keep stray fields below the tolerance level specified by the manufacturer and also to minimize mechanical vibration. Since vibration cannot be entirely avoided, a microscope design is employed in which the sample and the microscope column, including the recording medium, vibrate in unison, rather than in independent modes.

Drift in the accelerating voltage, or objective lens current, will result in changes in focal length that obviously can have adverse affects on the resolution achieved. Mechanical or thermal drift of the sample during exposure can also be serious. Again, the basic stage and microscope design are important. Although the lenses are water-cooled to dissipate the heat generated, it may be necessary to run the microscope with the lenses on for some time (even several hours) to achieve thermal equilibrium if optimum resolution is required. If this is not done, it may be found that the objective lens will slowly drift through focus.

In practice it is necessary to consider the combined effects of all the sources of loss of resolution. The latter fall into two classes; those that are independent of the exposure time and are additive such as spherical and chromatic aberrations and astigmatism, and those that result in a drift during exposure which may add or subtract in an unpredictable manner.

In order to resolve point details ≤ 10 Å in a photograph, it is necessary that suitable detail be present in the sample and that a suitable contrast mechanism be available. It is usually necessary that the sample be very thin, since we are looking at a projected image and have to consider the "signal to background" ratio. Even in a 100-Å-thick sample, a 10-Å-diameter sphere represents only 10% of the thickness. There may be an optimum thickness, dependent on the contrast mechanism. Contamination of the sample in the electron beam can cause a serious progressive loss in resolution. Cosslett[10] estimated from the Thomson-Whiddington empirical law for energy loss by incoherent scattering that the resolving power for amorphous carbonaceous specimens would be limited to about 10% of the specimen thickness. Thus the effect of contamination is understandable. Fortunately, contamination can be reduced to tolerable levels by cooling the sample surroundings, using a liquid nitrogen-cooled finger. Loss of resolution can similarly result from the presence of oxide layers on the surface of a metal sample.

3. Resolving Power

The discussion here follows that in Ref. 1. It has already been pointed out that there is no convenient way of correcting the spherical aberration of the objective lens, but the effect of this on the resolving power can be reduced by stopping down the lens aperture. A limit is imposed by the diffraction aberration Δr_d (referred to object space) which is given by[11]

$$\Delta r_d = \frac{0.61\lambda}{\alpha} \tag{8}$$

where λ is the electron wavelength, and α the semiaperture angle as before.

It will be seen from Equation 8 that Δr_d increases as α is reduced, whereas the spherical aberration Δr_s given by Equation 3 decreases. The optimum α is found by equating Δr_d with Δr_s and the minimum aberration by substituting the result in Equation 3:

$$\alpha_{opt} = A\lambda^{\frac{1}{4}}C_s^{-\frac{1}{4}} \tag{9}$$

$$\Delta r_{min} = B\lambda^{\frac{3}{4}}C_s^{\frac{1}{4}} \tag{10}$$

Hirsch et al.[12] point out that these relationships are approximate, since they are based on ray optics rather than wave optics, so that the constants A and B may be taken as unity for practical purposes. For a typical objective with $C_s = 1.6$ mm and $\lambda = 0.037$ Å (100 keV), the resolving power Δr_{min} is about 5.4 Å, while α_{opt} is about 6.9×10^{-3} radians. This corresponds roughly to a 20-μ objective aperture diameter for a lens of 1.6 mm focal length.

Equation 8, taken from physical light optics, assumes that the waves emitted by two point objects are incoherent and the objective aperture is filled with scattered waves. Neither assumption is really applicable when Bragg diffracted beams are formed.

D. Depth of Field and Depth of Focus

As in the case of light optics, the depth of field corresponds to the amount by which the objective lens may be off-focus on either side of the true object plane before blurring becomes comparable with the attained resolution.[5] The relation for the depth of field will now be derived with reference to Figure 1.

Consider a lens forming an image P_i of the axial object point P on a screen placed in the focal plane. A point Q, distant $D/2$ from P along the axis will be imaged at Q_i. This will appear as an out-of-focus disk of diameter χ_i on the screen; χ_i is the image of the cross section at the object plane, running through P of the ray pencil from Q. Because of diffraction and the lens aberrations, P_i

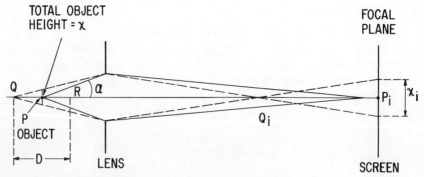

Figure 1   Schematic of the depth of field D. After Hall.[5]

will, in fact, be a diffuse disk rather than a point image. Therefore, although Q is out of focus on a solely geometrical argument, it will actually appear in satisfactory focus on the screen if

$$\chi_i \lesssim 2dM \tag{11}$$

where $2d$ is the diameter of the disk of confusion for P_i, referred to the object space, and M is the magnification. Now

$$\chi = \frac{\chi_i}{M} \tag{12}$$

Hence

$$\chi \lesssim 2d \tag{13}$$

where χ is the total object height (Figure 1).

On a similar argument, there will be a corresponding axial point R distant $D/2$ on the other side of P which will also be in satisfactory focus. The distance D between Q and R is defined as the *depth of field*.

Although greatly exaggerated in Figure 1, D is, in fact, very small compared to the distance between P and the lens, so that P, Q, and R subtend approximately the same angle α at the lens aperture. Hence

$$\frac{\chi}{2} = \frac{D}{2} \cdot \alpha$$

or

$$D = \frac{\chi}{\alpha} \tag{14}$$

Hence from Equation 13,

$$D = \frac{2d}{\alpha} \tag{15}$$

The *resolution limit* is the smallest distance between the centers of two points that can just be distinguished as separate in the image. On the Rayleigh criterion of classical optics,[5] we may consider this as given by $2d$ in Equation 15. Thus for a resolution of $2d = 10$ Å and $\alpha = 7 \times 10^{-3}$ radians, D is about 1400 Å, so that all of the detail in a foil of thickness $\leq$ 1400 Å is normally in focus.

The *depth of focus* is the distance in final image space, corresponding to D above in object space, that is, the distance D_i through which the screen can be moved without affecting the sharpness. If we define a satisfactory image again as one for which the disk of confusion, due to defocusing, is $\leq 2dM$ in

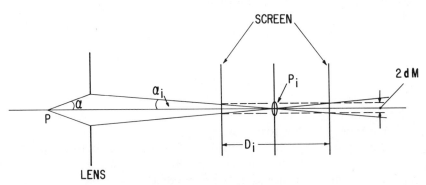

Figure 2 Schematic of the depth of focus D_i. After Hall.[5]

diameter, then, from Figure 2,

$$D_i = \frac{2dM}{\alpha_i} \tag{16}$$

Now

$$\alpha_i = \frac{\alpha}{M} \tag{17}$$

So

$$D_i = \frac{2dM^2}{\alpha} = DM^2 \tag{18}$$

Since M is the total magnification in the final image, the depth of focus is large; for example, for $M = 10^4$ and $D = 1400$ Å, $D_i = 14$ meters. Thus the image can be focused, even though the screen is tilted and the camera need not be placed in the same plane as the screen.

E. Fresnel Fringes

Fresnel diffraction fringes are commonly observed at edges of a foil or holes or around particles in an extraction replica in out-of-focus electron microscope images and are a useful guide to the instrumental performance. Only a simplified treatment of their origin will be given here. Fuller treatment and references are to be found in Refs. 5, 13, and 14.

The simplest case to consider is Fresnel diffraction at an opaque straight edge illustrated in Figure 3, after Heidenreich.[13] Rays diffracted by the edge interfere with other rays. The position of the fringe maxima may be calculated by path difference arguments:

$$\xi_{max} \approx \left[\frac{L(z_1 + L)}{z_1} (2n - 1)\lambda \right]^{1/2} \tag{19}$$

where n is an integer, λ is the wavelength, and the quantities L and z_1 are defined in Figure 3a.

The intensity distribution is shown in Figure 3b. It will be noted that the first maximum is the strongest and essentially defines the edge contour. The intensity falls off gradually within the edge. If the objective is focused on the edge, so that the distance L in Figure 3a is zero and the viewing plane coincides with the plane of the edge, the path difference near the edge is very small, so that the edge contrast is minimum at the best focus. Since exact focus is rarely achieved and some aberrations remain, a remnant of the first maximum at $n = 1$ may be visible. On defocusing Δf in either direction, so that $L = \Delta f$, fringes appear. Since, in practice, $L = \Delta f \ll z_1$, the Fresnel maxima are approximately located at

$$\xi_{max} \approx [\Delta f(2n - 1)\lambda]^{1/2} \tag{20}$$

In the final image plane, we multiply ξ_{max} by the magnification. This is essentially the same equation derived by Sommerfeld (see Ref. 5) for the Fresnel diffraction of light at a straight edge, when the source and observer (or screen) are both at finite distances from the edge.

Virtual fringes are obtained if the lens is defocused to the weak side, and a bright contour is observed. Conversely, if the lens is defocused to the strong side, a dark contour is observed. As discussed by Heiden-

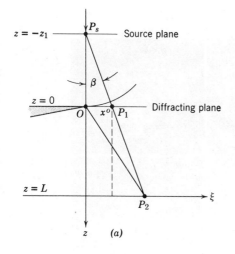

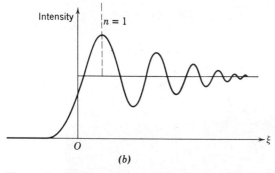

Figure 3 (*a*) Ray diagram for Fresnel diffraction at an opaque straight edge located at $z = 0$ with an electron source at $z = -z$, and a viewing plane $\xi - \eta$ at $z = L$. (*b*) Intensity distribution at position $z = L$ with maxima at positions given by Equation 19. From Heidenreich.[13]

reich,[13] the phase change on going through focus considerably modifies the intensities.

Since each point in the source effectively produces a separate Fresnel pattern, if the source is extended in a direction normal to the edge, the fringes tend to wash out as n (Figure 3*b* and Equation 20) increases, so that only the first fringe is commonly seen.

In the considerably out-of-focus condition, additional weak fringes may be observed inside the shadow of the opaque edge, as noted by Hall.[5]

In practice, whether at round particles in an extraction replica or at

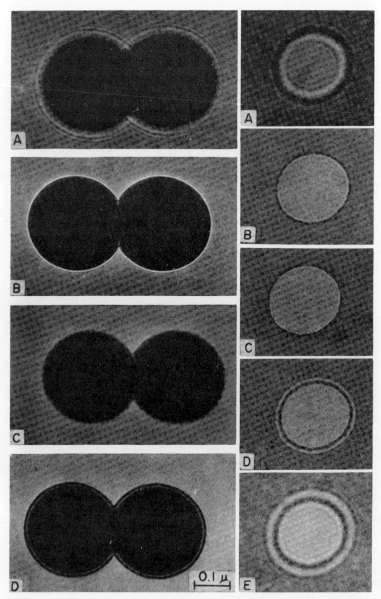

Figure 4 Fresnel fringes observed at polystyrene latex particles on the left and at holes in a collodion film on the right. Through-focus series with the lens current increasing from top to bottom. *C* is in exact focus. From Hall.[5] (Copyright 1953, by the McGraw-Hill Book Co., used with permission.)

holes in or at edges of thin foils, some transmission of electrons occurs through the edge leading to phase shifts and departures from the ideal case.[5,13] The fringes observed are illustrated in Figure 4. The symmetry and uniformity of the fringe width of the fringes, in this instance, indicates that the lens is well stigmated (see Section I.C.1b on Astigmatism).

Fresnel fringes provide a sensitive test of source coherence.[13] Unless a pointed filament is used which provides a highly coherent source of small half-angle, it is unlikely that higher order fringes will be visible. The coherence length determines the maximum Δf for fringe separation and the resolution determines the minimum.[13] If the two highest order fringes observed are called the nth and $(n+1)$th fringes, then from Equation 20:

$$(\Delta f \cdot \lambda)^{\frac{1}{2}}[\sqrt{2n+1} - \sqrt{2n-1}] \gtrsim \delta \qquad (21)$$

where δ is the resolving power. If only two fringes are observed, $n = 1$, giving

$$\delta < 0.73 \, (\Delta f \cdot \lambda)^{\frac{1}{2}} \qquad (22)$$

Thus the resolution may be checked at suitably high magnification, preferably using the overfocused condition which gives better contrast. The most suitable test object is a small hole in a carbon film. The astigmatism is first corrected, of course. More important, the Fresnel fringes provide a routine practical check of image quality, since they are sensitive to such factors as vibration, fluctuations in the high voltage or objective current, stage drift during the exposure, and the cleanliness of apertures. The fact that nearly 40 Fresnel fringes are visible in Figure 5, which is a 6-min exposure of a defocused image of a foil hole taken on an Elmiskop 101 instrument using a pointed filament, is a remarkable example of the performance obtainable with modern microscopes. It may be inferred that the drift rate was less than 0.05Å/sec.

F. Ray Paths

The ray paths and lens functions in the optical and electron microscope show a marked resemblance [see Figure 6, after Thomas[2]]. High resolution electron microscopes usually have an additional lens, called the intermediate lens, between the objective and projector lenses, giving three-stage magnification. A few have two intermediate stages that give four-stage magnification. The similarities between light and electron microscopes result from the fact that electrons can be focused by magnetic lenses just as light is focused by optical lens; however, the magnification of a magnetic lens can be varied by changing the excitation field. It

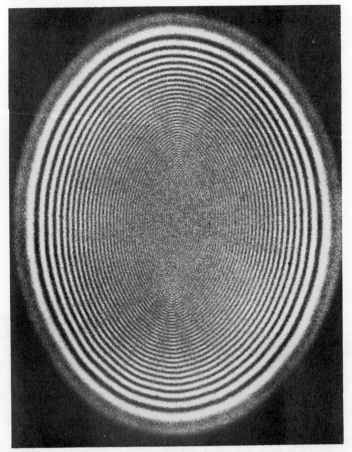

Figure 5 Fresnel fringes in defocused image of a foil hole taken on Elmiskop 101, using pointed filament, 100 keV, illumination aperture of 1×10^{-5} radians, objective aperture 1.3×10^{-2} μ, and electron optical magnification of 30,000$\times$. (Courtesy of Siemens America.)

is necessary to use a fluorescent screen or photographic emulsion to produce a visual representation of the electron image.

G. Bright and Dark Field

Bright and dark field micrographs of crystalline specimens can readily be made in the electron microscope. In bright field, the objective aperture serves to block out all electrons, except those in the transmitted beam (Figure 7), so that foil regions that are diffracting strongly appear dark

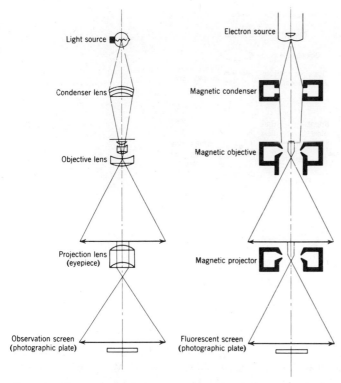

Figure 6 Comparison of the arrangement and the ray paths in a light microscope and a two-stage magnetic electron microscope. After Thomas.[2] (Courtesy of Siemens Corp.)

in the image. Since diffraction depends critically on orientation, the contrast may be changed by tilting the sample.

There are two principal ways of producing dark field. In one the objective aperture is simply displaced to interrupt the direct beam and allow one diffracted beam to pass through. The beam used is selected by inserting and moving the objective aperture, while viewing the diffraction pattern, and then refocusing the intermediate lens to give an image. The resolution of this image is inferior to that of the bright field image, principally because of spherical aberration. To minimize this, it is necessary that the diffracted beam used pass down the axis of the objective lens (Figure 7). This is accomplished in the second high resolution technique by deflecting the beam incident on the sample, using electrostatic or electromagnetic beam deflectors. By means of a deflector energizing

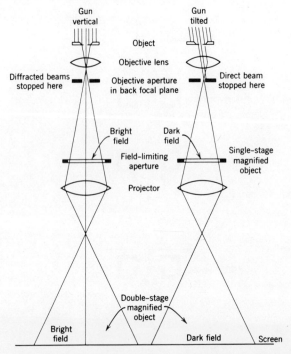

Figure 7 Ray paths for bright field imaging (on the left) and dark field imaging, using tilted illumination (on the right). After Thomas.[2] (Courtesy of Siemens Corp.)

switch, one can then move to and fro between the bright and dark field images. In older microscopes the same result can be accomplished by tilting the gun, a tedious and time-consuming procedure.

Dark field, electron microscopy procedures and applications are discussed in detail by Hirsch et al.,[12] including multiple dark field imaging procedures. Dark field microscopy is often helpful in interpreting bright field, diffraction contrast effects.

H. Selected Area Diffraction

The dark field image serves to give an area display of the intensity distribution of a particular diffracted beam, and a large specimen area may be scanned while viewing the image. A selected area diffraction pattern, on the other hand, permits the simultaneous display of several orders of the diffracted beams, integrated from an area (volume) of the sample, which may be as small as about $\frac{1}{4}$-μ diameter or quite large.

Usually, ¼ to ½ μ is the practical lower limit of the area selected, since the uncertainty in the area selected may amount to 0.1 to 0.2 μ [see discussion by Hirsch et al.[12]]. This technique permits determining the orientation of the area viewed and is also of considerable assistance in interpreting diffraction contrast effects in the corresponding bright field image, produced by refocusing the intermediate lens. The method of imaging the diffraction pattern is illustrated in Figure 8. A small field-

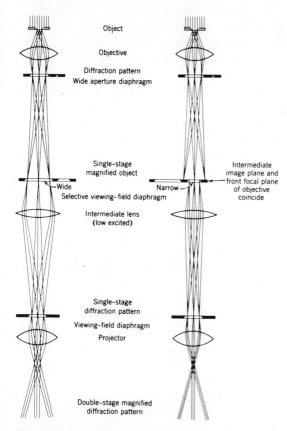

Figure 8 Ray paths for electron diffraction and the use (on the right) of a small area-selecting aperture in the plane of the first intermediate image. After Thomas.[2] (Courtesy of Siemens Corp.)

defining aperture is inserted in the image plane of the objective lens (first intermediate image). The intermediate lens is focused on this aperture, while the objective aperture is temporarily removed. The strength of the intermediate lens is then reduced, in order to view the diffraction pattern present in the back focal plane of the objective lens.

I. Calibration

1. Magnification

Magnification calibration is necessary if optimum precision is desired, since there are many sources of error and changes can occur as the microscope wears. The approximate magnification is usually obtainable from some kind of magnification meter built into the microscope.

Calibration is made against a test object of known dimensions, usually a carbon replica of a ruled diffraction grating. These are obtainable from standard electron microscope supply companies. Typical spacings are 28,800 and 54,800 lines per inch. Reliance is usually placed on the carbon replica which provides a faithful copy of the original standard ruled grating. However, it is a relatively quick and simple matter to determine the grating constant of an individual replica on an electron microscope grid, using an optical spectrometer with monochromatic light, or a laser and optical bench, to obtain a diffraction pattern from which the spot spacing is measured. Since the result is an average spacing, it is important that the microscope calibration be commenced at the low magnification end to include many rulings in the field from a good area of the replica. Calibration is then extended upward in magnification by ratioing pairs of fine points that may be recognized in the micrographs of the grating taken at lower magnification.

The objective current used in making the calibration, which determines the objective magnification, is an important variable and should be noted or standardized in subsequent use. The specimen position, and hence the objective current when focused on the specimen, varies from place to place if the specimen is buckled or tilted, varies from cartridge to cartridge, and even with the same cartridge varies with the firmness of seating and with time as the seating wears. It is desirable to record the objective current with each micrograph taken, and to calibrate the magnification for several specimen positions and hence the objective currents, so that a correction can be applied if desired. If a good quality current meter is added for this purpose, care should be taken that it cannot open circuit which might cause damage to the lens.

If the above precautions are observed, an accuracy of about ±2% is achievable for measurements made on the photographic plate. In a three-stage instrument, the magnification is usually calibrated against the intermediate lens current for one or more fixed projector settings. It is difficult to obtain better accuracy because of other sources of error such as lens hysteresis. One alternative approach is to use as an internal standard fragments of grating replicas which may be purchased in aqueous suspension. A drop is placed on the sample and allowed to dry. The older technique, using a suspension of latex spheres of known size, suffers from the disadvantages that the sphere size tends to increase as contamination occurs in the microscope and the apparent size is a function of the focusing.

At very high magnifications the lattice planes themselves, in a suitable crystal, can provide a test object. The spacing is known with high precision from x-ray measurements. Particular types of crystal have not yet become standardized for this purpose. We have used the {111} planes in (112) or (110) slices of high purity silicon or germanium for this purpose, which have spacings of 3.14 Å and 3.27 Å, respectively. These materials are rigid and not susceptible to plastic deformation at room temperature, so that slices may be cut and chemically thinned without residual damage. The dislocation density of standard semiconductor grade is low. The plane spacings are remarkably constant from place to place, which is apparently not true of some natural minerals that exhibit variations in plane spacings, presumably due to compositional inhomogeneity or variation in stoichiometry. The techniques involved have been discussed elsewhere.[12,15] (See also Section V.C.4c, on Resolution Limit for Periodic Structures.) Our experience indicates that a precision of ±2% should be realizable. The basal planes (spacing ~3.4 Å) in graphitized carbon powder have also been suggested as a test object.[16] Since 3-Å resolution is still difficult to achieve under less than ideal conditions, additional test objects with larger spacings in the 5 to 10-Å range are desirable.

2. Resolution

In high resolution instruments, two kinds of resolution evaluation are customary, one to check the so-called "point" resolution, the other to check the "line" resolution. The point resolution is usually defined as the smallest distance between the centers of two points that can be distinguished as separate in the image. Suitable test objects, obtained from electron microscope stocklists, usually consist of a carbon support film on a grid, onto which a very small quantity of a stable "amorphous" material has been evaporated. Carbon and platinum, simultaneously evaporated, is one combination that has been used, since it resists agglomeration in the electron beam. The contamination layer, which forms at an edge or hole in a specimen during exposure in the electron beam, has a granular structure and provides a good test object. In order to distinguish from electron noise, the pair of dots whose spacing is to be measured should be visible in two different exposures of the same field. Normally, pairs of dots in different orientations are measured to show that the resolution is present in several directions. The brochures of the leading high resolution microscopes contain micrographs that demonstrate resolutions of 3 to 5 Å measured in this way. Since the points being measured are typically 1 to 2 Å in diameter and are not claimed to be atoms, but are rather phase detail of an unknown nature, the physical significance of this kind of measurement is not at all clear. It would be much more satisfactory if atoms or rows of atoms parallel to the beam could be resolved with the correct known spacings.

Adequate test objects for the line resolution are more readily obtainable, since the lattice planes themselves may be imaged and checked to see that they have the correct directions (by selected area diffraction) and spacings (see previous section, I.I.1, on Magnification Calibration). A series of micrographs may be taken on the same test specimen, using different diffracted beams to resolve planes in different directions relative to the lens, to demonstrate that the line resolution is present in different directions. Alternatively, the specimen could be rotated between exposures. By using a graphitized carbon powder test specimen without an objective aperture, basal planes (~3.4-Å spacing) facing in different directions in different portions of the field may be simultaneously resolved, checking out many directions in one exposure. However, suitable micrographs are difficult to make on powder.

The use of a silicon (110) test specimen and a three-beam technique,[17] for example, permits two sets of {111} planes to be resolved simultaneously. Since the image now shows, not just two sets of lines, but dynamic contrast detail revealing the expected periodicities in the 100, 110, and two 111 directions in the foil plane, and the intersections of the two sets of {111} planes really define atom row positions, a single micrograph serves to check both the line and the point resolutions.

Micrographs showing lattice planes provide a severe instrumental check. The minimum spacing of planes or crossed lattice detail resolvable is a check of a resolution. However, the contrast and general quality of the image provide a stringent check on other factors such as the coherency of the illumination and the absence of electrical and mechanical variations.

3. Electron Diffraction

In order to determine the plane spacing d from the measured radius R of a spot about the center spot in a selected area, electron diffraction pattern, obtained in the electron microscope, it is necessary to assign a value to the product λL, usually referred to as the diffraction constant of the microscope, in the expression

$$d = \frac{\lambda L}{R} \qquad (23)$$

Here λ is the electron wavelength, and L the effective camera length employed which depends upon the lens magnification. Equation 23 is an approximate relation, based on the assumption that the diffraction pattern represents a plane section through the reciprocal lattice. This is a good approximation for small scattering angles. A more accurate expression is not required, since other errors due to distortion and stray fields in the microscope are dominant. Some workers prefer to use $2\lambda L$ as the "diffraction constant" and substitute the diffraction ring diameter D for the radius R in Equation 23.

The λL is determined by calibration with a standard specimen which gives a series of diffraction rings of known d values; it is averaged from the individual values obtained from each value of R and d. Suitable standards are fine grained evaporated films of aluminum, thallium chloride, or gold on a carbon substrate. An accuracy of 1 to 2% is possible by periodic calibration of the instrument at standard lens settings as used for the unknown. Since variations in specimen position lead to a variation in magnification, it is desirable to determine the diffraction constant for various objective currents. For a focused specimen, the diffraction constant is directly proportional to the focal length. If this correction is not made, it is better to use a fixed objective lens current, corresponding to the calibration, and tolerate a small amount of defocus.

Improved accuracy is possible if an internal standard is used. In the case of unknown particles in a matrix of known parameter, the latter may often be used as a standard for the former. Alternatively, a little standard material may be evaporated or otherwise deposited on the unknown sample. With correction for, or standardization of, the objective current, a standard specimen may be placed in the same specimen holder adjoining the unknown or included in the same batch of plates. Recording on plates avoids shrinkage that may appear on film. Measurements should be made on the plates, using a good measuring microscope, rather than on enlarged prints, where additional errors can occur, caused by the uncertainty of the enlargement magnification, aberrations in the enlarger lens, and nonuniform shrinkage of the paper. If the diffraction rings are not circular, the diffraction constant varies with the direction and an appropriate value should be used for a spot in the unknown pattern. Given a suitable sample tilted into an ideal orientation, with proper condenser settings and exposure, so that a sharp intense spot pattern is obtained, an accuracy of about 0.1% may be achievable. The sources of error are discussed in more detail by Hirsch et al.[12]

4. Rotation

In order to relate a selected area diffraction pattern to the detail in the corresponding micrograph, it is necessary to correct for the rotation. This rotation results from the fact that electrons follow a spiral path through a lens field, so that the image rotation varies with the field strength and the electron energy, there being less rotation the higher the energy for a given field strength. A correction is usually necessary, since, in a three-stage instrument, for example, the intermediate lens setting required to image the diffraction pattern is different from that required to image the structure. Using the diffraction intermediate lens setting as a reference point, the rotation is normally calibrated and plotted as a function of intermediate current and hence of magnification for fixed settings of the objective and projector lenses, for a given electron voltage.

The simplest calibration procedure is to use a replica of a standard single-ruled diffraction grating and take a series of micrographs of the same field, starting at low magnification and progressively increasing the intermediate lens current. At higher magnifications, where irregularities in the rulings become troublesome, a line may be drawn through two identifiable points in successive micrographs to determine the relative rotation. At very high magnifications, a single set of lattice planes in a suitable test object may be imaged and similarly used. Rotation calibration can thus be carried out simultaneously with the magnification calibration already discussed.

If a diffraction grating replica is not available, a crystal of MoO_3 may be employed for rotation calibration, as described by Thomas.[2] The edge of a crystal, which has a natural elongated tabular habit, is employed as a reference direction. A series of double exposures may be made, with the diffraction pattern superimposed on micrographs made with different intermediate lens settings. Errors can arise if the crystal is not laying flat on the grid.

Since the image is inverted at each stage, there is an additional net 180° rotation correction if an odd number of magnification stages is employed. It is necessary to identify suitably a particular corner of the plate in the microscope or on the print if a correction is made using prints, and then to print both the micrograph and the diffraction pattern with the emulsion the same way up. The use of an automatic device to "print" a number on a corner of the plate in the microscope helps to avoid errors of this kind if the number is visible on the print.

J. Stereoscopy

Stereoscopic pairs of electron micrographs may be obtained by tilting a replica at angles $\pm\alpha$, with reference to the perpendicular position, about an axis normal to the optical axis of the electron microscope between exposures of the same field. The angle α is commonly about 6° and is provided in a stereoscopic specimen cartridge which is standard equipment on most microscopes. When viewed through a stereoscopic viewer with the eyes perpendicular to the axis of tilt, an illusion of depth is obtained. Care is necessary in distinguishing height from depth, since the two are reversed if the micrographs are interchanged. It is convenient to record the images at lens settings such that the resultant image rotation sets the tilt axis parallel to one edge of the micrographs which is then noted. Since the cartridge tilt axis makes a fixed angle (usually a multiple of 45°) to the specimen traversing directions, the tilting direction may be identified with reference to the latter by a double-exposure technique, traversing the image in one direction a little way between exposures.

If the difference p in lateral separation of two image points perpendicular to the tilt axis, that is, the parallax, is measured, then:

$$\Delta z = \frac{p}{2M \sin \alpha} \tag{24}$$

where Δz is the vertical separation of the two points in the object, M the magnification, and $\pm\alpha$ the angle of tilt referred to above. If the two points selected are on opposite surfaces of the object, the thickness may be measured in this way. Nankivell[18] claimed that an accuracy of $\pm 5\%$ could be achieved with a 10° tilt, but this would necessitate the careful calibration of α for the stereoscopic holder employed [see discussion of errors by Wells[19]].

The technique is particularly useful for fractographs. Unfortunately, in the case of crystalline objects viewed by diffraction contrast, the contrast is usually changed by tilting, so that the stereoscopic effect is lost. In order to preserve the contrast, special diffracting positions must be chosen [see discussion by Hirsch et al.[12]]. This is facilitated if Kikuchi lines are present. The technique has been applied to view dislocation arrays and precipitates in crystals, and for diffraction contrast studies of defects such as vacancy and interstitial clusters, in which correct interpretation may require knowing the depth at which a defect is located in the foil.

II. INSTRUMENTATION

A. Electron Microscopes

Electron microscopes are commercially available from 10 or more suppliers around the world, most of whom manufacture several different models aimed at meeting various needs. High resolution microscopes are most commonly limited to 100 keV. A variety of smaller and cheaper medium resolution instruments are available. Since most of these attain maximum voltages of 75 keV or less, they are of limited use for transmission electron microscopy or electron diffraction on metallic foils but suitable for work on replicas or noncrystalline samples.

For high resolution, it is desirable to use as short an objective focal length as possible which means increasing the field strength and decreasing the bore diameter of the pole-piece, making it very difficult to introduce devices for such purposes as tilting, heating, and stretching the specimen because of the restricted space. Accordingly, high resolution microscopes tend to be designed either for biological work without provision for devices but with optimum resolution, or for metallurgical work with provision for devices, but some compromise on the resolution. Alternatively, the design may permit changing from a short focal length, high resolution stage to a longer focal length, lower resolution stage suitable for devices.

Two of the many 100 keV microscopes will be arbitrarily selected to illustrate the typical features of a high resolution microscope. The Siemens Elmiskop 101 (Figures 9 and 10) is an example of a three-stage instrument with separate power supplies (not shown), located remotely from the microscope. The Norelco EM300 (Figure 11) is a compact four-stage instrument with self-contained power supply, automatically controlled vacuum system, and fully transisterized electronics with plug-in, printed circuit boards. Both are capable of a point-to-point resolution of 3 Å under ideal conditions, including a low level of vibration.

The typical components of both instruments shown in Figures 9 to 11 follow. The electron beam is produced from a self-biased three electrode gun, employing a directly heated tungsten filament to produce a highly intense spot of small size. In the EM300 the gun has a separate airlock to facilitate changing the hairpin-shaped filament. Either hairpin or point filaments are standard in the Elmiskop 101, the point providing a greater current density, so that smaller illumination apertures can be

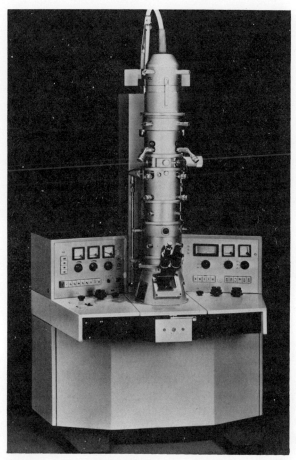

Figure 9   The Siemens Elmiskop 101, 100 keV, electron microscope. (Courtesy of Siemens America.)

used while maintaining the same image brightness, thus facilitating Fresnel fringe observations for correction of the astigmatism. The high voltage may be adjusted in 20 keV steps from 40 to 100 keV in the Elmiskop 101 and from 20 to 100 keV in the EM300 microscope.

The condenser consists of two independently controlled lenses, the first of which (condenser 1) demagnifies the spot, while condenser 2 focuses the spot onto or near the specimen plane. Condenser 1 controls the minimum size to which the spot can be focused on the specimen by condenser 2, while the latter varies the intensity, diameter, and aperture of the illuminating beam, and is provided with an electromagnetic

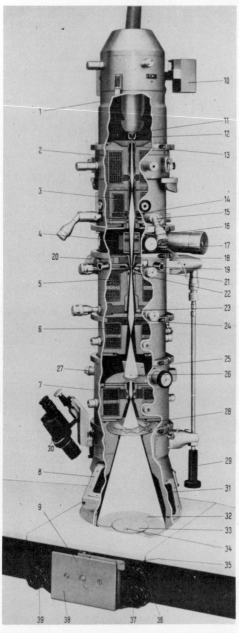

Figure 10 Cross section of the column of Elmiskop 101 electron microscope. 1, electron gun; 2, condenser 1; 3, condenser 2; 4, specimen airlock with beam deflecting system; 5, objective; 6, intermediate lens; 7, projector; 8, final image tube; 9, photographic chamber; 10, servomotor for cathode displacement; 11, cathode;

172

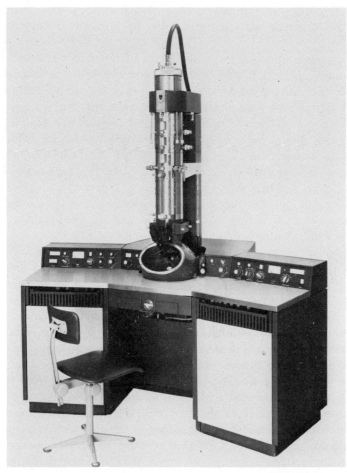

Figure 11 The Norelco EM300, 100 keV, electron microscope. (Courtesy of Philips Electronic Instruments.)

12, bias shield; 13, anode; 14, electromagnetic stigmator in condenser 2; 15, control for the apertures in condenser 2; 16, deflecting system; 17, automatic ventilation and evacuation of the airlock antechamber; 18, specimen cartridge; 19, specimen; 20, anticontamination device; 21, control for the objective apertures; 22, electromagnetic objective stigmator; 23, intermediate lens selector apertures; 24, electromagnetic intermediate lens stigmator; 25, mirror for intermediate image observation; 26, intermediate image screen; 27, drive for the intermediate image screen; 28, exposure shutter; 29, control for specimen stage adjustment; 30, binocular magnifier; 31, viewing window; 32, outer field screen; 33, central field screen; 34, small field screen; 35, lever for opening the photographic chamber door; 36, control for final image screen; 37, drive for film and plate transport; 38, door of the photographic chamber; and 39, airlock drive. (Courtesy of Siemens America.)

stigmator for eliminating astigmatism in the illuminating spot. The beam position can be centered electromagnetically to correct day-to-day variations.

The objective lens produces a magnified image of the specimen which is inserted through an airlock chamber. The objective lens operates at a fixed magnification for a given focal length and electron energy. Astigmatism can be corrected by an electromagnetic stigmator. An anticontamination device is provided to cool the surroundings of the sample. An electromagnetic beam deflector above the sample can be fed with direct current to change the angle of incidence of the beam on the sample, while the point of incidence remains fixed, permitting high resolution dark field work. When fed with alternating current, the device acts as a wobbler and increases the illumination aperture, thus decreasing the depth of focus and facilitating focusing at low and intermediate magnifications. The wobbler is switched off after focusing.

The first intermediate image, which is produced by the objective lens, is enlarged in the Elmiskop 101 by the variable intermediate lens and again by the projector lens, which is normally operated at fixed magnification, to produce the final image magnified by up to about $\frac{1}{4}$ million times. The projector lens magnification can be changed by changing the pole-piece. Is the EM300, an additional intermediate stage is provided by the diffraction lens which is situated between the objective and intermediate lenses. This is used as a variable reducing lens at low magnifications with the intermediate lens deenergized. At high magnifications it is used as a fixed enlarging lens with variable intermediate, extending the final electron optical range to about $\frac{1}{2}$ million diameters in normal working. In both microscopes, low magnification ranges are obtainable by deenergizing one or more lenses. Selected area diffraction (see Section I.H) and low angle (transmission) diffraction may be carried out in both instruments.

A number of choices of film and plate cameras are available for image recording, and a built-in exposure meter and electromagnetic shutter are provided. Alternatively, in the EM300 the image may be displayed on a TV monitor via a transparent phosphor-coated screen and a television camera located below the column. In the Elmiskop 101 either a ciné camera or TV camera may be used to look at the image on the normal phosphor screen through the front viewing port.

A built-in mechanical crane permits separating the column for maintenance. The gun and lenses are independently alignable. Built-in test units permit checking the current and voltage stabilities, as well as other important circuit conditions and the correct operation of the vacuum system. Safety interlocks protect against operator error or equipment failures.

A principal point of difference is in the design of the objective lens. In the Elmiskop 101, the specimen holder is lowered from above into the top of the objective lens, whereas in the EM300 it is inserted from the side into the split-immersion-type lens. The latter has an unusually short focal length of 1.6 mm, giving low aberrations (spherical aberration coefficient $C_s = 1.6$ mm, chromatic coefficient $C_c = 1.3$ mm) and a theoretical resolving limit of 2.3 Å. The former has a focal length of 2.7 mm ($C_s = 3.2$ mm, $C_c = 2.1$ mm). In both cases larger bore pole-pieces may be substituted with an increase in focal length and some reduction in resolution.

Either a fixed or a single-tilt specimen holder is at present available in the EM300 high resolution stage. However, using the somewhat lower resolution tilt or goniometer stages, with a larger bore pole-piece, a large variety of devices permitting double tilting, tilt and rotation, heating, cooling, stretching, and so forth, may be used. In the Elmiskop 101, which has only one stage, devices C, D, F, G, and H (Figure 12) may be used with the high resolution pole piece; whereas devices E, J, and K require a larger bore (lower resolution) pole-piece. General experience with devices in other microscopes indicates that the resolution obtainable tends to be limited in practice by other factors such as (a) the degree to which the sample is decoupled from external vibration; (b) freedom from drift which can be either thermal or mechanical in origin; (c) contamination due to such factors as dirt and oxidation which can cause charging-up or aberrations; and (d) stray fields from heater coils, and so forth. See Refs. 20 and 21 for reviews of devices.

In both instruments the contamination rate of the sample in the electron beam, due to the polymerization of organic vapors in the system, is greatly reduced by anticontamination devices of the type in which the sample surroundings are cooled preferentially to remove contaminants. The anti-contamination devices in Figure 12 are compatible with the use of all the stages and other devices mentioned. This represents a major advance over previous instruments for high resolution work.

Typical stabilities in these high resolution microscopes are high tension $\leq 5 \times 10^{-6}$ per min, objective lens $\leq 5 \times 10^{-6}$ per min, other lenses $\leq 2 \times 10^{-5}$ per min. Normal high vacuum is $\leq 5 \times 10^{-5}$ torr.

B. Accessories for Electron Microscopes

Although better results can usually be obtained by specialized instruments such as scanning electron microscopes, electron microprobes, and conventional electron microscopes, rather than by a single general purpose instrument, the cost tends to be prohibitive, unless sufficient work is available to keep them fully employed. Accordingly, a great deal of

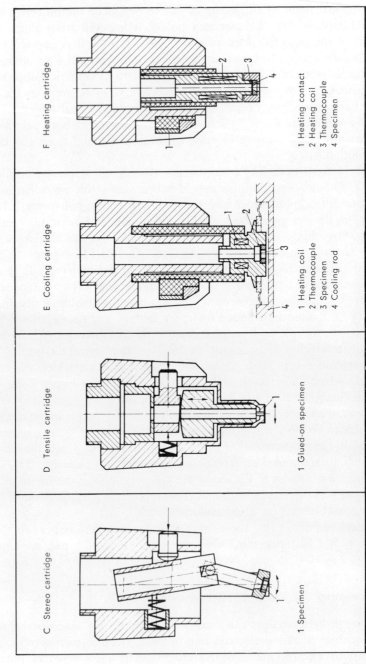

Figure 12 (*a*) Specimen cartridges for special procedures for the Elmiskop 101 electron microscope. (Courtesy of Siemens America.)

C Stereo cartridge

1 Specimen

D Tensile cartridge

1 Glued-on specimen

E Cooling cartridge

1 Heating coil
2 Thermocouple
3 Specimen
4 Cooling rod

F Heating cartridge

1 Heating contact
2 Heating coil
3 Thermocouple
4 Specimen

176

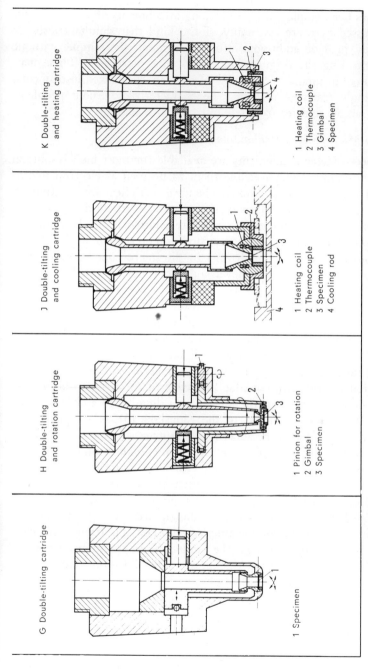

G Double-tilting cartridge

1 Specimen

H Double-tilting
and rotation cartridge

1 Pinion for rotation
2 Gimbal
3 Specimen

J Double-tilting
and cooling cartridge

1 Heating coil
2 Thermocouple
3 Specimen
4 Cooling rod

K Double-tilting
and heating cartridge

1 Heating coil
2 Thermocouple
3 Gimbal
4 Specimen

Figure 12 (*b*) Specimen cartridges for special procedures for the Elmiskop 101 electron microscope. (Courtesy of Siemens America.)

177

ingenuity has been employed in developing attachments of various kinds. Those discussed here are essentially of the kind that simultaneously or consecutively provide additional information about the sample, without seriously impairing the original function of the instrument. These may have a distinct advantage over the use of separate specialized instruments, in that a single (micro-) area may be characterized without problems inherent in sample transfer from one instrument to another.

1. Electron Microprobe Attachments

X-ray spectrometer attachments are available for most high resolution electron microscopes, which enable them to be used as electron microprobes and provide at least limited chemical analytical information on precipitates in extraction replicas or thin sections, particle dispersions, or biological sections. Characteristic x-rays are emitted in all directions, as a result of the impingement of the electron beam on the sample. Normally, these are blocked off to avoid hazard to the microscope operator. If a port is provided at the sample level and the sample is tilted at 15° or so to the beam, a portion of the emitted x-rays may be taken off horizontally into a spectrometer mounted outside the column at the stage level. Figure 13 shows a crystal spectrometer mounted on the Elmiskop 101. While a small takeoff angle is necessary to avoid undue foreshortening of the electron microscope image, it means that the x-ray signal is sensitive to topographical sample features. Furthermore, the x-ray corrections required are those for oblique incidence of the electron beam.

Using a double condenser, the sample area illuminated and hence analyzed may be $\leq$ 1-μ diameter. In the Elmiskop 101, 0.1-μ spot size is said to be attainable when the accelerating voltage is reduced to 40 keV. Since the specimen remains in the normal plane, high resolution microscopy and selected area diffraction may be carried out on the same area.

With suitable calibration, the x-ray intensity can be used to measure the thickness of a sample of known composition, although the result will be some kind of average over the illuminated area. With a normal transmission, electron microscope sample, a substantial portion of the incident beam is transmitted through the sample. Since transmission samples are typically 2000 Å or less in thickness, lateral diffusion of the x-rays is minimal and the geometrical resolution is principally controlled by the beam size. The correction procedures required to obtain quantitative results from the raw data obtained, even on samples of uniform thickness such as evaporated films, clearly differ from those discussed in connection with the analysis of a bulk sample in the electron microprobe. The most satisfactory way of obtaining quantitative results would be to ratio the x-ray intensities to those obtained on homogeneous standards of similar

Figure 13   Microprobe attachment, consisting of crystal spectrometer mounted on the Elmiskop 101 electron microscope. (Courtesy of Siemens America.)

known composition and of the same thickness as the sample to be analyzed. The problem is how to obtain such standards. The whole question of correction procedures for transmission samples appears to have been but little discussed. A further important question is how the x-ray intensity is affected in the vicinity of tilt contours, where Bragg diffraction of the incident electron beam is occurring, since this is necessary if diffraction contrast is to be obtained in transmission electron micrographs.

When the sample consists of an extraction replica or powder dispersion, it will be difficult to obtain more than qualitative analyses, due to the geometry. In the case of second phase particles which are smaller than the beam size, some average composition of the matrix plus particles will be measured. If the particles are larger than the beam size, averaging can be avoided, but the particles will tend to be opaque to the beam, so that electron diffraction patterns cannot be obtained and only shadow electron micrographs will be possible.

Since all x-ray wavelengths are normally analyzed using a single spectrometer, the crystals either have to be changed for different parts of the spectral range or a general purpose compromise adopted. One approach to this problem, used for example in the Norelco attachment for the EM300 microscope, is to use a bendable mica crystal whose radius of curvature is automatically adjusted as a continuous function of 2θ angle to focus the diffracted x-rays at the detector.

Nondispersive (energy) x-ray analysis may also be carried out by permitting the x-ray beam to impinge directly on a suitable detector, used in conjunction with the usual pulse height analysis equipment. Early microprobe attachments for the electron microscope were principally of this type and suffered from poor discrimination for neighboring elements in the periodic table. Focusing crystal spectrometer attachments were subsequently developed, giving improved resolution at the sacrifice of some sensitivity. Recent improvements in the discrimination of solid state detectors seem to indicate that nondispersive-type detectors may again shortly come into favor.

A gain in sensitivity should in general be achieved if a solid state detector can be mounted nearer to the sample than the crystal in a spectrometer. Ultimately, it may be possible to develop a hemispherical solid state detector, mounted coaxially with the beam, which would pick up the whole of the back-emitted x-rays, giving a gain in signal strength of at least 20 times over the optimum possible with present solid state detector configurations. Since the chief application of microprobe attachments is likely to be for qualitative analysis, it seems logical to use nondispersive analysis which permits a complete spectral display (see discussion under electron microprobe analysis in Chapter 9).

The typical electron beam current, and consequently the intensity of the emitted x-rays in the electron microscope, is comparable with that in the electron microprobe which is three to four orders of magnitude greater than that in the scanning electron microscope. Other factors being equal, much better sensitivity should be achievable using a microprobe attachment on a conventional electron microscope than on a scanning electron microscope. Of course there is some loss in intensity if the sample is thin.

Microprobe attachments can be used either for spectral analysis on a spot or for examining the distribution of one element in different parts of the sample. Since beam deflectors are normally provided in recent model microscopes, scanning x-ray images are also possible in principle. The electronic readout equipment needed for these various purposes is discussed in connection with the electron microprobe.

2. Reflection Electron Diffraction Attachments

Electron diffraction, carried out with an electron beam at glancing incidence—the so-called "reflection technique"—is particularly valuable for examining the surfaces of samples (bulk or otherwise) to a depth of about 100 Å with, say, a 60-keV beam. In many cases the diffraction pattern is attributable to electrons transmitted through small asperities on the surface.

For high precision it is desirable that there be no sources of stray fields between the sample and the camera, and shielding may be necessary against even weak external fields which may otherwise result in pattern distortion. The disadvantage of making reflection electron diffraction patterns in the electron microscope using a special holder in the normal stage position, rather than using a separate precision reflection electron diffraction instrument, is that even though the objective, intermediate, and projector lenses are switched off, some residual field remains and this is nonreproducible. Thus precision is lost, unless an internal standard is employed. Medium precision reflection diffraction may, nevertheless, be usefully conducted with such devices which are available for many microscopes. Some type of goniometer stage is desirable in order to perform the necessary sample manipulations. It may be possible to use the sample substrate as an internal standard or mount a standard alongside the sample. If the sample is thin enough, normal transmission electron diffraction and transmission electron microscopy may also be carried out.

The above disadvantage is overcome in an attachment such as that available for the EM300 microscope, in which the specimen is placed below the projector lens and the diffraction pattern (reflection or transmission) viewed on the fluorescent screen or recorded in the subscreen camera. Goniometer-type motions are provided for the manipulation of the sample, an airlock for sample charging, an auxiliary ion gun for neutralizing charges, and a hot stage for dynamic experiments up to 450°C. Since the device can be removed or attached in a few minutes using the projection chamber airlock, the microscope can be rapidly converted for normal use.

3. Scanning Electron Microscopy Attachments

With the addition of a suitable attachment, some electron microscopes that contain built-in, beam deflector systems can be converted for scanning electron microscopy. The necessary components, which will be clear from the chapter on Scanning Electron Microscopy, are some kind of detector and the electronics to provide a cathode ray display synchronized with the beam scan. A brief description of an attachment for the Philips (Norelco) EM200 or 300 electron microscope follows.

The Philips device is employed with the goniometer stage and involves the introduction into the column of a suitable electron detector. However, if a microprobe attachment is also installed, an x-ray signal may be used to form a scanning image. The attachment is designed for reflection or transmission scanning use with a resolution of approximately 250 Å; the normal detector is for primary electron energies. The detector must be removed from the column, in order to carry out normal transmission microscopy. In addition the special pole-piece required in the scanning mode needs to be interchanged, unless reduced resolution is acceptable.

The small spot size necessary to get the above resolution is achieved by using the upper half of the split immersion objective lens to focus a parallel beam of electrons from the double condenser in the sample plane. The lower half is then used to image the transmitted beam at infinity for transmission electron microscopy, and magnifications up to 100,000× may still be obtained using the intermediate and projector lenses.

Other features include twin cathode ray tube displays with different phosphors, contrast reversal; amplifier "bias" contrast control; area scan or line scan at 0, 45, or 90°; and variable frame and line sweep rates. A variety of holders permit transmission or opaque samples to be examined.

III. SPECIMEN PREPARATION

In view of the importance of specimen preparation for successful electron microscopical studies, a brief account will be given of some of the techniques of interest to the metallurgist or material scientist. No attempt will be made to survey the vast literature. Extensive reviews and references may be found in the book edited by Kay.[22]

A. Replica Methods

1. General Comments

Topographical features on bulk samples may be studied in great detail by making electron transparent replicas of the surface which are detached for examination in the electron microscope. Historically, the technique was developed before it was learned that metals could be thinned into electron transparent form. The replication of etched metal samples extended the resolution nearly 100-fold beyond that of the light microscope. Replication in conjunction with etching techniques to study internal structure has tended to be replaced by transmission electron microscopy which is more direct and capable of higher resolution. Replication remains useful, particularly for the examination of surface structure, for example, fractures, deformation markings, and worn surfaces. However, scanning electron microscopy, particularly as the resolution is further improved, will undoubtedly tend to replace replication for the examination of surfaces. Certain objects either because of their small size, fragility, or shape (e.g., reentrant angles or extreme roughness) can be almost impossible to replicate faithfully. Replication requires skill and experience, both in making high quality replicas and in interpreting micrographs and recognizing artifacts.

Replication has certain unique advantages. The technique can be nondestructive, and objects both large and remote from the microscope can be examined. Large selected areas, even square feet in extent, can be replicated and small regions subsequently selected for high magnification examination. Extraction replicas, by removing particles from the matrix, often provide an optimum sample for study, for example, for identification by diffraction. If only a moderate resolution is needed and the sample is straightforward, adequate replicas can be made relatively easily with a little practice. Replicas are normally small, easily stored, deteriorate only slowly with time, and thus provide a convenient semipermanent record.

Detailed replica techniques will not be described here but may be found in the literature; see Refs. 2, 22, and 23). There are innumerable variations. Replica techniques may be classified broadly into direct (one-stage) and indirect (usually two-stage, but three-stage are sometimes employed). Replicas are called *positive* when a hill on the specimen appears to be a hill on the replica, and *negative* when a hill appears to be a valley. It is customary to enhance the contrast of replicas by *shadow casting*, that is, by evaporating a heavy metal from an angle to give shadow effects in the final electron micrograph, equivalent to oblique light illumination. In a positive replica, a peak on the sample throws a dark shadow.

2. Plastic Replicas

Direct negative plastic replicas (Figure 14) may be made by applying a ½ to 2% solution of plastic in an organic solvent to the sample surface and draining off the excess. Formvar and collodion are the commonest plastics used. Repeated replication of the same area is possible. The first replica removes dirt from the surface and is usually discarded. Contrast results from variations in the thickness of the replica but can be enhanced by shadow casting (Figure 14). If the angle of shadow casting is noted, the elevation of specimen features may be determined from the shadow lengths. These materials have a coarser self-structure than carbon which tends to limit the resolution to approximately 200 Å at best. They are

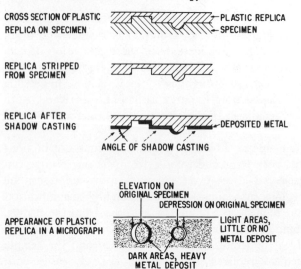

CROSS SECTION OF PLASTIC   PLASTIC REPLICA
REPLICA ON SPECIMEN   SPECIMEN

REPLICA STRIPPED
FROM SPECIMEN

REPLICA AFTER
SHADOW CASTING   DEPOSITED METAL
          ANGLE OF SHADOW CASTING

                    ELEVATION ON
                    ORIGINAL SPECIMEN
                        DEPRESSION ON ORIGINAL SPECIMEN
APPEARANCE OF PLASTIC   LIGHT AREAS,
REPLICA IN A MICROGRAPH   LITTLE OR NO
                                        METAL DEPOSIT
              DARK AREAS, HEAVY
              METAL DEPOSIT

Figure 14   Direct negative plastic replica technique. Prepared by E. F. Koch. (Courtesy of General Electric Company.)

weaker than carbon and easily damaged during stripping. They are also less stable than carbon in the electron beam and fairly easily damaged.

For good resolution of detail in the electron microscope, variations in thickness of a plastic replica due to object detail need to be a reasonable proportion of the total thickness (Figure 14). Significant variables, then, are the degree of etching, if an etched sample is employed, and the replica thickness. The etch employed would be considered a light etch for light microscopic work. No matter how good the replicating technique, the final result can be no better than the initial metallographic preparation, so that it is often desirable to employ electrolytic polishing.

Plastic replicas may also be used to back-up other replicas when there is danger of distortion or damage, either during stripping or during dissolution when swelling may occur. Many procedures have been devised, based on the fact that there are selective solvents for different plastics.

3. Carbon Replicas

Direct carbon replicas (Figures 15, 16, and 17) are made by evaporating carbon from a carbon arc onto the surface of the sample in a normal direction. Usually, one carbon rod is squared off and the other is turned down at the tip to a reduced section of controlled diameter and length. The whole of the reduced section is evaporated, giving a predetermined replica thickness for a given distance of the sample from the arc.

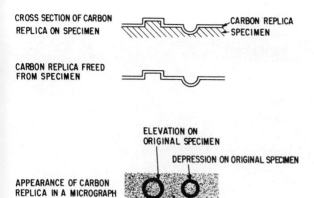

Figure 15 Direct carbon replica (unshadowed) technique. Prepared by E. F. Koch. (Courtesy of General Electric Company.)

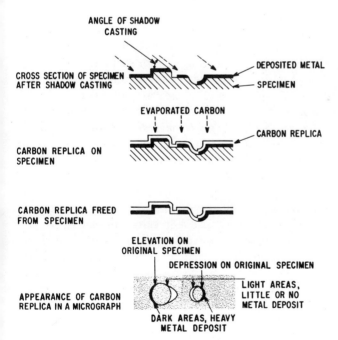

Figure 16 Preshadowed direct carbon positive replica technique. Prepared by E. F. Koch. (Courtesy of General Electric Company.)

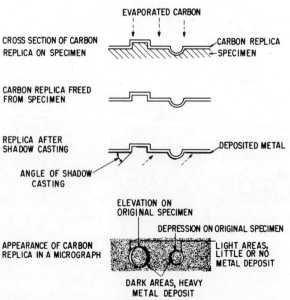

Figure 17. Postshadowed direct carbon negative replica technique. Prepared by E. F. Koch. (Courtesy of General Electric Company.)

Shadow casting may be carried out before or after stripping, depending on whether preshadowed positive (Figure 16) or postshadowed negative replicas (Figure 17) are required. As a comparison of Figures 16 and 17 shows, for the proper interpretation of replication electron micrographs, it is essential that adequate details be given of the replication and shadow-casting steps used, and of the sequence in which they were carried out. It is helpful to indicate the direction of shadow casting on a micrograph.

Direct carbon replicas, since they are stronger than plastic, may be made as thin as 100 to 200 Å and still remain intact on stripping. Because of their thinness and low self-structure, direct carbon replicas give the best resolution ($<$ 100 Å). While they may sometimes be floated off on a water surface, it is usually necessary to electropolish or etch-off which destroys the sample surface. By using indirect carbon replicas with cellulose acetate tape as a first stage, repeated replication is possible, but some resolution is lost (Figure 18). The tape, 1 to 5 mils thick, is either moistened on one side with a solvent to soften it, and then pressed onto the sample; or it is dipped in the solvent and simply hung over the sample. After being allowed to harden, it is dry-stripped and

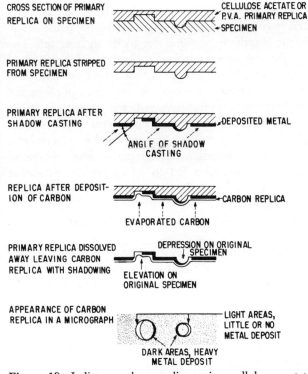

CROSS SECTION OF PRIMARY
REPLICA ON SPECIMEN

CELLULOSE ACETATE OR
P.V.A. PRIMARY REPLICA
SPECIMEN

PRIMARY REPLICA STRIPPED
FROM SPECIMEN

PRIMARY REPLICA AFTER
SHADOW CASTING

DEPOSITED METAL

ANGLE OF SHADOW
CASTING

REPLICA AFTER DEPOSIT-
ION OF CARBON

CARBON REPLICA

EVAPORATED CARBON

PRIMARY REPLICA DISSOLVED
AWAY LEAVING CARBON
REPLICA WITH SHADOWING

DEPRESSION ON ORIGINAL
SPECIMEN

ELEVATION ON
ORIGINAL SPECIMEN

APPEARANCE OF CARBON
REPLICA IN A MICROGRAPH

LIGHT AREAS,
LITTLE OR NO
METAL DEPOSIT

DARK AREAS, HEAVY
METAL DEPOSIT

Figure 18   Indirect carbon replica, using cellulose acetate or polyvinyl alcohol (PVA) primary replica. Prepared by E. F. Koch. (Courtesy of General Electric Company.)

shadow cast. Carbon is then evaporated on the sample, and the cellulose acetate is dissolved (Figure 18).

Since carbon replicas are produced by evaporation, the replica is of fairly uniform thickness; contrast results because it does not straighten out during stripping, so that it presents a greater path length to the electron beam in some places than in others. Without shadowing, hills and valleys are indistinguishable. (Figure 15). An advantage of the pre-shadowed, direct carbon positive replica (Figure 16) is that some contrast is obtained even if the replica were to straighten out completely on stripping.

4. Oxide Replicas

Aluminum may be anodized electrolytically to produce a film of alumina of constant controlled thickness on the surface; this film is then stripped either by electropolishing or by immersing in dilute mercuric chloride solution which is reduced to mercury at the metal/oxide interface by reaction with the aluminum, causing the oxide to float off. The oxide is scored into grid-sized squares before stripping. The origin of contrast is clearly similar to that in the case of carbon, and postshadowing is desirable to enhance the contrast. Like carbon, the oxide replicas are strong and stable in the electron beam but have rather more self-structure. Prior to the advent of the thin foil technique, aluminum oxide replicas were extensively employed in conjunction with chemical or electrolytic etching to study precipitation phenomena. Attempts to push the technique further and further in resolution led to its becoming somewhat discredited, partly because of artifacts and partly because of the difficulty of correctly interpreting etching effects at small particles. The thin foil technique is clearly superior in resolution and directness for internal structure studies.

Aluminum oxide replicas were also extensively employed with considerable success, prior to the advent of the thin foil technique, for the study of slip markings on the surface of deformed electropolished single crystals. The oxide replicas are somewhat more resistant to straightening than carbon films. Disadvantages are the limited applicability and the destruction of the metal surface in stripping. Evaporated silicon monoxide replicas were also used for deformation studies and were claimed to be less prone to distortion than carbon.

Although these techniques are clearly still powerful ones for the study of surface topography, the more direct technique of scanning electron microscopy is likely to replace them. The latter would permit dynamical studies of slip on testpieces of reasonable size, under known stress-strain conditions. Transmission microscopy on thin foils cut from deformed bulk specimens is the most direct way of studying the internal defects, for example, dislocation arrangements, resulting from deformation. The usefulness of dynamical experiments inside the electron microscope has been limited by the difficulty of defining the stress and strain, and the fact that the deformation behavior of thin foils tends to differ from that of bulk samples. It may prove possible to do better with the high voltage microscope which permits the use of thicker samples.

5. Fractography

Replication techniques have been extensively employed for fractography, and there is a substantial body of literature concerned with both the techniques and interpretation of microfractographs.[23-26] The most commonly used techniques are indirect carbon with cellulose acetate tape employed as a first stage, and the direct carbon technique. In the case of the latter, the sample may be rotated about a single axis during evaporation, in order to obtain a more uniform deposit. Silicone rubber has been used[23] for the first stage in the replication of very rough fractures. Being very flexible as well as strong, it resists damage during dry stripping even if reentrant angles are present.

6. Shadow Casting

Shadow casting is an important means of enhancing the contrast of replicas. The shadow length for a given angle of shadow casting is a measure of the elevation of the surface features. It is desirable to obtain the maximum contrast with the minimum thickness of shadowing material. The requirements of a good shadowing material are high scattering power, that is, high atomic number, ease of evaporation, low self-structure and stability in the electron beam. Some materials that are used are chromium, uranium, gold-4% palladium, platinum-palladium, carbon-platinum, and carbon. At least 5 to 15-Å thickness is necessary, or as much as 50 Å for lighter materials such as chromium and carbon. Carbon preshadowed direct carbon replicas are useful for high resolution work and easy to make, since the sample position is simply rotated between the two steps to change the evaporation angle from the carbon arc. The carbon shadow has particularly low self-structure and good stability. The scattering power can be improved by placing a small piece of platinum in a notch near the tip of the carbon rod to give a carbon-platinum shadow. Alternate procedures have been described by Bradley.[22]

7. Extraction Replicas

Extraction replicas, which were first introduced by Fisher[27] have proved particularly useful for inclusion identification and for studying their size distribution, morphology, and structure. One of the techniques that may be used is illustrated in Figure 19. Since it is necessary to etch or electropolish away the matrix, in order to free the inclusions, there is always a question as to whether extraction has modified the particles. In the case of coherent or semicoherent precipitates, coherency is lost by extraction. The extraction replica technique is of limited use, for example, in the case of aluminum alloys, in which most constituents tend to be electropositive to the matrix and preferentially attacked during attempts to remove them. Extracted particles which are too large for selected area electron diffraction may be examined by x-ray or electron microprobe techniques.

Since particles tend to be accumulated from a thin layer during etching and stripping, extraction replicas tend to give an exaggerated idea of the volume fraction of particles in a sample but give some idea of their

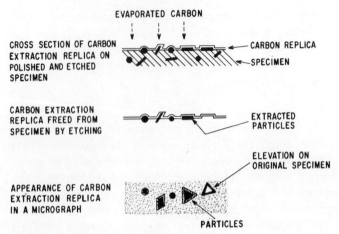

Figure 19   Direct carbon extraction replica technique.[21] Prepared by E. F. Koch. (Courtesy of General Electric Company.)

location, for example, at grain boundaries. If etching or extracting partially attacks the particles, the observed size distribution will be weighted. Incomplete extraction of all the particles may likewise weight the size distribution. Where possible, it is preferable to identify particles in situ by such techniques as selected area diffraction on thin films, or by electron microprobe, or microbeam x-ray examination of metallographic sections. However, the sensitivity of the extraction replica technique can be much greater because a concentrated sample is obtained. Furthermore, the contrast at particles may be increased in the absence of the matrix.

B. Preparation of Thin Foils from Bulk Samples

The principle methods of preparing thin electron transparent foils from bulk samples are (a) electrothinning, (b) ultramicrotomy, (c) chemical thinning, (d) cleavage, and (e) ion bombardment.

1. Electrothinning

Given sufficient ingenuity and effort, most electrically conducting materials can be successfully electrothinned. The sample is made the anode in an electrolytic cell and dissolved away until an appropriate thickness is reached, usually about 1000 Å for 100-keV examination. There is already a vast literature on this subject, since there are many permutations of materials, electrolytes, and techniques. The reader is referred to other sources, including Refs. 2, 12, 20, 28, and 29, for reviews.

Electrothinning is facilitated if foil of 0.001 to 0.010-in. thickness, preferably 0.003 to 0.005-in., is used as a starting point. This is thick enough to behave like bulk material in most experiments, provided that the grain size is kept well below the foil thickness. Preferred orientation may be avoided by proper processing but is often useful if ideal orientations are needed for diffraction contrast experiments.

If it is desired to examine the structure of large samples, methods must be found for removing thin slices, preferably in the thickness range mentioned above, without damaging the structure of interest, and for finishing by electrothinning. Methods commonly employed, depending on the material, include acid saws and other chemical machining methods, spark cutting, and sawing. Surface damage produced by the latter two techniques is commonly removed prior to final electrothinning by etching which is a relatively rapid process. The preparation of undeformed slices of soft single crystals can be extremely troublesome and time consuming, since even spark cutting tends to introduce dislocations.

If the material is inhomogeneous in composition or multiphase, and the composition fluctuation or particle size is larger than the final thickness desired, multiple breakthrough may occur during electrolytic thinning and it may be impossible to obtain good samples. Even if multiple breakthrough is avoided, surface etching effects may result in undesirable mass-thickness contrast effects which tend to mask the structure of interest in transmission micrographs. It is usually possible to improve the situation by careful choice of the electrolyte and thinning conditions.

Electrothinning techniques may be divided into static methods, which employ little or no stirring and are relatively slow; and dynamic methods in which the electrolyte, often in the form of a jet, is propelled against the surface of the sample, permitting the use of higher voltages and fairly rapid metal removal. The Bollman technique,[30] shown in Figure 20, is an example of the former type, while the Hugo and Phillips[31] technique (Figure 21) falls into the latter category. With the latter it is convenient to start with a precut disk of 0.003 to 0.005-in. thickness, which will fit into the electron microscope holder, and thin a spot at the center. The polished spot is lacquered, the disk reversed, and a single hole thinned through from the opposite side, using a footswitch to stop polishing the moment a hole forms. It is advantageous to thin in short pulses. The specimen is washed and the lacquer dissolved. The sample may be handled by the relatively thick surrounding area without damage to the thinned region. Thinning may be either electrolytic or chemical. A number of automatic electrolytic thinning devices are now available commercially. One of these employs a light source on one side of the

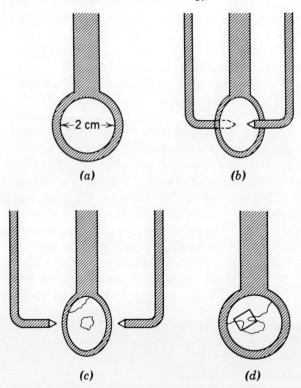

(a)   (b)

(c)   (d)

Figure 20   The Bollman electrothinning technique: (*a*) initial sample (the shaded region is masked with stop-off lacquer); (*b*) point cathodes moved close to sample to give central breakthrough; (*c*) cathodes moved apart to give peripheral breakthrough; (*d*) area from which foils are cut. From Thomas.[2]

sample and a photocell on the other to switch off the current automatically when breakthrough occurs.

2. Ultramicrotomy

Where electrothinning is impracticable, ultramicrotomy often provides a satisfactory method of obtaining electron transparent slices from bulk samples. The technique is a standard one for biological samples and such materials as wood, plant material pigments, textile fibers, bone, and cement. [There is an extensive literature which will not be reviewed here; a general account is given by Glauert and Phillips.[32]] It is less commonly realized that the technique is applicable to many metals and alloys,[32,33] materials such as carbon, polymers, composites, and fuel cell elements. Using a diamond knife ultramicrotome, good samples may be obtained

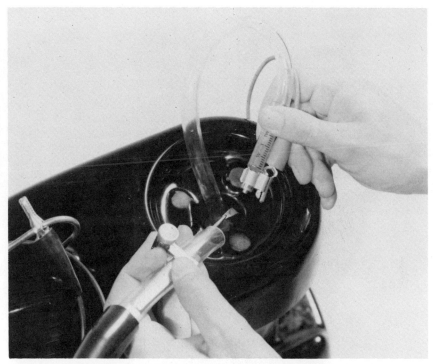

Figure 21 View of modified Disa-Electropol polishing table, showing sample held in tweezers ready to be thinned. From Hugo and Phillips.[31] (Courtesy of The Institute of Physics and The Physical Society.)

from most machinable materials, provided that some deformation is tolerable. Partially electron transparent, irregular fragments may often be obtained from unmachinable materials, particularly if tiny samples are cut in a suitable embedding material. These are useful for some purposes. Unfortunately, some metals such as iron, nickel, nickel-aluminum alloys, and zirconium, although they can be cut, give artifacts, probably caused by high friction on the knife, that obscure fine detail. This situation can sometimes be improved by annealing the slices in the electron beam.

A wide variety of ultramicrotomes is available commercially, of varying degrees of automation, convenience, and cost. A simple microtome of the mechanical advance type, which has been extensively employed in the author's laboratory for cutting metals and polymers, is shown in Figure 22. A single slice only a few hundred angstroms thick may be cut.

Figure 22   View of Porter-Blum ultramicrotome. From Phillips.[33]

A generally applicable procedure is to mount a sample in cold-setting plastic, usually an epoxy resin (obtainable from electron microscope stocklists). A half of a gelatin capsule provides a suitable mold, which can easily be removed after the mount has hardened. The plastic is pared away, as in sharpening a pencil, to expose the metal to be cut. If possible, only metal should be microtomed, since the plastic tends to charge up in the electron beam. Pointed rod samples may be cut without mounting. Foil samples may be cut while protruding slightly from a miniature vise, as an alternative to mounting. Even normally pliable materials such as rubber may be cut by using a cooling head.

The knife, which is typically a diamond wedge of 40 to 60° angle and 1 to 3 mm in edge length is mounted in the edge of a trough which is filled with water, so that the slice floats over the water surface as it is cut and can be picked up on a grid for examination. The slice width is typically 0.1 to 0.2 mm, and the slice length about 0.5 mm. A poor region of the edge is used for preliminary cutting to form a cut facet on the sample. Cutting occurs ideally by a simple shear process,[34,35] as shown in Figure 23. Under these conditions there should

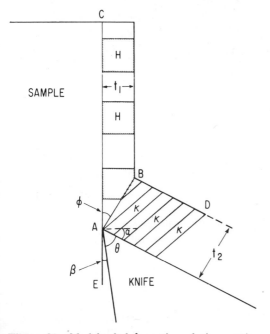

Figure 23 Model of deformation, during cutting a simple continuous chip. After Puspanen,[35] from Phillips.[21]

be no change in dimensions, for example, of pores or ductile inclusions, parallel to the knife edge, that is, across the width of the slice. A correction can be applied for the change parallel to the slice length.

Ultramicrotomy has a number of unique applications; for example,

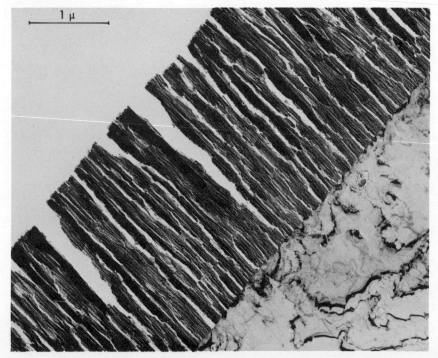

Figure 24 Transmission micrograph of anodized Al-2%Mg alloy, sectioned by ultramicrotomy using a diamond knife. From Hugo and Phillips.[36] (Courtesy of General Electric Research and Development Center.)

samples may be cut from powder or fine wires, or irregular samples of whatever size or shape. The distribution of porosity or internal cracks may be examined, whereas both electrolytic and chemical thinning would enlarge the holes. Electrically insulating materials may be examined; multiphase or composite samples may be prepared. Materials that, are reactive with water such as sodium may be prepared, although in this case a different liquid would be used in the trough. Sections may be taken in any desired plane to reveal features of interest. Thus sections can be taken normal to the plane of thin plastic filters to reveal the variation of hole shape through the thickness, or high strength carbon fibers a few microns in diameter may be sectioned, parallel and normal to the axis.

Since a fairly severe shear deformation is introduced, the technique is not applicable for the study of preexisting defects such as dislocations. Furthermore, low melting point materials such as aluminum, lead, tin, and zinc, particularly if of high purity, tend to recover or recrystallize

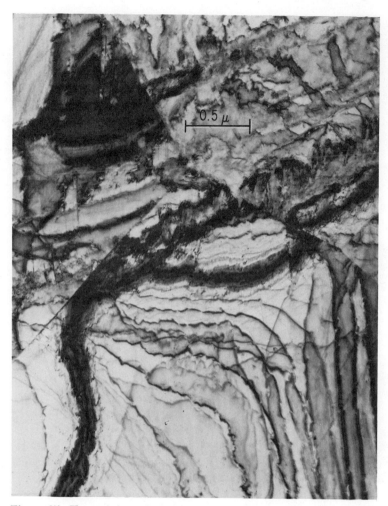

Figure 25 Transmission micrograph of annealed alpha brass alloy ultramicrotomed with a diamond knife showing grain boundaries. From Phillips.[21]

spontaneously after cutting. A disadvantage of the technique is the expense of the diamond knives. The life is poor if hard materials are cut. Figures 24, 25, 26, and 27 show examples of the results that may be obtained (see Refs. 32 through 34 for other examples).

3. Chemical Thinning

A number of materials such as silicon,[37] germanium,[37,38] and MgO[39] have been successfully prepared by chemical etching, although they

Figure 26 Transmission micrograph of annealed zirconium, ultramicrotomed and heated in the beam to cause recrystallization. From Phillips.[21]

would be difficult to prepare by almost any other technique, except ion bombardment which is very slow and tends to introduce damage and surface artifacts. Slices about 0.001 in. thick suitable for etching may be cut from hard materials of this kind, using a precision diamond cutoff wheel. The shallow damaged layer is removed during the subsequent etching. In the case of MgO, cleaved platelets may also be employed as a starting point. The technique has been successfully used to thin metals such as gold,[40] aluminum[41] and steel,[42] (see Appendix 1 in Ref. 12 for a fuller list), but electrothinning would now be preferred.

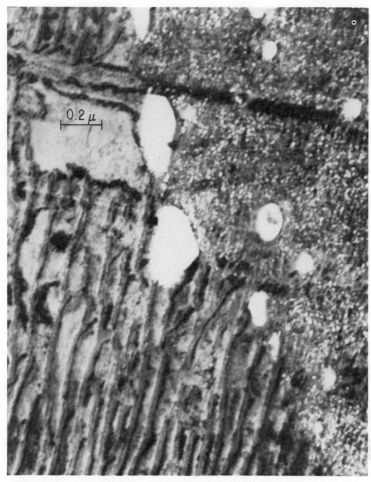

Figure 27   Transmission micrograph of ultramicrotomed, leaded alpha-beta brass, showing two-phase region and micropores resulting from heating to 880°C. From Phillips.[21]

It is somewhat difficult to find etchants capable of giving smooth-surfaced samples free from oxide. Multiphase or inhomogeneous samples are prone to differential attack. When it can be used, the technique may be simple, direct, and rapid, although some fairly elaborate chemical thinning cells have been devised. The apparatus shown in Figure 21 may be employed. In the case of materials transparent to light such as silicon and germanium the appearance of the appropriate interference color indicates that a desired thinness has been reached. The sample must then be rapidly removed and washed. Undoubtedly, a great many more chemical thinning procedures will be developed. The list of chemical polishes, published by Tegart,[28] is a useful starting point.

When large amounts of material need to be removed, electrolytic thinning is often preceded by chemical thinning. For this purpose the choice of reagent is not critical. If a jet technique is employed, the sample may be dimpled, in order to control the point of subsequent electrolytic breakthrough.

A principle advantage of the technique is that no deformation is introduced, so that it can be used in studying defects such as dislocations in materials that cannot be electrothinned.

4. Cleavage

If a material possesses an easy cleavage plane, use may be made of this property in preparing electron transparent samples. Excellent results have been obtained with layer-type materials such as graphite, molybdenite, mica, talc, and bismuth telluride (see Appendix 1 in Ref. 12 for a fuller list). The usual technique involves the successive cleavage of a single-crystal flake, using adhesive cellophane tape, the adhesive being removed with chloroform. The structure of a molybdenite sample prepared in this manner is shown in Figure 28. If the material is hard, slices may be separated from a crystal by a sharp hammer blow on a knife-edged chisel. These must subsequently be thinned.

Powder prepared by crushing brittle material with an easy cleavage plane tends to be platelike, and thin regions may be found, particularly if the fines are elutriated. If the material is deformable, a high density of dislocations may be introduced during the crushing.

In the case of a material such as silicon or germanium, a thin slice may be cut with a diamond wheel, chemically polished and fractured. Since the fracture tends to follow the cleavage, thin regions are usually found at the fractured edges.

The method is limited to materials, usually of hexagonal crystal structure, which have a single, easy cleavage plane. Clearly, the orientation of the sample obtained is also limited to the cleavage plane. The fact that an ideal orientation is obtained is often useful. A disadvantage of the technique is that cleavage steps are usually present on the sample and give additional contrast effects. Mica is somewhat unique, in that large atomically flat regions may be obtained by cleavage.

5. Ion Bombardment

Following pioneering work by Induni,[43] Castaing[44] (see also Ref. 2) designed an equipment in which a thin disk of aluminum was successfully

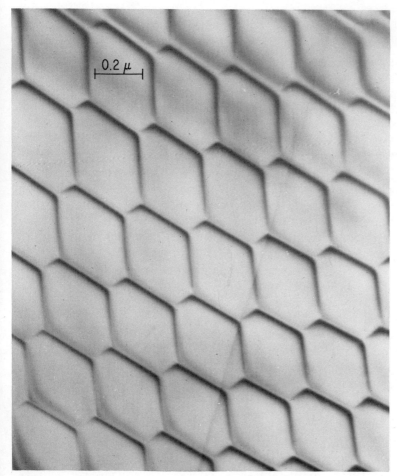

Figure 28 Transmission micrograph of cleaved molybdenite crystal, showing dislocation network in the basal plane. The network is probably pure screw in type. From Phillips.[21]

thinned by bombarding both sides simultaneously with 3-keV ions, produced from residual air introduced through a controlled leak into the continuously pumped system. The technique did not become popular, but recent improvements in design[45] awakened renewed interest and several commercial ion thinning units are now available, including twin-gun thinning units. The sample is rotated while under bombardment. Metals, ceramics, and minerals have been successfully thinned.[46]

The rate of material removal is extremely slow, typically about 1 to 3 microns/hr so that many hours are required to thin a 1-mil sample.

It is difficult to obtain smooth surfaces, if the starting surface is rough. Heating of the sample may be a problem. Ion bombardment may also produce internal structural damage, resulting in the presence of defects such as vacancy and interstitial clusters, and dislocation loops. In spite of these disadvantages, ion bombardment can be used to thin almost any material, so that it is useful for thinning materials such as oxides, carbides, nitrides, ceramics, and glasses for which other techniques may not be applicable. Another advantage is that clean surfaces may be obtained.

6. Miscellaneous

There are well-established techniques for preparing ceramic, mineral, and rock slices for petrographic examination in the transmission light microscope. Slices only 1 to 2 mils thick are cut with a diamond wheel, and then cemented onto a block and ground on abrasive laps. This process may sometimes be continued until a sample is transparent to 100 keV electrons, but hand-finishing is necessary and the method is tedious, with a poor yield. Silicon has been successfully prepared in this way by Dash and Phillips (see Ref. 21), as in Figure 29. Although some surface damage is evident, the results are satisfactory for some purposes.[47] The surface damage could be removed by chemical thinning.[37]

Evans and Phaal[48] obtained good results on diamonds by an interesting technique, in which a flake was heated at 1350°C in an atmosphere of carbon dioxide until most of the sample had oxidized. After cleaning in acid, a transparent diamond flake remained.

C. Naturally Thin Samples

An entirely different category of samples is the one in which we are interested in the structure of thin samples per se. We first discuss the preparation and applications of evaporated films and then briefly mention the sources of naturally thin samples.

1. Evaporated Films

Evaporated films have been much used for electron microscope studies.[2,12,41,49,50] A very wide variety of metals, and even alloy films, may be prepared by vacuum evaporation onto a suitable substrate. Details of methods of making evaporated films are to be found elsewhere.[51,52-54] Commonly, the metal is evaporated from a tungsten coil or boat, which has been heated to incandescence by the passage of a current, onto a relatively cool substrate, the whole being contained inside an evacuated bell jar. By suitable control of the evaporation conditions such as the evaporation rate, vacuum, choice of substrate, and substrate temperature,

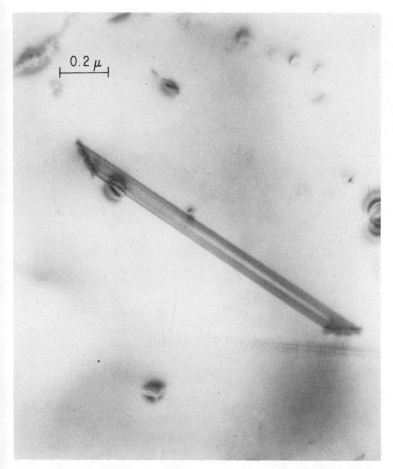

Figure 29 Transmission micrograph of a {111} slice of gold-diffused silicon, thinned by grinding. A section through a dislocation loop of about 2-μ diameter is seen in striped contrast. Unpublished work of W. C. Dash and V. A. Phillips from Ref. 21.

film structures ranging from fine grained, randomly oriented polycrystalline to highly oriented single crystalline may be obtained (Figure 30). The grain size of the former tends to be limited by the film thickness, and thus in electron transparent films is a factor of 10^2 to 10^3 smaller than is typical in bulk materials, giving rise to unusual mechanical and other properties.[53]

Control of the composition of alloy films made by evaporation has proved difficult. The three principle methods of depositing alloy films

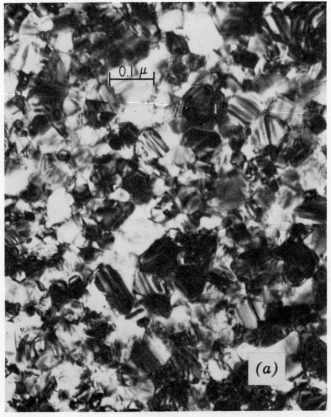

Figure 30 Transmission micrographs of evaporated silver films, grown on heated cleaved rock salt. (*a*) Fine grained, randomly oriented polycrystal, (*b*) (100) single crystal, showing stacking faults on {111} planes in striped contrast. From Phillips.[21]

that have been used are (*a*) direct evaporation of the alloy, (*b*) consecutive evaporation of the individual metals followed by a diffusion treatment, and (*c*) simultaneous evaporation of the component metals in the required ratio. Method *a* is limited by the fact that the composition of the deposit is affected by the vapor pressures of the alloy components, as well as by their proportion in the source material. Flash evaporation, in which alloy powder is slowly fed into a very hot boat, appears to overcome this difficulty.[53] Methods *b* and *c* can be used but require careful calibration or the use of sophisticated equipment for measuring

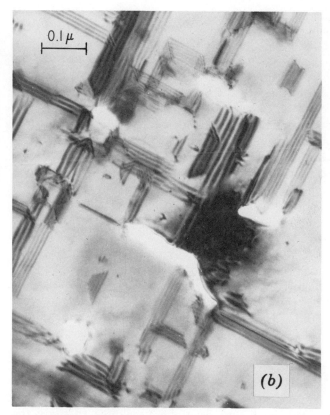

0.1 μ

(b)

Figure 30(b) Continued

the rate of evaporation. Unfortunately, it is difficult to analyze a film chemically, once it has been made.

It is necessary, of course, to separate the film from the substrate for examination in the electron microscope, unless the substrate is itself thin enough to be electron-transparent. If the substrate is water soluble, the film may be scored and floated off in grid-sized pieces. Films on glass substrates may also sometimes be floated off. Often it is necessary to etch away the substrate in a reagent that does not attack the film.

The phenomenon of epitaxy may be used to grow single-crystal films; a single-crystal substrate such as rock salt or mica is employed and the evaporation conditions carefully controlled.[54] In the case of mica, an intermediate epitaxially grown film of a different metal is sometimes employed to control the nucleation of the film of interest and to facilitate removal, for example, silver under gold.[50] Single-crystal films have been used for deformation studies in the electron microscope and may be stressed either by contamination in the electron beam,[55,56] or in a stretching device.[57]

Evaporated films lend themselves to studies of the early stages of the nucleation and growth of evaporated films on various substrates.[49,50,53] At the early stage of growth, when they are still discontinuous, a carbon film may be evaporated on top of the metal deposit and the two removed together for examination. Direct evaporation inside the electron microscope onto electron-transparent substrates such as molybdenum disulfide and carbon has given useful information on the mechanism of growth.[49,50] Whether single- or polycrystalline, evaporated films contain defects such as stacking faults (Figure 30b), twins, and dislocation loops which have been extensively studied.[49,50,55,58]

If a single-crystal film is grown epitaxially on a transparent substrate of a different material, so that both have identical orientation, and the composite is examined in the electron microscope, moiré fringes may be produced between the two.[49,50] In the case described, there would be a parallel moiré, arising from the periodic coincidence of lattice planes parallel to the microscope axis, giving fringes parallel to the planes but with a magnified period. Moiré fringes have proved useful in studying the nature and origin of defects in evaporated films.[50]

Films of alloys, capable of long range ordering, such as CuAu, made by evaporation, have been extensively employed[59,60] to study the mechanism of ordering. Evaporated films have also been used to study the mechanism of oxidation-type reactions; for example, Figure 31 is taken from a study[61] which showed that silver sulfide "patches" nucleate at the intersections of stacking faults and partial dislocations with the surfaces, and at hole edges in evaporated silver films.

2. Miscellaneous

Electron-transparent, platelike crystals can, in principle, be grown from the vapor,[53] liquid, or solid phase under appropriate conditions. Figure 32 shows the structure of a platelet of gold that was grown in air from the vapor phase in a gradient furnace. The gold source was maintained at 1030°C, and condensation occurred on an alumina substrate at 1000°C. Price[62,63] grew perfect crystals of pure zinc and cadmium from the vapor phase under low supersaturation and used them for direct studies of deformation inside the electron microscope. Techniques have been developed in the electronics field for depositing single-crystal or polycrystalline films, for example, of silicon on silicon by the decomposition of a gaseous compound such as SiH or SiCl₄ on the heated substrate.[53] Suito and

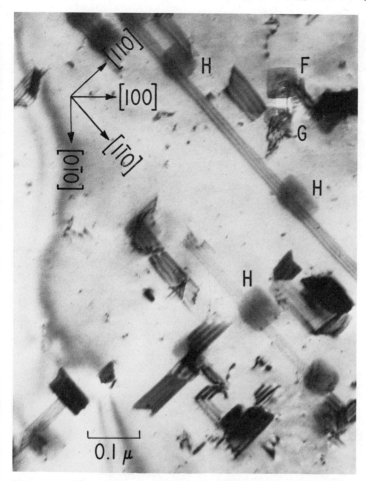

Figure 31   Transmission micrograph of single-crystal silver film, showing silver sulfide patches nucleated at stacking faults (H) and hole edges (F and G). From Phillips.[61] (Courtesy of the American Institute of Physics.)

Ugeda[64] prepared thin, single-crystal (111) flakes of gold 100 to 200 Å thick by reduction of a dilute solution of auric chloride. Use is frequently made in electron microscopy of small crystals removed from solid alloys by the extraction replica technique already described. These are useful for purposes of identification, and for structural and morphological studies.

Films prepared by electrodeposition have proved of interest for electron microscopic study of the structure of electrodeposits.[65,66] Thin films

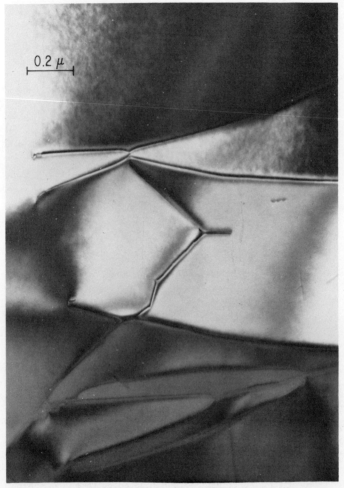

Figure 32 Transmission electron micrograph of gold platelet, grown from the vapor phase, showing dislocations. From Phillips.[21]

transparent to 100 keV electrons may be prepared by sputtering,[51,53] that is, by deposition on a substrate of particles knocked off a target cathode by gaseous ions. Such films usually have an extremely fine grain size but are of limited interest here. Thin oxide films, of alumina, for instance, may be made by anodization, already referred to under oxide replicas, but are of limited interest. Similar films can be prepared by the processes of normal oxidation and nitriding.

Samples of cast metal, thin enough for direct viewing, have been prepared by various "splatt-cooling" techniques. In one of these, drops of molten metal are thrown tangentially against a smooth cold surface such as copper[67,68] and extremely rapid cooling rates ($\sim 10^5$°C/sec) are achievable. Some alloy compositions yield amorphous films whose crystallization may then be observed directly in the electron microscope beam. Metastable solid solutions may be obtained from some alloys that would normally be two-phase, and the precipitation then studied in the electron microscope.[67] There has been a great deal of interest in this field. Alloy films similar to those obtained by splatt cooling may be obtained by evaporation onto a cold substrate; both amorphous and metastable structures have been observed by Mader.[53] Phillips[21] described a method in which polycrystalline cast tin films were made by squeezing molten tin between heated silica flats, and pointed out that glass films could be made similarly as an alternative to the usual technique which involves blowing thin glass bubbles.

Other techniques such as sublimation and crystallization from solution are described by Hirsch et al.[12]

D. Special Transmission Techniques

1. One-Sided Thinning

If a foil sample is electropolished and then deformed, so that slip steps are developed on the surfaces, the relationship between the internal dislocation arrangement and the steps on one surface may be revealed by electrothinning from the opposite surface only.[69-71] The steps give a contrast effect distinguishable from contrast at the dislocations. The technique has been applied to study fatigue[69,70] and tensile[72] deformation. The technique is also useful in studying etch pits in relation to dislocations.

2. Etched Transmission Samples

Compositional differences, due to segregation, rarely give rise to observable contrast effects in parallel-sided foils in transmission. It is well known in light metallography that segregation can often be revealed by chemical or electrolytic etching, since the rate of metal removal is a function of the composition. Since thickness differences give marked contrast effects in the electron microscope, etched foils can be used to study segregation. Phillips[73] observed ridges at grain boundaries (Figure 33) and mounds where dislocations intercepted the foil surfaces of iron samples, attributed to the segregation of carbon and nitrogen. The effects could be produced by electrothinning and subsequent chemical etching,

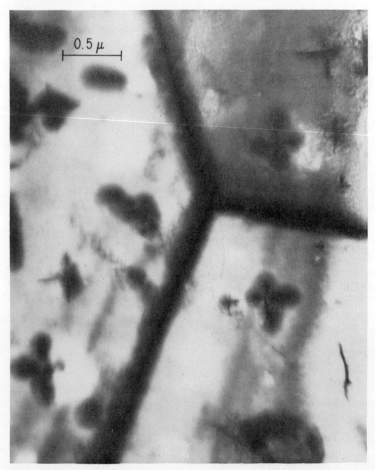

Figure 33 Transmission micrograph of electrothinned low carbon iron, showing ridges where grain boundaries intersect the foil surface and mounds where particles intersect the surfaces. From Phillips.[21]

or by electrothinning by a technique which also produced etching. The converse effect, namely, preferential attack at the segregated regions (grooving) was also noted[73] but is less useful, since it tends to result in holes.

Swann and Nutting[74] showed that stacking faults produced by deformation in certain copper alloys resisted etching which dissolved the matrix. This was attributed to the segregation of solute and suggested a mechanism of stress corrosion.

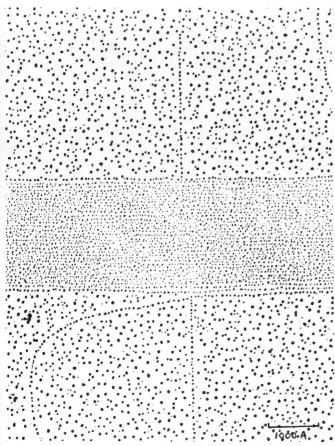

Figure 34 Electron micrograph of cleavage terrace on rock salt crossed by slip step. Many of the cleavage steps are 2.8 Å high. Gold nuclei extraction replica. After Bassett,[75] private communication.

3. Surface Decoration Technique

If a small quantity of gold, roughly equivalent to a continuous layer of 10-Å thickness, is evaporated onto the surface of a cleaved rock salt crystal under suitable conditions, small gold crystallites will nucleate, preferentially at cleavage steps. By transferring the crystallites to a carbon film, that is, by an extraction replica technique, a map of the surface topography can be obtained. Bassett[75] was the first to demonstrate that even monatomic steps 2.8 Å high could be revealed (Figure 34). Bethge[76] similarly decorated spiral evaporation markings on rock salt. Allpress and Saunders[77] studied the topography of thermally etched silver, using gold decoration. Henning[78] studied gold-decorated, cleaved graphite flakes which were thin enough to be examined directly in the electron microscope. He found that even dislocations with Burgers vectors parallel to the cleavage plane, were decorated. Robins, Rhodin, and Gerlach[79] studied gold-decorated cleaved MgO and claimed that point defects (impurity atoms), as well as monatomic steps, were decorated. Other successful decorating materials include platinum, tungsten, iridium, palladium, germanium, and Ni-60%Cr alloy. Figure 31 is an example of the silver sulfide decoration of silver. It should be pointed out that the shadow casting of replicas is a type of surface decoration technique.

We have seen that decoration is a valuable method of making visible small topographical features that might otherwise escape detection even with the electron microscope. A less obvious but important use has been to establish the time sequence of events, since decorated features can be distinguished from undecorated features that were developed after decoration. A more detailed review of decoration techniques is given by Stirland.[80]

IV. HIGH ENERGY ELECTRON DIFFRACTION

High energy electron diffraction (sometimes called HEED), as distinguished from low energy electron diffraction (LEED), can be carried out either in transmission or reflection. Transmission electron diffraction is commonly carried out in the electron microscope using normal specimen holders and employing 80 keV or more, in order to secure the advantages of penetration. Its growth in importance in recent years is derived from the vital role played in image contrast interpretation in transmission microscopy, in addition to the capability of diffracting from microareas down to a few tenths of a micron in diameter.

Reflection electron diffraction (sometimes abbreviated to RED) was formerly carried out exclusively in separate instruments, or electron diffraction cameras, predating the electron microscope. The same cameras can, of course, be used for transmission diffraction but without the imaging capability of the electron microscope. Most electron microscopes have special specimen holders, which permit this technique to be used, although with inferior precision, due to residual lens fields. These have not been extensively used, partly for this reason, and partly because, as a result of the glancing angle geometry, transmission imaging is not possible, so that selected area diffraction cannot be carried out.

Reflection electron diffraction is mainly used on bulk samples that

can be quite massive, necessitating large specimen chambers and manipu-
lating devices, commonly provided in separate electron diffraction instruments. It is commonly carried out at 50 to 60 keV, which permits the
examination of surface layers typically about 100 Å thick. The name
is somewhat of a misnomer, since most reflection electron diffraction
patterns result from transmission through surface asperities. Reflection
diffraction is commonly carried out in a vacuum of about 10^{-5} torr,
although high vacuum may be necessary for materials that react with
the atmosphere. Recently, attachments have been marketed for some electron microscopes that fit below the last lens and thus permit high precision
reflection (or transmission) electron diffraction to be carried out without
the necessity of a second instrument (see pages 180–181). In view of its
importance, we shall first discuss transmission electron diffraction. A
knowledge of elementary crystallography is assumed (see standard sources
such as Refs. 81 through 83).

A. Transmission Electron Diffraction

1. Reciprocal Lattice Concept

The reciprocal lattice is a lattice of points, each representing a plane in the
crystal and having the same (h, k, l) Miller indices. Higher order (fictitious)
planes, for example 220 and 330, are included, as well as the first order planes
such as 110. Each reciprocal lattice point lies on a line through the origin
normal to the corresponding planes in the crystal, at a distance from the origin
equal to the reciprocal of the interplanar spacing d_{hkl} (Figure 35). A similar
construction for other crystal planes leads to a three-dimensional array of
points that always form a space lattice. If a crystal has axial lengths a, b, c
and interaxial angles α, β, and γ, and a^*, b^*, and c^* are the lengths of the

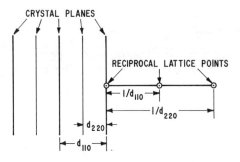

Figure 35 Relation between lattice planes in a crystal and their reciprocal lattice
points.

reciprocal lattice cell edges, then:

$$a^* = \left(\frac{bc}{V}\right) \sin \alpha \tag{25}$$

$$b^* = \left(\frac{ac}{V}\right) \sin \beta \tag{26}$$

$$c^* = \left(\frac{ab}{V}\right) \sin \gamma \tag{27}$$

where V is the volume of the unit cell in the crystal.

Following Thomas,[2] we can write these in vector notation as

$$\mathbf{a}^* = \frac{\mathbf{b} \times \mathbf{c}}{V} \tag{28}$$

and so forth, where

$$V = \mathbf{a} \cdot (\mathbf{b} \times \mathbf{c})$$

Since the vector product $(\mathbf{b} \times \mathbf{c})$, which has a magnitude $bc \sin \alpha$, is a vector normal to $\mathbf{b}$ and $\mathbf{c}$, $\mathbf{a}^*$ is normal to $\mathbf{b}$ and $\mathbf{c}$ and is equal to the reciprocal of the spacing of the planes in real space which contain b and c. The scalar or dot products satisfy the relations,

$$\mathbf{a}^* \cdot \mathbf{a} = \mathbf{b}^* \cdot \mathbf{b} = \mathbf{c}^* \cdot \mathbf{c} = 1 \tag{29}$$

and

$$\mathbf{a}^* \cdot \mathbf{b} = \mathbf{b}^* \cdot \mathbf{c} = \mathbf{c}^* \cdot \mathbf{a} = \mathbf{a}^* \cdot \mathbf{c} \ldots = 0 \tag{30}$$

If $\mathbf{a}$, $\mathbf{b}$, and $\mathbf{c}$ are unit translation vectors in real space, a lattice point at position $\mathbf{r}_n$ is defined by

$$\mathbf{r}_n = u\mathbf{a} + v\mathbf{b} + w\mathbf{c} \tag{31}$$

Similarly, a point at $\mathbf{r}^*$ in the reciprocal lattice with coordinates, corresponding to the reflecting planes hkl, is

$$\mathbf{r}^*_{hkl} = h\mathbf{a}^* + k\mathbf{b}^* + l\mathbf{c}^* \tag{32}$$

The reciprocal lattice vector $\mathbf{r}^*_{hkl}$, which is usually denoted by $\mathbf{g}$, is equal[12] to $1/d_{hkl}$, and is perpendicular to the crystal lattice planes with Miller indices hkl and spacing d_{hkl}.

2. Ewald Reflecting Sphere

Using the reciprocal lattice, a simple geometrical construction gives the condition for Bragg reflection for a given electron wavelength λ. If a sphere, the Ewald sphere, of radius $1/\lambda$ is drawn touching the origin of the reciprocal lattice with the incident beam through the origin as a diameter, then the reciprocal lattice points it touches correspond to planes in the Bragg diffracting position. This can be seen from Figure 36, where the beam is incident at the Bragg angle θ on the lattice planes of spacing d_{hkl}, for which the corresponding reciprocal lattice point r_{hkl}^* is shown. Clearly, r_{hkl}^* lies on the surface of the reflecting sphere of radius $1/\lambda$ if $g/(2/\lambda) = \lambda/2d_{hkl} = \sin \theta$, that is, the Bragg law $\lambda = 2d_{hkl} \sin \theta$ is satisfied (the order of reflection n is omitted, since in the reciprocal lattice fictitious planes of fractional d_{hkl} are included). In vector notation $\mathbf{g} = \mathbf{k}_1 - \mathbf{k}_0$, where $\mathbf{g}$ is the reciprocal lattice vector corresponding to the diffracting planes, $\mathbf{k}_0$ the incident wave vector (equal to $1/\lambda$), and $\mathbf{k}_1$ the diffracted wave vector. If the reflecting sphere is rotated while touching the origin of the reciprocal lattice, it sweeps out a limiting sphere of radius $2/\lambda$ (Figure 36). Thus all planes that can be made to reflect will have their reciprocal lattice points within the limiting sphere.

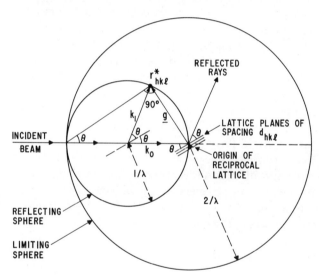

Figure 36 Ewald reflecting sphere construction for diffraction in the reciprocal lattice.

For 100-keV electrons, the wavelength λ is 0.037 Å, giving an Ewald sphere radius $1/\lambda$ of 27 Å, compared with typical reciprocal lattice vectors $(1/d_{hkl})$ of about 0.5 Å. The Ewald sphere is therefore nearly a plane section through the reciprocal lattice and nearly intersects many reciprocal points. The Bragg angle θ is of the order of $1°$, rather than as shown in Figure 36 which is typical of x-ray diffraction. In transmission electron diffraction using the electron microscope, the reciprocal lattice points are extended because of the thinness of the crystal, so that more intersect the Ewald sphere than would otherwise be the case. All reciprocal lattice points are extended to a length $2/\epsilon(hkl)$ in the $[hkl]$ direction, where $\epsilon(hkl)$ is the thickness of the crystal in this direction, so that the diffracted beam has a semiangular breadth λ/ϵ [130]. If the specimen is bent or buckled, so that different reflections occur in different areas, the overall diffraction pattern will show many more spots than that from one small area.

Not all the reciprocal lattice points are associated with Bragg reflections, because for some the structure factor, and therefore the intensity, is zero. There are simple rules for the permitted reflections. In the fcc lattice, h, k, and l are all odd or all even (zero is even); in the body-centered cubic (bcc) lattice $(h + k + l) = 2m$, where m is an integer; and, in the diamond lattice, either h, k, and l are all odd, or h, k, and l are all even and $(h + k + l) = 4m$. The reciprocal lattice points associated with Bragg reflections form a bcc reciprocal lattice for an fcc crystal, and vice versa. For a hexagonal crystal, a hexagonal lattice is obtained. The situation is more complex if more than one atom is associated with each lattice point, as in ionic crystals. [12]

The section of the reciprocal lattice normal to the incident beam cut by the Ewald sphere is, in fact, the observed electron diffraction pattern which appears in the back focal plane of the objective lens in the electron microscope. The portion allowed through the objective aperture is transformed by the lens into an image. The portions of the reciprocal lattice excluded are responsible for diffraction contrast, while those admitted control the degree to which the image resembles the object.

Since the incident beam lies along the zone $[HKL]$, the diffraction spots (hkl) are expected satisfying the relation:

$$Hk + Kk + Ll = 0 \tag{33}$$

Figure 36 indicates that the diffraction spots should correspond to the zero level, reciprocal lattice section, containing (000). The levels above and below should not normally appear for metals, unless the crystal is bent. If the plane of the specimen is normal to the incident beam, for each hkl point there will be a corresponding opposite reciprocal lattice point, so that the spots in the pattern are symmetrical about x and y axes, taking the beam along the z axis. An example of an electron diffraction pattern, obtained on a thin crystal of silicon at 100 keV, is shown in Figure 37. A rather complete set of indexed ideal diffraction patterns for the low indice planes of the common crystal structures is given in an appendix by Hirsch et al. [12] If the crystal is tilted relative to the

Figure 37 Transmission electron diffraction pattern, obtained on thin (112) crystal of silicon, showing cross-grating spot pattern and Kikuchi lines. (Courtesy of General Electric Company.)

incident beam, departures from symmetry can occur in the electron diffraction pattern.

The orientation of a crystal relative to the incident beam can be determined from diffraction patterns by indexing the spots and identifying the plane in the reciprocal lattice. This is a relatively straightforward matter if the material is known and d values are determined from the measured spot radii, using the camera constant. The d values can be compared with values tabulated in the ASTM index; this information, together with the measured angles between the spot zones [see tabulations for the cubic system by Barrett[81] and Hirsch et al.[12]] is often sufficient. It should be noted that the intensities often differ from those tabulated for x-rays. A set of standard stereographic projections is a useful aid in indexing and in performing rotations necessary to identify off-ideal orientations. The reader is referred to Refs. 2, 12, and 84 for more detail and alternate methods.

3. Structure Factor

The structure factor F_{hkl} relates the atom positions in space and the resulting diffraction pattern. It represents the relative intensities of the diffracted beams at the exact Bragg setting. For a perfect three-dimensional crystal, it consists of the sum of the amplitudes scattered by all the atoms in a unit cell, added with proper regard to phase. A kinematical structure factor may be derived, assuming only simple scattering; however, multiple scattering is common and requires a dynamical structure factor. The structure factor is required, in order to determine an unknown crystal structure from a diffraction pattern, making use of the intensities as well as spot positions. This turns out to be even more difficult than in the case of x-ray patterns (see Refs. 12 and 13).

For some reflections the structure factor is zero, so that they are forbidden (for geometrical rules see page 216). However, violations of the structure factor rule can result from double diffraction, which also can change the relative intensities of the allowed reflections.

4. Double Diffraction

We can regard diffracted beams inside a crystal as acting as primary beams for further diffraction. Certain forbidden reflections can then occur, which would not be expected on the kinematical theory which assumes diffraction is weak, but would be accounted for on the dynamical theory. In addition, spot intensities can be modified. Fortunately, the simple concept of double diffraction gives a reliable prediction of which forbidden spots can occur, without resort to dynamical theory. We simply translate the primary diffraction pattern, without rotation, to bring the center spot into successive coincidence with each of the strongest spots in the primary pattern. Any new spots thus introduced are possible double-diffraction spots. If reflections $(h_1, k_1, l_1,)$ and $(h_2 k_2 l_2)$ occur, then reflections $(h_1 \pm h_2, k_1 \pm k_2, l_1 \pm l_2)$ can also occur.

Much stronger double-diffraction effects can occur if a crystal contains regions of different orientation, for example, twins, epitaxial layers such as oxide, or two phases (see examples and discussion in Refs. 12 and

84). Reflections that would be forbidden in a disordered alloy by the structure factor can occur in the ordered condition.

5. Twinning

Twins have a special orientation relative to the matrix which can formally be obtained by rotating the matrix orientation 180° about the twin axis. The plane perpendicular to this axis is known as the twinning plane, which for fcc crystals is the {111} plane so that four sets of twins are possible. The observed diffraction pattern is a composite of the orientations present; if the twins are thin, additional streaking effects parallel to the twin axis can be expected. See Hirsch et al.[12] for detailed analysis.

6. Crystal Shape Effects

The crystals used for transmission electron diffraction are usually in the form of a thin plate with the beam parallel to the small dimension. The reciprocal lattice points are then elongated normal to the plate and becomes spikes symmetrical about the reciprocal points, which may now intersect the Ewald reflection sphere and give observable but reduced intensity where no spot would otherwise be seen. In crystals containing diffracting platelike precipitates, such as GP zones, which may be only one plane thick, the plates may be at any angle to the incident beam, so that a variety of effects may be observed in the diffraction pattern [see Hirsch et al.[12] for detailed analysis, also Heidenreich[13]]. Plates only one plane thick give rise to continuous streaks through the reciprocal lattice points; thicker plates give shorter spikes. In practice the spike length is approximately inversely proportional to the plate thickness, until the thickness reaches the extinction distance which gives the minimum possible spike length.[12] The nature of the diffraction pattern depends again on the geometry of the spike intersection with the reflecting sphere and either streaking or displacement of the diffraction spots may occur.

Spikes perpendicular to the faces of a polyhedral crystal are expected to extend from the reciprocal lattice points, with relative intensities roughly proportional to the respective facet areas.[12] If the crystal is needle-shaped, all the spikes will lie perpendicular to the needle axis. If it is round in cross section, the reciprocal spikes will merge into a disk with a diameter inversely proportional to the needle diameter for small diameters. In the limiting case of a linear crystal such as a GP zone, consisting of a single line of molecules, or a single-polymer chain, a continuous sheet of intensity would occur in the reciprocal lattice normal to the needle axis.

The detailed electron diffraction effects expected from crystal shape effects can be calculated.[12,13]

7. Effects Due to Elastic Strain

In the absence of double diffraction, the broadening or elongation of reciprocal lattice points, due to elastic strain, can usually be distinguished from crystal shape effects by the fact that the latter affect all reciprocal lattice points including the origin, whereas effects due to elastic strain are zero for the origin and depend strongly on the indices of the point.[12] The shape of the diffraction spots is not affected by a uniform elastic bending about an axis in the plane of a foil, but additional spots may occur. However, tangential spreading of the diffraction spots is expected if the axis is parallel to the incident beam. A twist $d\phi$ gives an extension $|\mathbf{g}|d\phi$ in reciprocal space in the plane perpendicular to the beam.

A radial broadening of the diffraction spots proportional to $|\mathbf{g}|$ is produced by a nonuniform elastic strain, in distinction to small particle size effects that affect all reciprocal lattice points equally.[12] If the range of elastic strain is ϵ, the reciprocal lattice points are extended by an amount $|\mathbf{g}|/\epsilon$ along $\mathbf{g}$.[12]

8. Kikuchi Lines

Single-crystal, electron diffraction patterns from thicker regions of rather perfect crystals free from buckling often show a complex pattern of lines, known as Kikuchi lines,[85] in addition to spots (see Figure 37). Their origin involves both elastic and inelastic scattering, requiring dynamical theory for a complete explanation; however, the main geometrical features are adequately given by the oversimplified treatment of Kikuchi.[85] He assumes that they arise from the subsequent elastic scattering of electrons which have first suffered an inelastic collision involving a small energy loss. The inelastically scattered electrons are concentrated about the incident beam, and their intensity decreases monotonically with an increase in the angle of scattering. These electrons form a diffuse background in a normal diffraction pattern.

Some of the electrons inelastically scattered at a point such as P (Figure 38) in the direction PD impinge on a crystal plane such as AB at the correct Bragg angle for diffraction and are scattered in the direction CE. In addition, some of the electrons originally traveling in the direction CE will likewise impinge on the planes AB at the Bragg angle and be diffracted in the direction PD; however, these will be fewer in number, since the scattered intensity falls off with increasing angle to the incident beam. Thus pairs of bright and dark Kikuchi lines result, with an angular separation of 2θ, symmetrically located on either side of the intersection between the planes AB and the photographic emulsion. The bright, or deficiency line, representing a decrease in background intensity, is always nearer the center spot. When all possible directions for reflections from the planes *AB* are considered, it may be seen that the directions along which either an excess or deficiency of electrons results correspond to two cones of semiangle $90° - \theta$ which intersect the plate placed normal to the incident beam in hyperbolae. These are nearly straight lines, since θ is small.

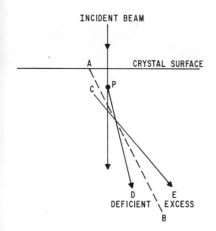

Figure 38   Simple geometrical illustration of the origin of Kikuchi lines by planes *AB*.

Typical patterns from thick crystals show several pairs of Kikuchi lines, corresponding to various orders of (*hkl*). Since the lines midway between each pair represent the extension of the (*hkl*) planes, the crystal orientation may be determined from two or more intersections or Kikuchi poles. When the crystal is tilted or rotated, the Kikuchi lines move as though rigidly attached to the crystal and can be used to measure the orientation change with considerably greater accuracy than by the use of the spot pattern. The latter does not move, when the crystal is moved, although the spot intensities change. This is because the reciprocal lattice spikes intersect the Ewald sphere and give rise to the same spot pattern over a small angular range of tilt of the crystal.

An idealized example of Kikuchi line patterns for incident beam directions near [100] in an fcc crystal is shown in Figure 39, after Howie,[86] no distinction being made between the bright and dark line contrast in the pairs. It will be noted that each pair is associated with a particular Bragg reflection, giving rise to a spot **g** in the diffraction pattern. The pair are separated by a distance equal to the magnitude of **g** and run normal to **g**. In Figure 39(*b*), where one Kikuchi line passes through the center spot and the other line of the pair passes through the 020 spot, exact 020 Bragg reflection is achieved. The

Kikuchi line displacements in a pattern straddling a low angle boundary enable the boundary rotation or tilt to be determined with high precision.

Kikuchi line plots can be prepared by a reciprocal lattice construction [see Hirsch et al.[12]]. Methods of indexing Kikuchi patterns are described elsewhere.[86-89] If two-beam diffracting conditions are employed, Kikuchi poles

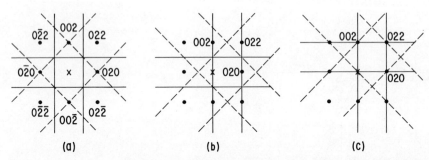

(a)   (b)   (c)

Figure 39   Schematic showing Kikuchi line positions near (200) orientation of fcc crystal: (a) beam exactly parallel to [200], (b) crystal tilted by Bragg angle θ about [002] direction into exact reflecting position for 020 reflection, (c) crystal tilted so that 020, 002, and 022 reflections are simultaneously in exact reflecting position. After Howie.[86]

(intersections of Kikuchi lines) may be absent. In this case, it is possible to determine the orientation by comparison with standard Kikuchi projections, called Kikuchi maps. These have been published for diamond cubic[90] bcc[91] and hcp[91] crystals. Applications of Kikuchi line analysis have been discussed by Von Heimendahl, Bell, and Thomas,[88] and Thomas.[92] The Burgers vector determination is considered in Ref. 90. The use of a goniometer stage is desirable.

One important use of Kikuchi line patterns is in determining the magnitude of the deviation parameter s from the Bragg reflecting position, shown in Figure 40 [see Refs. 84 and 88; also Howie[86]]. The parameter s is taken as positive or negative, depending on whether the reciprocal lattice point G lies inside or outside the Ewald reflecting sphere. The situation shown is equivalent to that in Figure 39a, where the reflecting sphere is equidistant from points 020 and 0$\bar{2}$0, and s from the geometry is $-g^2\lambda/2$. Then s is zero, when the crystal is in the exact 020 reflecting position (Figure 39b), so that one Kikuchi line passes through the center spot and the other through the 020 spot. The parameter s is $-g^2\lambda$ for the 0$\bar{2}$0 reflection in Figure 39b. Interpolation is used to obtain intermediate values of s.

9. Departure from Bragg Reflecting Condition

The diffraction intensity is maximum in the exact Bragg reflecting position and falls off as the crystal is tilted out of this position, analogous to light diffracted from an optical grating. The angular halfwidth β of the reflection

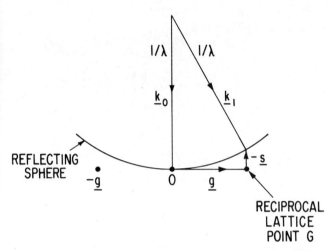

Figure 40 Ewald sphere construction in the reciprocal lattice, showing incident and scattered wave vectors k_0 and k_1, the deviation parameter s, and vector g normal to the diffracting plane. λ is the wavelength.

after Guinier[83] is then:

$$\beta = \frac{K\lambda}{\epsilon \cos \theta} \tag{34}$$

where K is a constant varying with the indices but not the order of the reflection (for spherical crystals, K is ~ 0.9 for all reflections; see Ref. 93), λ the wavelength, θ the Bragg angle, and $\epsilon = Nd_{hkl}$ where N is the number of reflecting planes of spacing d_{hkl}, that is, ϵ depends on the extent of the crystalline lattice.

B. Reflection Electron Diffraction

The subject of reflection electron diffraction has been reviewed at length by Bauer,[94] who also discussed the kinematic and dynamic theory.

In its simplest form (Figure 41), a camera for reflection electron diffraction consists of an electron gun, a single electromagnetic lens used to focus the beam at a glancing angle on a specimen mounted on a goniometer head, and a screen or plate camera mounted normal to the undeviated beam, the whole being mounted in a chamber capable of

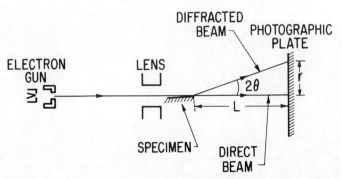

Figure 41   Schematic showing simple arrangement for reflection electron diffraction.

evacuation to 10^{-5} torr or better. If L is the camera length, that is, the distance between the point of impact of the electron beam on the specimen and the photographic emulsion, and r is the radius measured from the center spot to the diffraction spot (or ring), we have

$$d_{hkl} = \left[\frac{\lambda L}{r}\right]\left(\frac{1 + 3r^2}{8L^2}\right) \approx \frac{\lambda L}{r} \qquad (35)$$

where d_{hkl} is the spacing of the diffracting $\{hkl\}$ planes and λ is the electron wavelength (usually in the range 0.12 to 0.04 Å, corresponding to an accelerating voltage range of 10 to 100 keV). The camera constant λL can be determined by calibration with a standard substance such as evaporated aluminum of accurately known parameter. With suitable screening to protect against the influence of stray external fields, accuracies of 1 part in 10,000 are realizable.

The beam size is typically of the order of 1 mm and at glancing angle intercepts a region on the specimen several millimeters long. Under these conditions the penetration normal to the sample surface with a 50-keV beam is typically of the order of 100 Å. Since, unlike x-rays, electrons are strongly scattered in the forward direction, the sensitivity is excellent, and under favorable conditions thin epitaxial layers in the form of patches equivalent to a uniform film thickness of a monolayer may be detected.[95] Since diffraction patterns usually result from transmission through asperities, a slight roughness of the surface is advantageous and no pattern may be obtained if the surface examined is too smooth. The patterns obtained are essentially the same as in transmission electron diffraction, except that a part of the pattern is obliterated by the shadow of the specimen. The angle of incidence of the beam on the specimen may be determined from the distance between the shadow edge on the plate and the undeviated spot. Identification may be made from ring patterns, comparing the d values with those of known substances determined with x-rays which are listed in the ASTM index or other standard sources such as Wyckoff[96] or Pearson.[97] It should be noted that the structure factors for electrons differ from those for x-rays, so that the intensities may be different. Identifications from spot patterns are difficult, since only a single cut through the reciprocal lattice is obtained, so that supplementary information is usually needed. In view of the complexities and variety of surface reactions possible, even when a good ring pattern is obtained, it may not correspond to a listed substance. In view of the complexities and the present state of the theory,[94] there is considerable difficulty in determining other than simple unknown structures.

The theoretical interpretation of true reflection diffraction patterns, obtained from smooth specimens, differs from that for patterns of the usual kind, discussed above, that result from transmission through asperities [see Bauer[94]]. Appreciable refraction effects, due to the inner potential, may occur from smooth surfaces, resulting in deflection of the diffracted beams toward the shadow edge. Kikuchi patterns may be obtained from reasonably flat and rather perfect single crystals.[94]

Reflection electron diffraction may also be carried out using attachments to an electron microscope, as discussed on page 180. Substantially smaller beam sizes may then be obtained. Recently, there has been a trend toward the use of high vacuo (10^{-9} to 10^{-10} torr), employing bakable stainless steel chambers, equipped with an ion gun for cleaning the specimen surface and an electron spray gun (200 to 1000 eV) for removing surface charges from nonconducting specimens.[94,98] Additional ports may be provided to permit the insertion of devices for cleaving in situ, heating, cooling, or stretching the sample; coating by evaporation; and admission of gases for controlled oxidation studies.[94] More elaborate equipment, employing reflection electron diffraction as the principle mode of operation, with alternate modes, is also available commercially.[99] (Also see Ref. 94.)

The applications of reflection electron diffraction are similar to those in transmission electron diffraction but relate to very thin surface layers on bulk samples. Thus the technique lends itself to studies of surface texture, grain size, abrasion, oxidation and corrosion, identification of surface layers or small particles protruding from a surface (e.g., after appropriate etching), epitaxy, adsorption, sputtered layers, surface faceting, and surface deformation. Reference should be made to Bauer[94] for more details and bibliography. The recent availability of high vacuo RED equipment opens up many possibilities of using electronic readout of intensities for in situ experiments of many kinds, in such areas as the growth of thin films or the oxidation of clean surfaces. Developments in electron microscopes or recent years have tended to overshadow the usefulness of RED. The availability of high resolution RED attachments for some electron microscopes may reawaken interest.

V. CONTRAST THEORY

Following Heidenreich,[13] it should be pointed out that there are two basic types of contrast in electron images, namely, phase contrast and amplitude contrast. *Phase contrast* results from the recombination by the objective lens of the direct and scattered rays in the image and applies over a wide range of object distances. It is particularly important in the case of periodic structures and, in the range of interatomic distances, is the dominant type. *Amplitude contrast* is produced by inserting a small aperture in the back focal plane of the objective lens, and includes both mass thickness and diffraction contrast. The object distances involved are much greater than the interatomic distances.

The concept of mass thickness contrast is familiar from the discussion on replica techniques, in which contrast is partly due to the variations from point to point in the thickness traversed by the electron beam, and partly to mass difference introduced by shadowing with heavy atoms. Similar effects are produced in tissue slices by staining with heavy metal atoms. Thickness variations are usually present in powder samples and tend to be the major source of contrast. Mass thickness contrast effects are frequently obtained with thin foil samples, due to either variations in thickness or composition.

In the last 15 years a whole field has grown around the study of crystals and their defects by the technique of transmission electron microscopy, mainly employing diffraction contrast. The term "diffraction contrast" is normally used to denote electron images of crystalline materials, in which an objective aperture is used to block all except the zero order diffraction spectra from the image (bright field image). Alternatively, dark field images can be formed instead, as already discussed, using only a single nonzero order beam. The aperture thus removes nearly all information concerning atomic distances in the object. This information can, however, be obtained from the electron diffraction patterns. It follows that the distance between diffraction contrast detail does not correspond necessarily to real object distances. The dynamical theory of electron diffraction has been developed to the point at which much of the fine contrast detail can be explained. The mathematics are complex and approximations are often necessary, so that the interpretation is subject to the limitations of the theory, as well as to errors resulting from lens aberrations. The main features of diffraction contrast images are often adequately explained by the simpler kinematic theory. In order to interpret the diffraction contrast image correctly, it is usually necessary to record a selected area, electron diffraction pattern. It will not be possible to do more here than indicate some of the principal results of the theory; the reader is referred to Refs. 12, 13, 100, and 101 for detailed accounts.

A. Electron Scattering from Amorphous Solids

Heidenreich[13] considers an idealized case, in which an amorphous object illuminated by a monoenergetic parallel beam of incident electrons is imaged by an objective lens with no aberrations. The electrons scattered by such an object form a diffuse diffraction pattern at the objective aperture plane which is roughly coincident with the back focal plane of the objective lens. The distribution of the scattered electrons in the aperture plane then determines the intensity distribution in the image plane. The intensity scattered by a point in the object falls off monotonically with the increasing scattering angle if there is no wave interference.

We are interested in the total electron scattering outside an angle which will correspond to the objective aperture. The total cross section for scattering outside the aperture is given after Heidenreich by

$$Q = \frac{N_0 \, \sigma_{atom}}{M} \, \rho \qquad (36)$$

If t is the object thickness, the product Qt determines the scattering. Here $N_0 = 6 \times 10^{23}$ is Avogadro's number; M is the molecular weight; ρ is the density, so that ρt is the mass thickness; and σ_{atom} is the total cross section per atom for scattering outside the aperture, which is assumed independent of the presence of neighboring atoms. Some values of σ_{atom} for beryllium, chromium, germanium, platinum, and uranium are 0.5, 6, 8, 10, and 13 cm^2 $\times$ 10^{18}, respectively.[13] If the atoms are organized in groups, this mass thickness interpretation is applicable only as long as the crystallite size is below the microscope resolution. For larger crystallites, image intensity variations must be interpreted by diffraction contrast.

The intensity I, scattering outside the objective aperture, falls off from the incident intensity I_0 with the usual exponential decay,

$$I = I_0\, e^{-Qt} \tag{37}$$

where t is the thickness and Q is the total cross section given by Equation 36. Since the dimensions of Q are reciprocal distance, Q is also the number of scattering acts per unit distance traveled; Qt can be taken as 1 below a critical mass thickness which is approximately 3×10^{-5} gram/cm^2 at 65 keV for many substances.[13] Thus, for carbon, Q is about 8×10^4 cm^{-1}. The mean free path for scattering is simply the critical mass thickness divided by the density. For thicker samples, in which multiple scattering occurs, the situation is more complex than Equation 37 indicates and the intensity falls off more gradually with mass thickness than expected. The cross section Q for a variety of substances and for 65 keV electrons can be taken as about $4 \times 10^4 \rho$,[13] where ρ is the density. Mass thickness contrast tends to decrease with an increase in the accelerating voltage, since the scattered intensity tends to concentrate in the forward direction, so that scattering outside the aperture decreases.

Equations 36 and 37 indicate how, for replicas, point-to-point differences in effective thickness, and differences in the amount of shadowing material, produce amplitude contrast. The subject is considered in much more detail by Heidenreich,[13] who concludes from the simple single-scattering approach that the contrast due to a particle on a substrate film should be independent of the substrate thickness. This ceases to hold for thick substrates because a background "electron fog" results from multiple scattering and spherical aberration effects.

There is a considerable advantage in using dark field (with tilted illumination to maintain the beam axial). The problem is now the reverse of that in bright field, since we consider the intensity scattered into the objective aperture. Heidenreich[13] shows that the dark field contrast $\Delta I/\bar{I}$ for an amorphous particle of thickness δt and total cross section Q_p supported on an amorphous substrate film of average thickness $\bar{t}$ and average cross section $\bar{Q}$ is

$$\frac{\Delta I}{\bar{I}} \approx \frac{\delta t}{\bar{t}} \frac{\overline{D_p(\beta_0)}}{\overline{D(\beta_0)}} \frac{N p}{N} \tag{38}$$

assuming Q_p, δt and $\bar{Q}t$ are much less than unity; where N_p and N are the number of atoms per cc in the particle and substrate, respectively. The radial intensity in the scattering pattern is assumed nearly constant over the diameter of the objective aperture and has average values $\overline{D_p(\beta_0)}$, due to the particle, and $\overline{D(\beta_0)}$, due to the substrate, where β_0 is the tilt angle of the beam incident on the sample. The value $\overline{D(\beta_0)}$ is also known as the differential cross section.

If we assume that, just to be visible, $\Delta I/\bar{I}$ is approximately 0.05, then the minimum particle thickness δt_{min}, visible in dark field, is

$$\delta t_{min} \approx 0.05\bar{t} \cdot \frac{D(\beta_0)}{D_p(\beta_0)} \cdot \frac{N}{N_p} \tag{39}$$

By comparison, for bright field, it can be shown that[13]

$$\delta t_{min} \approx \frac{0.05}{Q_p} = 0.05\Lambda_p \tag{40}$$

where Q_p is the total cross section for the particle for scattering outside the aperture, and Λ_p is the mean free path for the scattering in the particle.

For a carbon protuberance on a carbon film, Equation 39 is simply $\delta t_{min} \approx 0.05\bar{t}$; therefore, for a 100-Å film, the minimum protuberance thickness detectable at 65 keV is about 5 Å using dark field, compared with 70 Å using bright field.

To summarize, the single-scattering approximation implicit in Equation 38 represents a close approximation to image contrast for objects having a thickness not exceeding the mean free path for scattering, provided that crystalline regions are smaller in extent than the resolution. However, the mass thickness interpretation is still fairly satisfactory, when the crystallite size exceeds the resolution, provided that there are many randomly oriented crystallites through the thickness. For amorphous materials the bright field image intensity is controlled by the total scattering cross section and the thickness, whereas the dark field image intensity is determined by the differential cross section and the density.

B. Electron Scattering from Quasiamorphous Solids

Quasiamorphous solids may be defined[13] as those in which, although the atoms are randomly arranged, certain interatomic distances are more frequent than others. Interference among the wavelets, scattered by the atoms, now results in a nonuniform distribution of electrons in the back focal plane of the objective lens and typically give rise to diffraction patterns consisting of diffuse rings, whose radii are reciprocal to interatomic distances in the object. The radial intensity distribution in the scattering pattern represents the radial atom density around an arbitrary point (atom) in the sample. This radial distribution function can be calculated from the observed scattering pattern by Fourier integral methods.[13]

The same treatment can be applied to crystalline materials, provided that there are a large number of randomly oriented crystallites included in the incident beam. However, if the diffraction pattern consists of sharp rings, the mass thickness interpretation becomes invalid and must be replaced by one based on diffraction contrast. A distinction may be made in the image because diffraction contrast effects are tilt-sensitive whereas the mass thickness contrast effects are insensitive to tilting.

Since the scattered electrons carry information on the atom periodicities, in principle, it should be possible to image the atoms if the objective aperture is large enough to admit the scattered beams, or no aperture is used and the sample is a monolayer. In practice this has not been achieved because the resolution is usually inadequate; since the lens is not aberration-free, the bigger the aperture, the worse are the aberrations introduced. Finally, there may be the problem of insufficient contrast. The possibilities are indicated by the fact that macromolecules of virus particles, packed in a crystalline array, have been successfully imaged[102] by what is essentially mass thickness plus phase contrast.[13] The molecular spacing was about 320 Å, so that most of the diffraction pattern was included in a cone of 5×10^{-4} radian and passed through the aperture, enabling a complete image of the globular molecules to be obtained. The constituent atoms of the virus particles were not resolved and would give a quasiamorphous scattering pattern.

To summarize, the presence of radial maxima in the scattering pattern indicates that certain periodicities of scattering features are present in the object (phase detail) resulting in interference, the scattered wavelets reinforcing (being in phase) in certain directions. For a periodic distance r in the object, the angle of scattering β is simply about λ/r, where λ is the wavelength. For the phase information to be available for image formation, it is necessary that $\beta_{obj} > \beta$, where β_{obj} is the solid half-angle, defined by the objective aperture. If $\beta_{obj} > \beta$ for even the shortest r distance r_0 in the object, and the resolution is adequate, both the amplitude and phase contrast mechanisms can operate, giving a *complete image*. Conversely, if $r_0 < \lambda/\beta_{obj}$, r_0 cannot appear in the image, that is, it is a *partial image*. This is presently the case with atomic detail in real crystals.

C. Electron Scattering from Crystalline Solids

We are principally concerned with contrast effects in single crystal regions or between adjoining crystals which extend through the thickness of the foil. This is because, in order to be reasonably transparent to electrons, foils prepared from polycrystalline metal samples must usually be $< \sim 2000$ Å in thickness, whereas grain sizes are usually > 5 μ.

Furthermore, at reasonable magnifications >10,000×, the field size is normally less than the grain size. Complexities from overlapping crystals can be considered separately and arise when inclined boundaries run through the foil; when there are particles such as precipitates within the foil; or in special cases such as heavily rolled, fine grained material, in which the grain thickness can be extremely small normal to the foil plane.

If the foil is tilted, so that Bragg diffraction of electrons occurs and a small aperture is placed in the back focal plane of the objective, *diffraction contrast* can be obtained. This happens because either diffracted electrons are intercepted by the aperture (bright field), so that strongly diffracting regions appear dark, or because a selected diffracted beam is passed through the aperture (dark field), so that only regions diffracting strongly in that direction appear bright. Contrast will normally be observed between adjoining crystals, since they tend to diffract differently as a result of their different orientation. A flat, uniformly thick perfect crystal should show uniform diffraction contrast. If defects are present, such as stacking faults and dislocations, these may cause localized changes in diffraction contrast. Buckling or thickness variations may also give diffraction contrast effects.

The problem then is to interpret observed diffraction contrast effects and to be able to recognize and characterize various kinds of defects. Other kinds of contrast already discussed, such as mass thickness contrast, may be simultaneously present. Diffraction contrast effects may usually be recognized by the fact that they are very tilt-sensitive, that is, they can be changed greatly by tilting the foil. For the proper interpretation of images, it is usually necessary to record a selected area, electron diffraction pattern from the area photographed or a known portion of it. The recording of appropriate dark field images, in addition to a bright field image, is also extremely valuable in some cases.

Most electron microscopic studies of crystals to date have employed the method of diffraction contrast, in which one beam is used to form the image. However, two or more beams may be allowed to contribute to the image in the technique of *phase contrast* or *direct resolution* microscopy. The direct resolution of lattices, at least in the form of partial images, is then possible. This technique will undoubtedly become more commonly used as the resolution of microscopes is improved, although the instrumental requirements will remain exacting. Moiré patterns which may be regarded as magnified pseudolattice images, are already commonly observed.

A number of different theories of diffraction contrast have been developed (see Refs. 12, 13, 100, and 101 for detailed accounts). The qualitative kinematical theory assumes that the intensity of the diffracted beam is small, compared with the direct beam, and that the amplitude of the incident beam is unaltered as it passes through the crystal. In order even approximately to be true, the planes must not be near the Bragg angle. The effect of all but one diffracted beam is neglected. In practice, however, at least one diffracted beam is usually intense in regions photographed, and so dynamical effects must be taken into account for quantitative interpretation. By the appropriate tilting of the foil and by imaging very small areas, the number of strongly diffracting beams may be restricted to one, so that a simplified two-beam dynamical theory can be applied.

The dynamical theory takes account of the depletion of the incident beam, due to (elastic) scattering, and also the further diffraction of the diffracted beam as it passes through additional planes of atoms. Account can also be taken of "absorption," the name used to describe the depletion of the waves in given directions due to large angle inelastic scattering, as a result of collisions with individual electrons in the crystal. This is done by introducing a small complex lattice potential into the theory. The theory neglects the fact that a large proportion of the electrons undergo small angle inelastic collisions but this does not appear to invalidate it.

1. General Theory of Contrast without Lattice Resolution (Diffraction Contrast)

The bright field image can be considered to be a magnified view of the intensity distribution of electrons leaving the bottom surface of crystal parallel to the direction of the incident beam. If absorption is neglected, this is found by subtracting the sum of the calculated intensities of the diffracted beams from the incident intensity. The two-beam case, where only one strong diffracted beam occurs, is the simplest to compute and is often approximated in practice over a small region of a crystal, appropriately tilted.

a.   Concept of the Wave Function

As pointed out by Makin,[101] a basic concept of the theory is the wave function $\phi = \exp(2\pi i \mathbf{k} \cdot \mathbf{r})$ which defines a plane wave. This may be seen as follows.

Both the real and imaginary parts of the expression, $\exp(2\pi i x/\lambda) = \cos(2\pi x/\lambda)$ and $i \sin(2\pi x/\lambda)$, where λ is the wavelength, represent plane waves. We consider only the real part which is a cosine wave. Hence $\exp(2\pi i x/\lambda)$

represents a plane wave of unit amplitude travelling in the x direction. A plane wave 90° out of phase with this can be represented by $i \exp(2\pi i x/\lambda)$. Since the direction of interest may not be parallel to the direction of the wave, vector notation is introduced by letting $1/\lambda = |\mathbf{k}|$, where the direction of the vector $\mathbf{k}$ is the direction of the wave. As seen from Figure 42, the product

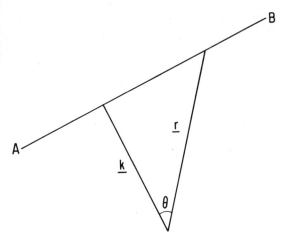

Figure 42 *AB* represents the surface of a stationary plane wave, that is, a coherent wave with plane surfaces of constant amplitude and phase. Vector k is in the direction of the wave, and vector r in the direction of interest.

$\mathbf{k} \cdot \mathbf{r} = kr \cos \theta$ is constant over the plane AB, so that the wave equation $\psi = \exp(2\pi i \mathbf{k} \cdot \mathbf{r})$ defines a plane wave.

2. Dynamical Two-Beam Theory for a Perfect Crystal

In the dynamical theory, the motion of an electron in the periodic potential of the crystal lattice is described by a wave equation that takes account of the multiple reflection of waves inside the crystal.

In the *column approximation*, it is assumed that the wave emitted at any point on the bottom surface of the crystal is caused only by diffraction within a narrow column of material above this point. The wave function $\psi(\mathbf{r})$ of electrons travelling in a column is

$$\psi(\mathbf{r}) = \phi_0(z) \exp(2\pi i \mathbf{k} \cdot \mathbf{r}) + \phi_g(z) \exp(2\pi i \mathbf{k}' \cdot \mathbf{r}) \qquad (41)$$

where ϕ_0 and ϕ_g are the amplitudes, and $\mathbf{k}$ and $\mathbf{k}'$ the wave vectors in vacuo of the direct and diffracted beams, respectively, that have magnitudes $1/\lambda$ and $1/\lambda'$

(Figure 43). In contrast with the kinematical theory, where the amplitude ϕ_0 of the incident wave is assumed to remain constant as it passes through the crystal, in the dynamical theory both the wave amplitudes ϕ_0 and ϕ_g vary with the depth z in the crystal. Changes $d\phi_0$ and $d\phi_g$ in ϕ_0 and ϕ_g as a result of scattering,

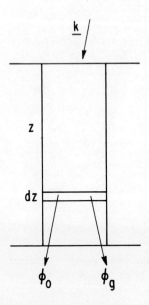

Figure 43 Schematic to illustrate column approximation. After Makin.[101]

when the electron passes through a small slab dz in the column of the crystal (Figure 43) are considered.

It may be shown[101] that

$$\psi(\mathbf{r}) = \psi^{(1)}[C_0^{(1)} \exp(2\pi i \mathbf{k}^{(1)} \cdot \mathbf{r}) + C_g^{(1)} \exp(2\pi i (\mathbf{k}^{(1)} + \mathbf{g}) \cdot \mathbf{r})]$$
$$+ \psi^{(2)}[C_0^{(2)} \exp(2\pi i \mathbf{k}^{(2)} \cdot \mathbf{r}) + C_g^{(2)} \exp(2\pi i (\mathbf{k}^{(2)} + \mathbf{g}) \cdot \mathbf{r})] \qquad (42)$$

where $\mathbf{g}$ is the appropriate reciprocal lattice vector, and $C_0^{(1)}$, $C_g^{(1)}$, $C_0^{(2)}$, and $C_g^{(2)}$ constants. The first $\psi^{(1)}$ term is known as the first Bloch wave and the $\psi^{(2)}$ term as the second Bloch wave. The Bloch wave functions appear naturally as the steady state solutions for the wave function $\psi(\mathbf{r})$ of the electron in the periodic crystal potential. The values of C are governed solely by the extinction distance ξ_g and by the deviation s from the Bragg reflecting position. Values of $\psi^{(1)}$ and $\psi^{(2)}$ are determined by the boundary conditions at the top of the crystal, where values of ϕ_0 and ϕ_g are known.

At the exact Bragg reflecting position, it can be shown[12] that the first Bloch wave reduces to

$$b^{(1)}(\mathbf{k}^{(1)},\mathbf{r}) = i\sqrt{2}\sin\left[\pi\,\mathbf{g}\cdot\mathbf{r}\right]\exp\left[2\pi i\left(\mathbf{k}^{(1)}+\frac{\mathbf{g}}{2}\right)\cdot\mathbf{r}\right] \qquad (43)$$

and the second Bloch wave to

$$b^{(2)}(\mathbf{k}^{(2)},\mathbf{r}) = \sqrt{2}\cos\left[\pi\,\mathbf{g}\cdot\mathbf{r}\right]\exp\left[2\pi i\left(\mathbf{k}^{(2)}+\frac{\mathbf{g}}{2}\right)\cdot\mathbf{r}\right] \qquad (44)$$

These represent plane waves, that is, a current flow in the z direction parallel to the diffraction planes (Figure 44). The intensities $|b^{(1)}|^2$ and $|b^{(2)}|^2$ of the two Bloch waves are proportional to sin and cos functions, respectively, so that they are modulated across the atomic planes but are $\pi/2$ out of phase (Figure 44). For the first wave the current is concentrated between the atoms,

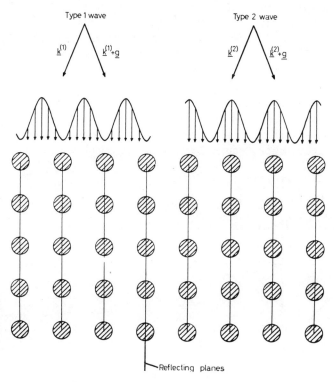

Figure 44 Schematic diagram showing possible standing wave fields at the Bragg reflecting position in a simple cubic lattice. Absorbing regions of atoms are shaded. The type 2 wave (right) is absorbed more than the type 1 wave (left). After Hashimoto, Howie, and Whelan,[103] from Howie.[86] By permission of The Royal Society.)

whereas in the second it is concentrated in the vicinity of the atoms. This fundamental difference persists to a considerable extent, even away from the exact Bragg reflecting position. The second wave is scattered more by the atoms than the first wave, accounting for the *"anomalous absorption effect,"* that is, for the depletion of the waves in given directions, due to large angle inelastic scattering. This effect can be formally accounted for by adding an imaginary term to the crystal potential which results in a modified extinction distance.

The fact that Bloch waves of the first type, which avoid the atoms and are therefore well transmitted, can be excited in crystals makes it possible to apply transmission microscopy to reasonably thick crystals of certain orientations. In practice the wave $\psi^{(1)}$ can penetrate to the order of ten extinction distances, whereas $\psi^{(2)}$ will only penetrate one to two extinction distances.

Qualitatively, we can argue that, since the first Bloch wave is concentrated between the atoms and therefore has a higher average potential energy than the second Bloch wave, it must have a lower kinetic energy, in order to maintain the total energy constant, $k^{(1)} < k^{(2)}$ (corresponding to a larger wavelength). The boundary condition at the top of the crystal means that the wave vector components parallel to the surface must be identical; consequently, $k^{(1)}$ and $k^{(2)}$ differ only normal to the crystal surface (z direction). Because of the difference in wavelength, there is a beating effect between the two waves, resulting in oscillations of total transmitted or diffracted intensity (Figure 45). The periodicity of these oscillations in the z direction, in the Bragg reflecting position, is called the extinction distance ξ_g. This may be calculated from the

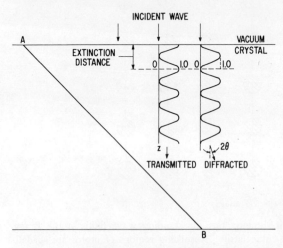

Figure 45 Oscillations of transmitted and diffracted intensities with depth in crystal. These give rise to characteristic Pendellösung fringes, when they meet plane *AB*, which may be the bottom surface of a wedge-shaped crystal, a fault, or a grain or interphase boundary.

following relation:

$$\xi_g = \frac{\pi V_c \cos \theta}{\lambda F_g} \tag{45}$$

where V_c is the volume of the unit cell, λ the relativistic wavelength, θ the Bragg angle, and F_g the relativistically corrected structure factor for the reflection **g**.

The extinction distance is an important quantity, since, as shown below, it controls the periodicity of the thickness fringes and stacking fault fringes; it also controls the periodicity of oscillation effects in dislocation images, the dislocation image width, and the visibility of precipitates. Some approximate values of the extinction distance for 100 keV electrons are given in Table 2 [a fuller list is given by Hirsch et al.[12] and by Howie[86]].

Table 2   Approximate Values of the Extinction Distance ξ_g (in Angströms) for 100 keV Electrons Relativistically Corrected

Crystal	Z	Reflections				
FCC		(111)	(200)	(220)	(311)	(331)
Al	13	560	670	1060	1300	1880
Ni	28	240	280	410	500	750
Cu	29	240	280	420	510	750
Ag	47	220	260	360	430	610
Au	79	160	180	250	290	410
BCC		(110)	(200)	(211)	(220)	(310)
Fe	26	270	400	500	610	710
Nb	41	260	370	460	540	620
HCP		(0002)	$(\bar{1}100)$	$(\bar{1}101)$	$(11\bar{2}0)$	$(\bar{2}201)$
Mg	12	810	1510	1000	1410	2020
Co	27	250	470	310	430	620
Zn	30	260	550	350	500	700
Diamond cubic		(111)	(220)	(311)	(400)	(331)
C	6	480	670	1250	1220	1970
Si	14	600	760	1350	1270	2050
Ge	32	430	450	760	660	1030
Ionic crystals		(111)	(200)	(220)	(311)	(222)
LiF		1720	650	940	2200	1210
MgO		2730	460	660	1800	850

After Hirsch et al.[12]

238   Conventional Electron Microscopy

It will be noted that extinction distances are typically of the order of a few hundred angstroms. They increase with a decrease in the atomic number Z, or with an increase in the order of reflection. They increase with a decrease in the incident wavelength (Equation 45), that is, increase with an increase in the acceleration voltage. The values given in Table 2 may be multiplied by the correction factors of 0.68, 0.81, 0.92, 1.42, and 1.72 for voltages of 40, 60, 80, 300, and 1000 keV, respectively [see Hirsch et al.[12] for a fuller list].

In practice it is difficult to excite a single Bragg reflection $\mathbf{g}$, without, to some extent, exciting additional systematic reflections $n\mathbf{g}$, corresponding to spots in the same zone. If this occurs, the values of extinction distance shown in Table 2 are reduced by an amount that is greatest for low order reflections in high atomic number elements. For the 111 reflection in aluminum and gold, the values given should be multiplied by 0.90 and 0.75, respectively. For the 200 and 220 reflections in gold, the corrections are 0.8 and 0.9, respectively. The corrections may be even larger if low order reflections in other zones are excited.

a. EXTINCTION CONTOURS

We now look at the more quantitative aspects of the two-beam dynamical theory, from which the concept of extinction contours is derived. In the absence of absorption effects (e.g., in thin crystals) the intensity I of the transmitted wave on the dynamical theory, for a crystal of thickness t, is for unit incident amplitude:

$$I = 1 - \frac{1}{1 + w^2} \sin^2 \frac{\pi(1 + w^2)^{\frac{1}{2}}t}{\xi_g} \tag{46}$$

where w is a parameter showing a deviation of the incident beam from the Bragg angle given by

$$w = s\xi_g \tag{47}$$

where ξ_g is the extinction distance for the reflection operating and s is the distance of the reciprocal lattice point from the reflecting sphere, measured normal to the crystal surface ($s \approx |\mathbf{g}|\Delta\theta$, where $\Delta\theta$ is the departure from the exact Bragg angle). The intensity of the diffracted beam is simply $(1 - I)$. Substituting for w in Equation 46, and putting $[s^2 + (1/\xi_g^2)]^{\frac{1}{2}} = s_{\text{eff}}$, we find:

$$I = 1 - \left(\frac{\pi}{\xi_g}\right)^2 \frac{\sin^2(\pi t s_{\text{eff}})}{(\pi s_{\text{eff}})^2} \tag{48}$$

Hence for zero crystal thickness ($t = 0$), the intensity I of a bright field image is 1; as t increases, I oscillates with a periodicity given by

$$t_e = \frac{1}{s_{\text{eff}}} = \frac{\xi_g}{(1 + s^2\xi_g{}^2)^{\frac{1}{2}}} \tag{49}$$

(1) *Thickness Extinction Contours.* Thus, when s is constant and t varies, we get *thickness fringes* (also known as *thickness extinction contours*)—note that the periodicity in the exact Bragg position ($s = 0$) is simply ξ_g, the extinction distance for the operating reflection (Table 2). When the beam is incident at the Bragg angle, crystals of thickness $(m + \frac{1}{2})\xi_g$, where m is an integer, will have zero transmitted intensity and maximum diffracted intensity, while crystals of thickness $m\xi_g$ have maximum transmitted and zero diffracted intensity (Figure 45).

A wedge-shaped crystal will thus show a set of fringes parallel to the edge. If the orientation is identical over the field of view, that is, s is constant (a condition that rarely holds exactly because of buckling), the fringes denote contours of equal thickness. The thickness can be determined by counting the fringes from the thin edge inward. Five or six can usually be seen. As the thickness increases, the fringes fade, as a result of the progressive weakening of the Bloch wave $\psi^{(2)}$ by anomalous absorption. In the bright field, a bright fringe, that is, high transmitted intensity, represents a thickness corresponding to an integral number of extinction distances. In the dark field, the contrast is reversed. Both the visibility and the spacing of the fringes decrease as the crystal is tilted away from the reflecting position in either direction. The effective extinction distance is then less than the values given in Table 2, and the thickness increment is then t_e from Equation 49. A further decrease may occur as the result of many-beam reflection. The presence of a surface feature such as a slip step on the crystal crossing a thickness contour will give rise to a fringe displacement. This effect has been used by Phillips[104] to measure the step height.

Even without a change in foil thickness, a structure discontinuity such as a grain boundary on an inclined plane is sufficient to produce a wedge effect.

(2) *Tilt Extinction Contours.* We see from Equation 49 that, for a constant thickness, the intensity oscillates as s varies, giving rise to *tilt extinction contours (bend contours)*. If the foil is buckled, these contours correspond to regions of the foil at constant inclination to the electron beam, as the reciprocal lattice points sweeps to and fro through the reflecting sphere. In practice, the situation is more complex, since as indicated by Equation 46, the intensity distribution around a reciprocal lattice point shows subsidary maxima which correspond to weaker sub-

sidiary *tilt extinction contours*. This is apparent in the intensity profiles computed in Figure 46 by Hashimoto et al.[103] It will be noted that the dark field fringe pattern is symmetrical about $s = 0$, whether or not

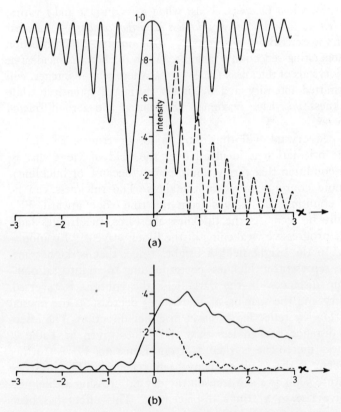

(a)

(b)

Figure 46 (*a*) Rocking curves, computed on the two-beam theory for a crystal of thickness $t = 4\ \xi_g$, showing the bright field (continuous curve) and dark field (dashed curve) intensities versus $x(x = s\ \xi_g)$ across a tilt contour. (*b*) The effect of including absorption $(\xi_g/\xi_{g'} = 0.10)$ that is, thick foil. The dark field curve is symmetrical about $s = 0$. After Hashimoto, Howie, and Whelan,[103] from Howie.[86] (By permission of The Royal Society.)

absorption is included, corresponding to thicker regions of the crystal. The asymmetry at the contour corresponds to a region of low intensity, that is, poor transmission, inside the reflecting position, where Bloch wave $\psi^{(2)}$ predominates, and a region of good transmission outside the reflecting position, where Block wave $\psi^{(1)}$ predominates. This shows that wave $\psi^{(1)}$ is well transmitted, whereas wave $\psi^{(2)}$ is scattered more strongly and is therefore more rapidly attenuated. Since two low order tilt contours for reflection $+g$ and $-g$, frequently occur back to back in bent crystals, they appear in the bright field as a dark band (sometimes called an *absorption band*), with regions of excellent transmission outside the reflecting position on either side. In a flat crystal the highly transmitting orientation can be developed by tilting, so that the center spot in the diffraction pattern is located just outside the two Kikuchi lines, corresponding to a strong low order reflection.

Clearly, in bright field. extinction contours, due to other reflections, may also be present, as seen in Figure 46 from Howie.[86] Using Equation 49 and the measured spacing of the subsidiary fringes in a bright field pattern, we can compute the foil thickness.[105] In this way, Howie[86] calculated the thickness in the area $ABCDE$ in Figure 47 as about 500 Å.

The effect of absorption in thicker regions (Figure 46) is to decrease the intensity of the subsidiary fringes in both the bright and dark field images and make the main fringe more diffuse. This is because the standing wave $\psi^{(2)}$ is weakened by absorption, so that beating with wave $\psi^{(1)}$ is less pronounced.

3. Kinematical Theory of Contrast Effects

The main features of diffraction contrast effects from defects in crystals may be understood on the basis of the simpler kinematical theory.[12,106] The results of the dynamical theory become asymptotic to this for large-enough deviations from the Bragg angle or for sufficiently thin crystals. The column approximation already referred to is used, and only one diffracted beam is considered.

a. PERFECT CRYSTAL

The term perfect crystal refers to a thin crystal free of defects such as dislocations, stacking faults, and so forth, which may contain thickness variations and long wavelength strains due to bending. Considering the dark field image, we find that the amplitude A, diffracted in the direction of the diffracted beam by a column of perfect crystal one unit cell diameter thick such as that in Figure 43, is

$$A = \sum_j F_j \exp\left[-2\pi i(\mathbf{g} + \mathbf{s}) \cdot \mathbf{r}_j\right] \qquad (50)$$

where F_j is the scattering factor for the contents of the unit cell situated at r_j (a lattice vector) in the column, $\mathbf{g}$ the reciprocal lattice vector corresponding

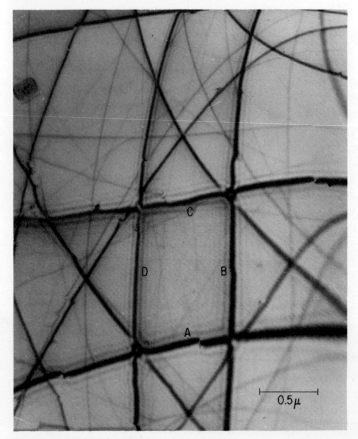

Figure 47   Bright field electron micrograph of saucer-shaped aluminum foil near ($\bar{2}15$) orientation, showing four [113]-type tilt contours A through D. Private communication from Howie, after Howie.[86]

to the operating reflection, and s a small vector giving the deviation of the reciprocal lattice point from the reflecting sphere. Since the Bragg angle is only a degree or two, the column can be taken as normal to the crystal surface, as in Figure 43, instead of parallel to the diffracted beam, without appreciable error. Since the product $g \cdot r_j$ is an integer, we have for a perfect crystal:

$$A = \sum_j F_j \exp\left(-2\pi i s \cdot r_j\right) \tag{51}$$

The unit cells are assumed similar, so that the constant F_j's can be moved outside the summation sign. The origin is taken at the midpoint of the column halfway through the thickness and the summation replaced by an integral. Hence

$$A \propto \frac{\sin \pi ts}{\pi s} \tag{52}$$

or

$$I \propto \frac{\sin^2 \pi ts}{(\pi s)^2} \tag{53}$$

where t is the length of the column, and s the projection of $\mathbf{s}$ onto t. Also, I is the kinematical intensity distribution around a reciprocal lattice point which gives rise to a spike in reciprocal space in the t direction.

Equation 52 gives the amplitude, scattered in a particular direction from a column one unit cell wide, that is, the amplitude a long way from the crystal effectively at infinity, as in the diffraction pattern. The image contrast depends upon the intensity distribution on the lower surface of the crystal. Diffraction contrast then results from aperturing out either the diffracted or transmitted beam. We now consider the latter, that is, the dark field image, and the consequences of a finite column size.

Although the rays diffracted from an isolated column would be spread over a range of directions, in a crystal the rays at the bottom surface are concentrated in a small range of angles around the principle ray, corresponding to the Bragg reflection, as a result of interference with the rays diffracted from neighboring columns. The total intensity for a column of finite width was computed by Hirsch, Howie, and Whelan[106] by calculating the average intensity per unit area of cross section and integrating the intensity scattered over the reflecting sphere. This gave an expression identical with Equation 53, with the constant of proportionality evaluated. In their later treatment,[12] by a somewhat different method, Hirsch and his co-workers arrive at a relation for the total diffracted amplitude emerging from the crystal, when the reflecting planes are normal to the surface, of:

$$A = \frac{i\pi}{\xi_g} \int_0^t \exp\left(-2\pi isz\right) dz \tag{54}$$

$$= \frac{i\pi}{\xi_g} \frac{\sin \pi ts}{\pi s} \exp\left(-\pi ist\right) \tag{55}$$

where ξ_g is the extinction distance for the operating reflection. The intensity, therefore, varies as $[\sin^2(\pi ts)]/(\pi s)^2]$, that is, oscillates sinusoidally with the depth in the foil for fixed s, in agreement with the result already given from the dynamical theory. The oscillations have a depth periodicity $\Delta z = s^{-1}$, that is, the fringe spacing decreases as the deviation from the reflecting position increases. This can be seen, as we shall now show, in rather simpler fashion from an amplitude phase diagram.

The amplitude phase diagram[106] is a convenient graphical way of representing the amplitude of the diffracted wave given by Equation 55. The amplitude scattered by a column element dz, relative to the column midpoint 0 taken as the origin of zero amplitude, can be considered as a vector dz with a phase angle $2\pi sz$ (Figure 48). For a perfect crystal

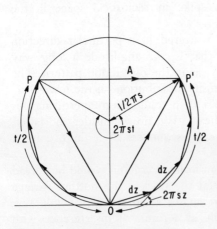

Figure 48   Amplitude phase diagram for a perfect crystal. The circle has a radius $(2\pi s)^{-1}$. The amplitude A, diffracted by the column of the crystal of thickness t, is PP', where the arc POP' is equal to t. The intensity is $(PP')^2$. The value of PP' will oscillate as s or t is varied. After Hirsch et al.[106]

the angle between successive vectors is constant; since successive vectors are equal, they are chords of a circle with radius $1/2\pi s$. Arcs PO and OP' of length $t/2$ represent the top and bottom halves of the column. Also, PO and OP' represent the resultant diffracted vectors from the two halves that give the resultant PP' which represents the diffracted wave of amplitude A. The wave PP' subtends an angle $2\pi st$ at the center. The amplitude A is PP which is twice the radius, multiplied by the sine of half the angle subtended at the center. Hence

$$A = \frac{1}{\pi s} \sin \pi st$$

which is the same result as Equation 52. As the thickness changes, P and P' move around the circle, causing the amplitude A to oscillate with thickness, in accord with Equation 52. The average A is proportional to $1/s$, and the average diffracted intensity I to $1/s^2$. The depth periodicity of the intensity oscillation is $1/s$. The intensity of the transmitted beam is $1 - I$ on the principle of conservation of flux, so that the bright and dark field images are complimentary. If s is increased, the circle radius is decreased, and both the intensity maxima and the periodicity of the depth oscillations are decreased. For a given thickness, tilting the crystal to change s changes the circle radius; hence oscillations in intensity as a function of angle are expected. This is more easily seen in Figure

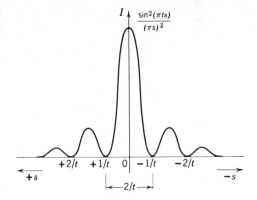

Figure 49   Kinematical intensity distribution about a reciprocal lattice point. From Thomas,[2] after Hirsch et al.[106]

49 which shows the intensity about a reciprocal lattice point as a function of s, calculated from Equation 53.[2] The main bend contour occurs at $s = 0$, and the subsidiary maxima give rise to subsidiary bend contours in a buckled foil. This treatment breaks down for $s = 0$, unless t is very small; one must then use the dynamical treatment (Figure 46).

b. IMPERFECT CRYSTAL

If a defect is present that results in the displacement of an atom at depth z in a column from its ideal position by a vector $\mathbf{R}$, an additional phase factor $\exp(-i\alpha)$ where $\alpha = 2\pi\mathbf{g} \cdot \mathbf{R}$ is introduced, giving a diffracted amplitude A:[12]

$$A = \frac{i\pi}{\xi_g} \int_0^t \exp(-2\pi i\mathbf{g} \cdot \mathbf{R}) \exp(-2\pi isz) \, dz \qquad (56)$$

It may be noted that, if $\mathbf{R}$ and $\mathbf{g}$ are parallel, contrast is a maximum, whereas there is no contrast if the displacement $\mathbf{R}$ is normal to $\mathbf{g}$. The latter is the basis of determination of the Burgers vectors.

(*1*) *Stacking Fault.* A stacking fault such as *AB* in Figure 50, passing through the crystal on an inclined plane, represents the simplest type of planar imperfection because the atom displacements are all given directly by the displacement vector **R**. The crystal above and below *AB* is perfect, but the

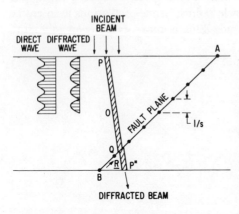

Figure 50   Schematic showing intensity oscillations of periodicity 1/*s* of the direct and diffracted waves in the crystal on the kinematic theory. A column *PP″* is displaced by vector **R** at *Q*, where it crosses the inclined stacking fault *AB*. The black dots correspond to intensity minima (fringes) in the dark field image of the fault. After Hirsch et al.[106]

portion below is translated, relative to that above, by the vector **R** parallel to *AB*. The phase factor that determines the contrast is thus zero above *AB* but changes abruptly to the value $\alpha = 2\pi \mathbf{g} \cdot \mathbf{R}$ below *AB*, where **g** represents the reciprocal lattice vector, corresponding to the operating reflecting plane. When $\mathbf{g} \cdot \mathbf{R}$ has an integral value, the fault is invisible, since the Bragg reflecting planes above and below the fault are in registry with one another. If **R** is a lattice translation vector, the fault is invisible for *all* reflections. However, if **R** is not a lattice translation vector, there may be particular values of **g** for which the product $\mathbf{g} \cdot \mathbf{R}$ is integral. This forms the basis of the method of determining **R** by looking for reflections in which the fault is invisible. Faint contrast is expected[12] if $\mathbf{g} \cdot \mathbf{R}$ differs from an integer by 0.02 or more. Since the contrast is unchanged by adding a further lattice translation vector to **R**, the value of **R** so determined suffers from this uncertainty which may in particular cases be removed by studing the contrast behavior of the partial dislocations bounding the fault.

In the important case of stacking faults on the close-packed {111} planes in fcc crystals, **R** can take the values $\pm\frac{1}{3}$ [111]. A displacement $\mathbf{R} = \frac{1}{3}$ [111], equivalent to moving the lower crystal (Figure 50) in an upward direction, corresponds to an *intrinsic* fault which is equivalent crystallographically to removal of a {111} layer of atoms, followed by collapse of the crystals on either side. A displacement $\mathbf{R} = -\frac{1}{3}$ [111] corresponds to an *extrinsic* fault, equivalent to the insertion of an extra layer of atoms. Faults can also be produced by shear parallel to the fault plane with displacement vectors such as $\pm\frac{1}{6}$ [$\bar{1}\bar{1}2$] and may have different partial dislocations but differ from $\pm\frac{1}{3}$ [111] only by a lattice translation vector, so that they can be classed similarly as intrinsic or extrinsic. Techniques for distinguishing the latter depend on dynamical contrast effects.[12]

The intensity distribution at a fault can be determined[12] from Equation 56. However, the contrast effects can be more simply understood by means of an amplitude phase diagram. For the case discussed above, $\mathbf{R} = \pm\frac{1}{3}$ [111], and for a reflection $\mathbf{g} = h, k, l$, the phase angle α is

$$\alpha = 2\pi\mathbf{g} \cdot \mathbf{R} = \pm \frac{2\pi}{3} (h + k + l) = \pm \frac{2n\pi}{3} \qquad (57)$$

where n is an integer. The possible values of α are $\pm 120°$ or $0°$, depending on the indices of the reflection. The amplitude phase diagram for $\alpha = -120°$ is shown in Figure 51, after Hirsch et al.[106] The circle POP' represents the

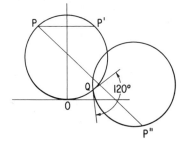

Figure 51 Amplitude phase diagram for column PQP'' in Figure 50, crossing a stacking fault at Q in an fcc crystal ($\alpha = -120°$). The resultant amplitude is represented by the vector PP''. After Hirsch et al.[106]

diagram for the perfect crystal of radius $1/2\pi s$, as in Figure 48. Zero is the midpoint of the column. The periodicity of the amplitude oscillations inside the crystal is $1/s$, corresponding to the circumference of the circle; POQ represents the diagram for the upper part of the column shown in Figure 50. At Q where the column crosses the fault, an abrupt change of phase, taken to be $-120°$, occurs, and the diagram thereafter follows the second circle of the same radius $1/2\pi s$ which passes through Q and makes an angle of $120°$ with the first circle. The amplitude from the top part of the column is PQ, while that from the bottom is QP''; PP'' is the resultant diffracted amplitude which in general differs from the amplitude PP' from an unfaulted region of the crystal. The point Q changes according to the position of the column, and P'' and PP'' change accordingly. As Q varies, PP'' will oscillate, since Q and P'' occupy the same positions for points on the fault which differ in depth by $1/s$ (Figure 50). The contrast, therefore, takes the form of fringes parallel to the intersection of the fault with the surface as in Figure 30b.

If the fault plane is parallel to the foil surface, the column length above and below the fault is independent of position, so that the position of Q is fixed in Figure 51 with PP'' constant at a value which may be larger, smaller, or equal to PP'. The fault will show a uniform contrast or amplitude, depending on its depth in the foil. This situation is commonly encountered in hexagonal lamellar materials which both cleave and fault on the basal plane.

Typical fault contrast effects may be observed in any case in which a constant nonlattice displacement exists between two crystals. They are sometimes seen[12] at very thin platelike precipitates, where they are called "displacement fringes" and are parallel to the intersection of the precipitate plane with the foil surface. They can occur at antiphase domain boundaries in ordered alloys [see review by Marcinkowski[107]].

The contrast at overlapping stacking faults has been discussed on the kinematical theory with the use of the amplitude phase diagram by Amelinckx.[100]

For a complete discussion of the detailed contrast effects observable at stacking faults near the exact Bragg reflecting position and in thick crystals, in which absorption effects assume importance, reference may be made to the dynamical theory.[12,13,100] The dynamical effects may be made use of for determining the nature of a fault, for distinguishing the top and bottom fault intersections with the foil surfaces, and for distinguishing stacking faults from thin twins. Distinction from thickness fringes, and fringes at grain boundaries, is also possible, although this is usually fairly obvious on other grounds.

(2) Dislocations. Diffraction contrast due to dislocations is more complex than that due to stacking faults, since a mixed situation of phase shift and atom displacement field is involved. The intensity distribution around such defects cannot be dealt with rigorously by either the kinematical or dynamical theories, as pointed out by Heidenreich,[13] but the geometry can be well approximated.

The simple concept is that dislocations in thin crystals cause Bragg diffraction contrast effects because of bending of the reflecting planes near the dislocation, which introduces a local change in the deviation parameter s, representing the departure from the reflecting position. The presence of an extra half-plane of atoms at an edge dislocation results in the bending's being in the opposite sense on either side, so that the crystal is locally rotated toward the reflecting position on one side and away on the other. Planes not displaced by the dislocation, that is, having no component of the Burgers vector normal to them, diffract as in a perfect crystal. The diffracted intensities from those undergoing displacements are modified. Maximum contrast is obtained when the displacements are normal to the reflecting plane. The defect will tend to vanish if the crystal is tilted, so that only undistorted planes diffract. This principle forms the basis of the common methods of determining the Burgers vector, described in detail elsewhere.[2,12,13,20,100] Since it is difficult to arrange in the bright field for only a single reflection to contribute to the image contrast, systematic dark field study employing appropriate reflections is commonly employed. Since the strain field, rather than the dislocation line, is responsible for the contrast, the dislocation image is usually asymmetrical and need not coincide with the dislocation line. Far from being a line, the image width is often about 100 Å. Furthermore, multiple images may occur if more than one type of reflection operates simultaneously.

The contrast effects produced by dislocations depend, as in the case of stacking faults, on the quantity exp $(2\pi i g \cdot R)$, except that the atomic displacement vector R is a continuously varying function of position. The principle atom displacements are parallel with and proportional in magnitude to the Burgers vector b. This may be defined as the relative displacement across the slip plane, produced by the passage of the dislocation. For an edge dislocation, b is perpendicular to the line direction; for a screw dislocation it is parallel. Clearly, for a dislocation loop some portions may have edge character, some screw character, and some mixed character. The contrast depends mainly on the quantity $g \cdot b$ which is an integer for a perfect dislocation, for which b is a lattice vector, but need not be integral for partial dislocations bounding stacking faults.

(3) *Screw Dislocations.* Pure screw dislocations are simpler in that the line directions are parallel to the Burgers vector b, so that $g \cdot b$ completely determines their visibility. Thus the dislocation is invisible when $g \cdot b = 0$; appears as a dark line (in bright field) for $g \cdot b = 1$, whose width is about $\xi_g/5$, that is, about 30 to 100 Å for low order reflections; and for $g \cdot b = 2$ is wider and may consist of two peaks when s is small. Two images may also be obtained when two reflections g_1 and g_2 operate, and may both be on the same side or on opposite sides of the true dislocation position, depending on whether the signs of $(g_1 b)s$ and $(g_2 b)s$ are the same or different. The position of the image, relative to the dislocation, is reversed by changing the sign of $(g \cdot b)s$, so that a discontinuity is observed in which a dislocation crosses an extinction contour (s reversed in sign) or an absorption band (g reversed). The image is usually within the image width $(\sim\xi_g/5)$ of the dislocation position.

The intensity profiles of a screw dislocation parallel to the surface in the kinematical theory are shown in Figure 52, from Hirsch et al.[106] The intensities are obtained from these curves[12] by multiplying the ordinates by

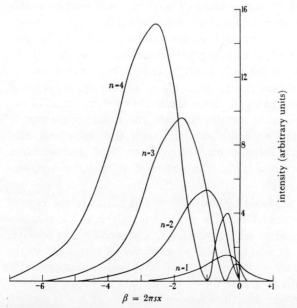

$$\beta = 2\pi s x$$

Figure 52   Intensity profiles of images of a screw dislocation parallel to the foil surface for various values of n ($n\pi$ is the phase difference between waves scattered immediately above and below the dislocation). The abscissa β is $2\pi s x$, which for constant s is proportional to the distance x from the dislocation located at $\beta = 0$. From Hirsch et al.[106] (By permission of The Royal Society.)

$(2\xi_g s)^{-2}$. Since the extinction distance ξ_g is likely to increase with n, the relative peak heights need adjustment, but in any case they are not reliably estimated by the kinematic theory. The increase in image width with n is qualitatively correct, as is the fact that the image displacement is of the order of its width. The multiple images predicted for $n = 3$ and 4 have not been experimentally confirmed.[12]

When the screw dislocation is inclined to the foil plane at an angle ψ, the theory indicates[12] that the dislocation width and amplitude decrease by a factor cos ψ, so that the dislocation becomes less visible. Since different parts of the dislocation are now at different depths in the foil, various intensity oscillation effects analogous to stacking fault fringes are observed, particularly near the Bragg reflecting position. These require the dynamical theory for detailed explanation. If the dislocation is nearly normal to the foil, instead of vanishing, since **g · b** is approximately zero, bright and dark dot contrast pairs lying on a line normal to **g** are observed because of stress relaxation at the foil surfaces,[108] which results, in the bending of the reflecting planes in opposite senses on opposite sides of the dislocation.

(4) Edge and Mixed Dislocations. Mixed dislocations can always be resolved into edge and screw components for purposes of calculation, so that only pure screws and edges need be considered. Edge dislocations have displacements R_2 normal to the slip plane, in addition to the displacements R_1 parallel to b, considered in the case of screws above. Usually, R_1 is dominant, so that the main image features are quite similar to those of screw dislocations. Unlike a screw dislocation that vanishes when $g \cdot b = 0$, a faint symmetrical image of an edge dislocation may remain because of the displacements R_2, although displacements R_1 produce no contrast. Image profiles for edge and mixed dislocations on the kinematic approximation have been computed by Gevers.[109,110]

(5). Partial Dislocations. Partial dislocations lie at the boundary of a stacking fault and have a Burgers vector b that is not a lattice vector. The product $g \cdot b$ need not be an integer; for example, it may take the values 0, $\pm \frac{1}{3}$, $\pm \frac{2}{3}$, ± 1, $\pm \frac{4}{3}$, and so forth, in an fcc crystal, for which the Shockley partials have $b = a/6[211]$. Usually, screw partials with $g \cdot b = +\frac{2}{3}$, ± 1 or $\pm \frac{4}{3}$ are visible; those with $g \cdot b = 0$ or $\frac{1}{3}$ are invisible. Frank edge partials for which $b = a/3[111]$ also occur in the fcc system and can take $g \cdot b$ values, as for the Shockley partial above. Since the Frank partial is of edge type, the contrast from R displacements may help distinguish it. The dark field image profiles of partial dislocations have been computed by Gevers.[111]

(6) Precipitates. The general problem of diffraction contrast effects at precipitates is complex, since there are many variables such as the precipitate shape, depth in the foil, crystal structure, orientation, composition, and lattice mismatch from the matrix which may produce strain in both the particle and matrix, and is often a function of the foil temperature and treatment. Composition and lattice defect gradients may exist near the interface.

Grain boundary fringes may be visible at an inclined precipitate/matrix boundary. Moiré fringes are often observed at precipitates. Displacement (fault) fringes may be observed if a platelike precipitate produces a relative displacement of the matrix on either side of it.

(a) COHERENT PRECIPITATES. Contrast effects at coherent precipitates, particularly those of spherical shape, are the simplest to discuss. Coherent precipitates are those in which both the precipitate and the matrix atoms lie on the same lattice, and the mismatch, if any, is accommodated elastically. The presence of small precipitates of this kind is often detected by *strain field contrast* which can occur if there are atomic displacements normal to the diffracting planes.

For an isotropic misfitting spherical inclusion in an infinite isotropic matrix, the displacements R_r are purely radial and are given after Mott and Nabarro[112] by

$$R_r = \frac{\epsilon r_0^3}{r^2} \qquad \text{outside the particle} \qquad (58)$$

$$R_r = \epsilon r \qquad \text{inside the particle} \qquad (59)$$

where r_0 is the particle radius, and ϵ the constrained or in situ strain (the same quantity measured by x-ray diffraction on the alloy). The strain ϵ is given by:[12]

$$\epsilon = \frac{3K\delta}{3K + 2E/(1 + \nu)} \qquad (60)$$

where K is the bulk modulus of the inclusion E and ν are Young's modulus and Poisson's ratio for the matrix, and δ is the conventional misfit:

$$\delta = \frac{2(a_1 - a_2)}{(a_1 + a_2)} \qquad (61)$$

where a_1 and a_2 are the unit cell sides of the precipitate and matrix, respectively. If $\nu = \frac{1}{3}$ and the particle and matrix have similar moduli, Equation 60 indicates that $\epsilon \approx (\frac{2}{3})\delta$. Equation 60 arises because the strain in the matrix is a pure shear, while that in the particle is a dilation. This is a simplification, since most materials are elastically anisotropic.

The bending of the lattice planes around a particle of this type is illustrated schematically in Figure 53, from Hirsch et al.[12] The distortion viewed in the z direction would be identical. Since all displacements are radial, planes passing through the center of the particle undergo no displacement (Figure 53), so that it is immediately apparent that there is a plane of no contrast[113] normal to g. There are two distinct sources of diffraction contrast from such a particle, which may be called *precipitate contrast* and *matrix contrast*.[12] The former arises because the amplitude and phase of waves in a column of crystal passing through the particle are altered by its presence. The latter is caused entirely by the strain displacement field in the matrix and can give diffraction contrast in a manner analogous to dislocations, although the situation is more complex. We consider matrix contrast first.

The *matrix contrast* (*strain field contrast*) effects depend essentially on the quantity $\mathbf{g} \cdot \mathbf{R}$, where $\mathbf{g}$ is the reciprocal lattice vector for the Bragg diffracting planes, and $\mathbf{R}$, which varies with position, is the displacement in the matrix. For spherical precipitates, considered in the kinematical theory by Phillips and Livingston[113] and in the dynamical theory by Ashby and Brown,[114] the displacements $\mathbf{R}$ are all radial, their direction (inward or outward) depending on whether the inclusion contracts or expands the matrix. A void is equivalent to the former. If we consider a column parallel to the diffracting planes, passing through the center of the particle in Figure 53, the displacement **R** normal to the planes, that is, parallel to **g**, is zero, so that $\mathbf{g} \cdot \mathbf{R} = 0$ and a line of no contrast results in the image. If two strong reflections operate, two lines of no contrast are observed.[113]

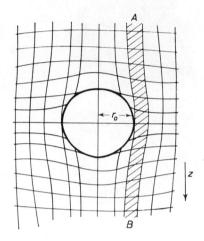

Figure 53 The bending of matrix lattice planes around a spherical misfitting inclusion which expands the matrix. The vertical lines can be taken as the diffracting planes which in practice are slightly inclined to z, the axis normal to the crystal surface. From Hirsch et al.[12]

Maximum contrast is expected for a column passing close to the surface of a small particle, as shown in Figure 53. The phase angle α is then

$$\alpha = 2\pi\mathbf{g} \cdot \mathbf{R} = \frac{2\pi\epsilon g r_0^4}{(r_0^2 + z^2)^{3/2}} \qquad (62)$$

where r_0 is the particle radius and ϵ the strain parameter given by Equation 60, and z is measured from the point at which the column touches the particle. Intensity profiles may be computed or determined from phase amplitude diagrams.[12] Figure 54 shows the two principal types of contrast, seen in the kinematic region after Phillips and Livingston,[115] and drawn from the computed intensity levels.[113] Ashby and Brown's[113] dynamical two-beam calculations confirm the contrast features above and permit quantitative interpretation of the intensity in terms of the strain ϵ. It appears that, for sufficiently small particles, the condition for visibility is

$$\frac{\epsilon g r_0^2}{\xi_g} \geq 10^{-2} \qquad (63)$$

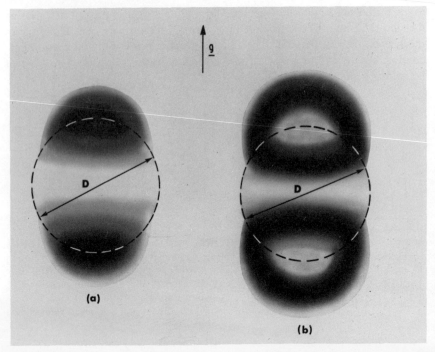

Figure 54 Drawing of two main types of (kinematic) bright field contrast at the spherical cobalt particle in the interior of a copper crystal. The dashed circle represents the particle boundary. Note the symmetrical twin contrast lobes with "line of no contrast" normal to the diffraction vector g. Closed-D contrast lobes (a) correspond to large s, open-D contrast lobes (b) to intermediate s. After Phillips and Livingston.[115]

Figure 55 from Howie[86] may be used as a guide in selecting the reflection used; the values of the parameter $g\xi_g$ are tabulated by Howie (see Table 2 for values of ξ_g). For very small precipitates ($r_0 < \sim 30$ Å), low order reflections, giving short extinction distances, are favorable. For large precipitates ($r_0 > \sim 100$ Å) and small mismatch, a high order reflection is best.

Care is necessary in determining the particle size, since the particle itself, shown by the dashed line in Figure 54, may be invisible. The diameter is best measured between the lobe corners indicated by the arrows, taking care to avoid undue contrast in printing. The image width parallel to g (Figure 54) is mainly determined by the quantity $\epsilon g r_0^3/\xi_g^2$ and may be used to measure the strain when $s = 0$ [see Ashby and Brown[114] for details].

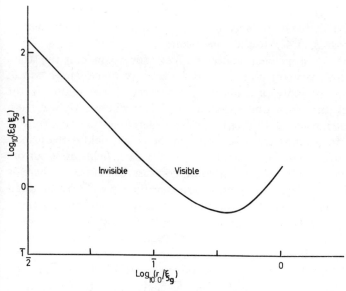

Figure 55 Visibility limit of spherical inclusions. The curve represents the minimum particle radius r_0 for which it should be visible for a particular reflection of reciprocal lattice vector g and in situ mismatch strain ϵ; ξ_g is the extinction distance. From Howie.[86] (By permission of Blackwell Scientific Publications.)

Anomalously wide asymmetrical images are expected,[114] caused by strain relaxation, if the particle is within $\xi_g/2$ of a foil surface. In a bright field image the sense of the asymmetry is opposite at the two foil surfaces, but in a dark field image it depends only on the sign of g and thus may be used to determine the sign of the strain.[114] For an interstitial type of inclusion, that is, one that expands the matrix, the contrast in a positive print is dark on the side of positive g.

Similar methods, based on the dynamical theory, may be used to measure the sign and magnitude of ϵ for platelike inclusions[114] and tetrahedra,[116] and the Burgers vector of small loops nearly normal to the foil.[114,116] The dynamical diffraction contrast from cuboidal inclusions was considered by Sass, Mura, and Cohen,[117] who also measured the sign and magnitude of ϵ, as well as Weatherly.[118] The twin lobe contrast with a line of no contrast at cuboidal particles resembles that from spherical particles; however, a distinction is possible by the occurrence of a notch in the line of no contrast[117] and/or asymmetric image[117,119] if the reflecting plane is not a symmetry plane of the cube. Tanner[120] considered the kinematic diffraction contrast from thin disk-shaped {100} zones in a copper-2% beryllium alloy; by using elastic anisotropy of the matrix he was able qualitatively to account for contrast striations parallel to {110} traces, that is, at 45° to the zone planes.

In the absence of strain field contrast effects, for example, if the strain ϵ is small or zero, coherent particles may be seen by *precipitate contrast*, arising from one or more of a number of sources: (*a*) structure factor contrast[114,121] if the extinction distance in the precipitate differs from that in the matrix; (*b*) if the particles are ordered and the matrix disordered, uniform contrast can arise either in the bright field if the principle reflections are from the superlattice[119] or in dark field, using an ordered reflection[119,122-124]; (*c*) orientation contrast can arise if there is some parameter difference, for example, so that the crystal may be tilted in such a way that either the precipitate or matrix is diffracting more strongly; (*d*) displacement fringe contrast[125,126] can occur if the atom displacements produced by the particle are such that there is an abrupt phase change at the interface, (*e*) moiré fringe contrast may occur, and (*f*) interface fringe contrast can arise to curvature of the interface or bending of the lattice planes near the interface, analogous to the behavior near a void. These six effects are discussed in some detail by Hirsch et al.[12] with additional references to the literature. Dynamic contrast effects from small voids have been discussed by Van Landuyt, Gevers, and Amelinckx[127] and Hashimoto.[128]

The dynamical theory of diffraction contrast effects (orientation contrast) for nondiffracting *incoherent inclusions* (no matrix distortion) was considered by Guyot.[129]

4. Theory of Contrast with Lattice Resolution

In the lattice resolution technique, sometimes called direct resolution microscopy, one or more diffracted beams are allowed to pass through the objective aperture, in addition to the transmitted beam, and are recombined in the image. If the increased aberrations can be tolerated, no aperture need be employed; if the lens were aberration-free; an improved image with maximum information content would result, according to the Abbé diffraction theory. In practice, factors such as spherical aberration may introduce phase differences between the different beams that detract from the image.

The contrast in the image is determined by the way in which the beams are combined (or interfere) in the image, since the beams are coherent with one another. Periodic contrast results, with periodicities related to those in the crystal. Detail finer than the resolution limit of the microscope cannot, of course, be reproduced and this factor alone has so far limited the rapid growth of direct resolution microscopy. With the new high resolution microscopes available, now capable of resolving the lattices of metals, the field should assume vastly increased importance over the next decade.

The fringe patterns obtained may be complete or more often are partial images, depending on the amount of information admitted to the image and on the effective resolution. The contrast details in partial images, for example if a single set of planes is imaged, are simpler than complete images and adequate for many purposes. In some circumstances harmonic or subharmonic fringes may occur. In a simple two-beam case, one fringe may result for each lattice plane, so that there is a one to one correspondence between the number of planes and the number of fringes, and the correct plane periodicity is seen.

Detailed theoretical consideration with adequate knowledge of the experimental conditions would be necessary to decide the exact relationship between the fringe position and the atom (plane) positions. Dynamical theory indicates that, in general, the two are not coincident. Since phase shifts due to lens aberrations are usually general rather than local, that is, tend to move all the fringes equally, this is often of little practical consequence. Localized fringe shifts relate to atom plane distortions, associated with the strain fields around lattice defects such as dislocations. Spurious effects can result from local changes in thickness or deviation parameter s, so that care is necessary in interpretation. Projected images must tend to average distortions which vary through the foil thickness. Strain relaxation effects around defects may occur as a result of the proximity of the foil surfaces. A great deal of work evidently needs to be done before image interpretation is as firmly based as in the case of diffraction contrast.

A one to one correspondence between image and object points is apparent in a ray diagram for the objective lens (Figure 56). Rays coming

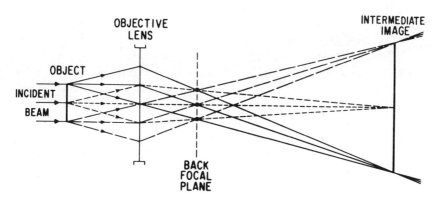

Figure 56 Schematic ray diagram for objective lens which is assumed aberration-free. After Phillips and Hugo.[15]

from each object point, whether representing undeviated or scattered electrons, recombine at a corresponding point in the conjugate image plane. If the object structure is crystalline or has a regular periodicity, the deviated rays in the figure can be taken to correspond to diffracted beams in specific directions. Rays in a particular direction from different object points converge at the same point in the back focal plane, giving rise to a diffraction pattern. The rays entering the image can be controlled by the selection of the size and position of the objective aperture inserted in the back focal plane.

In a very simple approach, after Menter,[130] the periodic image is regarded as a pattern formed by interference between the transmitted and diffracted beams contributing to the image. According to interference theory, a single diffracted beam at an angle 2α to the transmitted beam will give fringes spaced $s' = \lambda/2\alpha$, where λ is the wavelength. From Bragg's law, the first order diffracted beam from a set of planes of spacing d, makes an angle α of $\lambda/2d$ (taking sin $\alpha \approx \alpha$ for small α) with the undeviated beam, hence $s' = d$. Thus the fringe spacing corresponds to the spacing of the diffracting planes. Menter concluded that the beam necessary to image the projection of a set of crystal planes was the Bragg reflection from the planes and was able to confirm this on crystals of copper and platinum phthalocyanine. In a bent crystal, the periodic image of a particular set of planes appeared only in the extinction contour, corresponding to diffraction from that set of planes, and appeared only if the particular beam was admitted through the objective aperture. If two diffracted beams were admitted, a complex two-dimensional lattice image was formed,[131] whose periodicity in any direction was related to that of the crystal planes perpendicular to that direction.

a. DYNAMICAL TWO-BEAM THEORY FOR A PERFECT CRYSTAL

The dynamical theory of lattice resolution was discussed by Cowley;[132] Cowley and Moodie;[133] Hashimoto, Mannami, and Naiki;[134] Heidenreich;[13] and by Amelinckx.[100] Unfortunately, in spite of its importance, the dynamical theory for the beam tilt case does not appear to have been discussed. An approximate kinematic treatment was given by Komoda,[135] who pointed out that the (lattice) fringe position varied with the focal plane in the untilted two-beam case but was independent of the focal plane in the two-beam symmetrical tilt case. If confirmed by dynamical considerations, this is an important result, since the experimentalist is concerned not only with the spacing of the fringes, which is correctly predicted by the existing treatments, but also with establishing the exact position of a fringe relative to a plane of atoms. The kinematical two-beam relations derived by Hirsch et al.[12] appear qualitatively to predict correctly the effect of varying thickness or deviation parameters, but break down when $s = 0$ which is the very condition aimed at experimentally, unlike most diffraction contrast microscopy.

We shall follow the dynamical approach of Amelinckx[100] for the nontilt case and consider the intensity distribution for the well-focused image, which is equivalent to that at the exit surface of the foil, as a function of the position vector r on the exit surface. Amelinckx shows directly from the two-beam dynamical equation for the wave function that, for a plane-parallel plate crystal, the image (resolution permitting) consists of a set of parallel lines perpendicular to the diffraction vector with period d_{hkl} of the planes diffracting. This is the same result obtained by Hashimoto et al.[134]. The intensity is given by

$$I = I_t + I_s + 2(I_s I_t)^{1/2} \sin (2\pi g x + \chi) \tag{64}$$

where the x axis is parallel to the reciprocal lattice vector g and the y axis perpendicular to it ($\mathbf{g} \cdot \mathbf{r} = nx$), where I_t and I_s are the transmitted and scattered intensities, respectively.

where

$$\sin \chi = \frac{-2s}{\sigma^2 \xi_g} \sin^2 \pi \sigma t \tag{65}$$

$$\cos \chi = \frac{-2}{\sigma \xi_g} \sin \pi \sigma t \cos \pi \sigma t \tag{66}$$

$$\sigma = \frac{1}{\xi_g} [1 + (s \xi_g)^2]^{1/2} \tag{67}$$

$$\tan \chi = \frac{s}{\sigma} \tan \pi \sigma t \tag{68}$$

and

$$I_s = \frac{\sin^2 \pi \sigma t}{(\sigma \xi_g)^2} \tag{69}$$

Here ξ_g is the extinction distance, and t the crystal thickness. If we neglect absorption and take the incident intensity as unity, that is, put $I_t = 1 - I_s$, Equation 64 reduces to:[100]

$$I = 1 + 2[I_s(1 - I_s)]^{1/2} \sin (2\pi g x + \chi) \tag{70}$$

This relation indicates[100] that the image consists of a sinusoidal fluctuation, superimposed on a uniform background of unit intensity. If the two beams (transmitted and diffracted) are equal in intensity, that is, $I_s = \frac{1}{2}$, maximum fluctuation intensity (unity) is obtained, corresponding to a contrast of 2.

The condition for the two beams to be equal in intensity is obtained by equating the following two relations for the amplitudes ϕ_s and ϕ_t of the scattered and transmitted beams:[136]

$$\phi_s = \sin\left(\pi z/\xi_g\right) \tag{71}$$

$$\phi_t = \cos\left(\pi z/\xi_g\right) \tag{72}$$

where ξ_g is the extinction distance for the operating reflection. The beams are equal, when the specimen thickness $t = z = \xi_g/4$, $3\xi_g/4$, and $5\xi_g/4$. Since absorption increases with thickness, the optimum position should correspond to $\xi_g/4$.

It is apparent from Equation 68 that, in the exact Bragg reflecting position, where the deviation parameter $s = 0$, the phase shift term χ which enters into Equation 70 is zero. Equation 70 for the image intensity then reduces to

$$I = 1 + \sin\frac{2\pi t}{\xi_g}\sin\left(2\pi gx\right) \tag{73}$$

The first intensity minimum is then at a point shifted by an amount $d_{hkl}/4$ ($d_{hkl} = 1/g$) from the origin along the x axis. Hashimoto et al.[134] point out that, if the crystal is simple and the x origin coincides with the center of symmetry, the projected mass thickness would have a maximum at the origin, since the lattice points are generally located at the center of symmetry. However, the intensity minimum of the image is shifted by a quarter of a plane (fringe) spacing, so that it does not directly indicate the mass thickness. If the crystal does not have a center of symmetry, there is a further phase shift of the fringes. We are ignoring at present fringe displacements due to unequal phase shifts, introduced into the two beams as a result of lens aberrations, that is, we assume an ideal lens.

It is apparent from Equation 73 that, for this parallel-sided plate crystal, no fringes will be seen if the term $\sin 2\pi t/\xi_g = 0$, that is, if $(2t/\xi_g)$ is integral. Reversal of the sign of the term $\sin 2\pi t/\xi_g$ causes a shift of half a fringe and corresponds, for example, to a thickness change from $\xi_g/2$ to ξ_g. The behavior of the image in a wedge-shaped crystal, when the fringes are normal to the edge, is also as indicated above.

If $s \neq 0$, then $\chi \neq 0$ in Equation 70, and an image shift results that also depends on the thickness; however, the correct lattice periodicity is maintained. Variations of s (i.e., crystal bending) or t (i.e., thickness variations) with the change in x across the field, can give rise to a number of fringe shift effects which are purely electron optical (see Refs. 12, 100, and 134 for details). The details vary with the orientation of the fringes, relative to the axis of bending, or relative to the edge if the crystal is shaped like a wedge.

The phase shift term χ in Equation 70 is modified if lens aberrations are present and axial illumination imaging techniques are used, since the two beams then pass asymmetrically through the objective lens and are affected in different degree by the aberrations. The additional phase delay term χ_s for spherical aberration is then:[12]

$$\chi_s = \frac{\pi C_s \alpha^4}{2\lambda} \qquad (74)$$

where C_s is the spherical aberration constant for the lens, and α the angle between the diffracted beam and the optical axis, that is, twice the Bragg angle for the planes imaged. In addition there is a defocus term $\chi_{\Delta f}$ that is positive for underfocus and negative for overfocused images given by[12]

$$\chi_{\Delta f} = \frac{\pi \Delta f \alpha^2}{\lambda} \qquad (75)$$

where Δf is the amount of the objective defocus.

Chromatic aberration, that is, a change in the wavelength of the incident beam, likewise gives a change Δf in the focal length of the objective. The phase shift χ_c, due to a change ΔV in the accelerating voltage V, is[12]

$$\chi_c = 2\pi C_c \alpha^2 \frac{\Delta V}{\lambda V} \qquad (76)$$

where C_c is the chromatic aberration coefficient of the lens.

b. IMPERFECT CRYSTAL

(1) *Dislocations.* Following Bassett, Menter, and Pashley[137] the effect of a dislocation AB, passing through the crystal on the direct lattice image, may be seen by treating the image as if it simply represents a two-dimensional projection of planes of atoms (Figure 57). The projection of the slip plane (shaded) is shown superimposed on the projection of the (hkl) planes. Then $A'B'$ is the projection of the dislocation line. The introduction of the dislocation of Burgers vector $\mathbf{b}$ into the perfect lattice shifts the fringes on one side of the image relative to those opposite, since there is a component of $\mathbf{b}$ normal to the (hkl) planes, causing N fringes to terminate on the dislocation. The unit vector normal to the planes is then $d\mathbf{g}$, where d is the plane spacing and $\mathbf{g}$ the corresponding reciprocal lattice vector. The relative fringe shift is $d(\mathbf{g} \cdot \mathbf{b})$ that is, $(\mathbf{g} \cdot \mathbf{b})$ fringe spacings, so that we have:[12]

$$N = \mathbf{g} \cdot \mathbf{b} \qquad (77)$$

Sine $\mathbf{g} = [uvw]$ we have:[12]

$$N = hu + kv + lw \qquad (78)$$

which is the same expression derived by Bassett et al.[137]

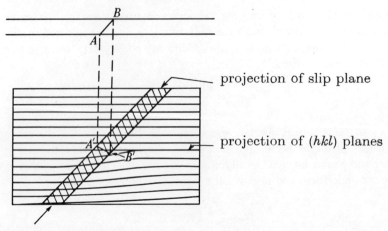

projection of slip plane

projection of (hkl) planes

Figure 57   Schematic showing effect of a dislocation line AB passing through the crystal on the directly resolved image of planes (hkl). From Bassett, Menter, and Pashley.[137] (By permission of The Royal Society.)

The N is always integral for a perfect, that is, unsplit, dislocation and a permissible (hkl) reflection. Thus, if $[uvw] = [\frac{1}{2}\frac{1}{2}0]$, the permitted $\{hkl\}$ reflections for an *fcc* crystal have their indices either all even or all odd. Hence $N = \frac{1}{2}(h + k)$ is always integral. Slip will leave perfect matching on lines away from $A'B'$ (Figure 57). The number of terminating fringes may be determined by making a Burgers circuit around $A'B'$.

If only a single set of planes is imaged and the direction of the dislocation is unknown, there is insufficient information to determine the dislocation character. The terminating half-fringes do not necessarily correspond with real extra half-planes in the crystal, since the image is a two-dimensional lattice projection. Thus a screw dislocation can also give rise to extra half-planes in a lattice image. The terminating fringes are, however, a direct measure of the Burgers vector component of the dislocation normal to the planes imaged.

An example of a dislocation in germanium is shown in Figure 58 from Phillips and Hugo.[15] The $\{111\}$ planes of listed spacing 3.266 Å were imaged by the two-beam tilt technique. The authors argued that the line direction was probably normal to the foil, so that this would be an edge-type dislocation of the usual $(a/2)\langle110\rangle$ type which can be produced in the diamond cubic lattice by insertion of an extra half $\{111\}$ plane.[138]

A dislocation should be invisible if $\mathbf{g} \cdot \mathbf{b} = 0$ for then $N = 0$. This is the same criterion as that which applies for visibility by normal diffraction contrast, which is understandable, since in both cases the contrast depends on there being distortion normal to the diffracting planes.

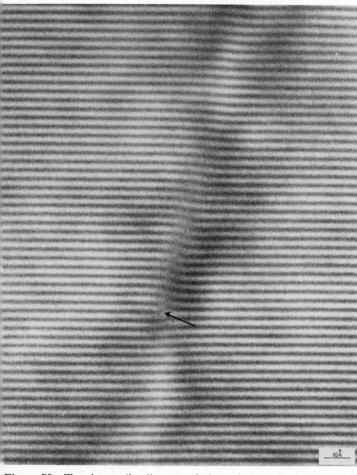

Figure 58 Two-beam tilt, direct resolution micrograph, showing dislocation, probably of edge type, in germanium. Note extra half-fringe (arrowed) and distortion around the dislocation. From Phillips and Hugo.[15]

The number N of extra half-fringes for a given dislocation depends on the particular planes imaged. The Burgers vector is completely determined if the components in three noncoplanar directions are determined. This can be done in principle from a single four-beam image or from a series of two-beam images with appropriate tilting. Usually, the possible types of Burgers vector will be known, or the line direction will be known, so that a conclusive determination is possible from less lattice fringe information.

An example of a dislocation in a crossed {111} lattice image of an (011)

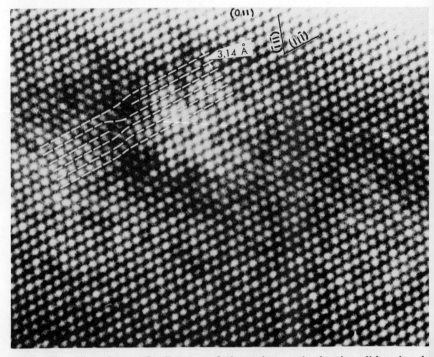

Figure 59 Three-beam tilt direct resolution micrograph showing dislocation in crossed {111} lattice image in (011) crystal of silicon. The white lines were added to make the defect more visible. From Phillips and Hugo.[139] (Courtesy of General Electric Company.)

silicon crystal, produced by the three-beam tilt technique, is shown in Figure 59 after Phillips and Hugo.[139] The defect is interpreted (white lines) as a pair of extra half (11$\bar{1}$) fringes of opposite sense spaced only a few angstroms. No extra half-fringe is present in the ($\bar{1}$1$\bar{1}$) set, but some strain is evident, indicative of a minor component of Burgers vector normal to the ($\bar{1}$1$\bar{1}$) planes. Assuming the usual (a/2) ⟨110⟩ Burgers vector, all but two vectors can be ruled out, leaving (a/2) [10$\bar{1}$] laying in ($\bar{1}$1$\bar{1}$) or (a/2) [110] laying in ($\bar{1}$11). The defect appears to consist of a positive-negative noncoplanar pair of one of these.

Mannami[140] extended the two-beam dynamical theory to interpret the direct resolution of an edge dislocation perpendicular to the crystal surface. His conclusions generally support the account given above. His conclusion that the strain around a dislocation would not be visible in a lattice image is at variance with the results of recent experiments mentioned below.

Although Hirsch et al.[12] discount the possibility of obtaining useful information about the strain field of a dislocation from lattice images, recent experiments tend to indicate otherwise.[15,141] In two-beam tilt lattice micrographs, diffraction contrast can be obtained from diffracted beams intercepted by the objective aperture. The lattice planes show displacements inside the "dot" diffraction contrast region, typical of a dislocation viewed end on[15] (Figure 58). Thus, qualitatively, one can determine the extent of the frozen-in strain around a dislocation or a group or wall of dislocations in a material such as germanium.[15]

(2) *Grain Boundaries.* If the lattice planes could be imaged simultaneously on both sides of a grain boundary, useful information on the nature of the boundary could in principle be obtained. The requirement that the planes be parallel or nearly parallel to the microscope axis clearly limits the type of high angle boundary which can be examined. Observations so far appear to be limited to low angle boundaries and twin boundaries.

Parsons and Hoelke[141,142] imaged the {111} lattice planes of spacing 2.34 Å, crossing a low angle tilt boundary in a (110) crystal of aluminum, using the two-beam tilt technique. The observed plane rotation of 2.8° at the boundary correlated well with that expected from the measured mean dislocation separation of about 55 Å. Figure 60, from Phillips and Hugo,[15] shows the {111} planes of spacing 3.27 Å crossing a grown-in, low angle boundary in germanium. The "strained region" is about 25 Å wide, from which it was inferred that the dislocation spacing was about 10 Å. The plane rotation of 1.2° across the boundary was ascribed to the net edge component and correlated with the two extra half-fringes, counted by making a Burgers circuit between RS. Similar studies on high angle boundaries have yet to be made.

(3) *Twins.* An abrupt change in direct of the lattice planes should be visible at a coherent twin interface. Certainly, at least one diffracted beam from the twin and one from the untwinned crystal must be admitted to the image, in addition to the undeviated beam. Twinning appears to have been observed first by Komoda et al.[143] in copper phthalocyanide, where the spacing changed abruptly from that corresponding to the (001) planes to that corresponding to the (20$\bar{1}$) planes. Komoda[144] observed mirror reflection of the {111} fringes across the boundaries in multiply twinned evaporated gold particles. Figure 61 shows the {111} planes of spacing 3.14 Å crossing grown-in twins in a (110) crystal of epitaxial silicon from Phillips and Hugo.[145] Classical "mirror reflection" of the planes at the {111} twin interfaces is shown. The twin

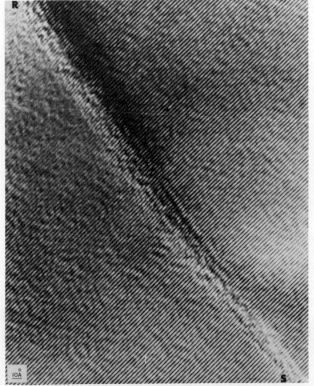

Figure 60   Two-beam tilt, direct resolution micrograph of grown-in, low angle boundary in (112) germanium crystal. The {111} planes are resolved. From Phillips and Hugo.[15]

thicknesses range from about 10 to 25 Å, so that they would appear only as a line by diffraction contrast.

(4) *Structure of Fibers.* Fibers are assuming great importance as reinforcements in composite materials. It is well known from x-ray diffraction work that carbon fibers can have structures ranging from amorphous to crystalline, which presumably can be related to their mechanical and other properties. Figure 62, from the work of Hugo, Phillips, and Roberts,[146] shows the internal (002) plane structure near the periphery of a fluted Thornel 50 high modulus carbon fiber. The micrograph was obtained by the multibeam technique, using axial illumination. The authors point out that normal grain boundaries or subboundaries are not observed; instead, the lattice planes maintain continuity over long dis-

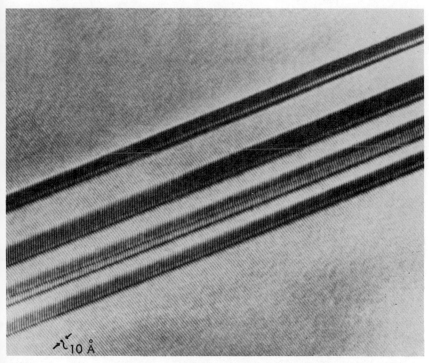

Figure 61 Direct resolution micrograph of epitaxially grown silicon film, taken with tilted illumination, showing the deviation of {111} planes across microtwins. From Phillips and Hugo.[145] (Courtesy of General Electric Company.)

tances, in spite of substantial changes in the direction of the planes, "flowing" around intruding material and leaving what appear to be elongated micropores.

The technique is, in principle, applicable to other kinds of fiber that show small or large regions of crystallinity. Remarkable pictures have been obtained by Yada,[147] showing the lattice planes in crystalline fibers of chrysotile asbestos.

c. RESOLUTION LIMIT FOR PERIODIC STRUCTURES

In a kinematical treatment, where the fringes in images of periodic structures can often be regarded as formed by interference between two beams, the resolution limits from various lens aberrations may be considered after Hirsch et al.[12] in terms of the unwanted phase shifts discussed on page 261. A phase shift χ between the diffracted and transmitted beams causes a fringe displacement of $\chi d/2\pi$, where d is the fringe spacing. It is argued that recognizable

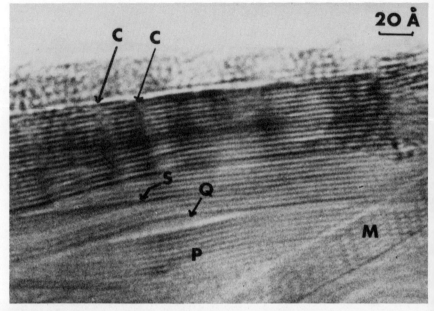

Figure 62   Direct resolution micrograph of carbon fiber, showing (002) planes of about 3.4 Å spacing, roughly parallel to the fiber axis. The principle features are the gradual curvature (*P*) and splitting (*S*) of packets of planes, moiré (*M*), possible micropore (*Q*), and contrast reversal (*C*), attributable to variation in the deviation parameter s. From Hugo, Phillips, and Roberts.[146] (By permission of *Nature*.)

fringes will still result, provided that the total spread of phase does not exceed π.

With axial illumination, the phase shift due to spherical aberration given by Equation 74 can be compensated by defocusing, according to Equation (75) as pointed out by Menter.[126] However, if there is a spread in phase across the diffracted beam, due to beam divergence, only partial compensation is possible. The resolution limit d_{min} is then:[12]

$$d_{min} = [2C_s\lambda^2(\delta + \Delta\theta)]^{1/4} \tag{79}$$

where δ is the incident beam divergence; $\Delta\theta$ ($\approx \lambda/w$) the divergence due to the finite width which is diffracting; and w the width of the crystal perpendicular to the planes diffracting, that is, the width of the appropriate extinction contour. Substituting $C_s = 0.32$ cm, $\lambda = 0.04$ Å, $\delta = 10^{-4}$, and $w = 1000$ Å yields $d_{min} \approx 2$ Å.

If chromatic aberration is caused by ripple and drift in the high voltage supply, the resultant phase shift χ_c (Equation 76) cannot be compensated by a change in focus, and a similar resolution limit can be derived:[12]

$$d_{\min} = \left(2C_c d \frac{\Delta V}{V}\right)^{1/2} \tag{80}$$

For typical values of $C_c = 0.13$ cm, $\lambda = 0.04$ Å, $\Delta V = 0.5$ V, and $V = 100$ keV, $d_{\min}$ is approximately 2.3 Å.

Fluctuation and drift in the objective lens current will change the focal length, analogous to change in high voltage. If ΔI is the total spread in the objective lens current I, we have:[12]

$$d_{\min} = \left(4C_c \lambda \frac{\Delta I}{I}\right)^{1/2} \tag{81}$$

Using typical values of $C_c = 0.13$ cm., $\lambda = 0.04$ Å, and $\Delta I/I = 2.5 \times 10^{-6}$, we find that $d_{\min}$ is approximately 2.3 Å.

The cumulative effect of these aberrations, together with mechanical (stage) drift and vibration is hard to predict. Electrical and mechanical drift effects can be minimized by using short exposures; they may add or subtract. The effects of spherical aberration, electrical ripple, and mechanical vibration are expected to be additive.

d. Resolution with Tilted Illumination

The two-beam technique with tilted illumination,[137,148,149] so that the microscope axis bisects the angle between the diffracted and transmitted beams, is expected to give a substantial gain in resolution of periodic structures. Equal and opposite phase shifts are introduced into the two beams because of spherical aberration, so that defocusing is no longer necessary to compensate. The resolution limit $d_{\min}$, due to beam divergence coupled with spherical aberration (Equation 79), is reduced[12] by a factor of $(4)^{-1/3} = 0.63$, since the beam inclination is now equal to the Bragg angle, instead of being twice it. Phase shifts from chromatic aberration in the source, or variations in the objective lens current, cancel.[149,150]

The tilted illumination technique was generalized to the multibeam case by Komoda,[151,152] who placed all the beams symmetrically around the optic axis. He concluded that the chromatic aberration phase shift was then negligible. The uniform spherical aberration shifts would also tend to cancel, leaving only those parts associated with beam divergence.

5. Indirect Lattice Resolution—Multiply Periodic Structures

Even though the microscope resolution is insufficient to image the lattice directly, multiple periodicities may be imaged if present. In order to produce an image, it is again necessary that the diffracted beam representing the multiple periodicity, for example, a superlattice reflection, enter the objective aperture for recombination with the undeviated beam. Some sources of these multiple periodicities are moiré effects from overlapping crystals, long range ordering giving rise to antiphase domains, graphite with intercollated heavy atoms, and staining centers in biological samples. The common feature is the fact that new reciprocal lattice vectors g and g′ result, given by:[13]

$$\mathbf{g} = \mathbf{g}_1 - \mathbf{g}_2 \qquad (82)$$

$$\mathbf{g}' = \mathbf{g}_1 + \mathbf{g}_2 \qquad (83)$$

where $\mathbf{g}_1$ is the lattice vector of the original system and $\mathbf{g}_2$ is the multiple period vector. In effect we combine two simple harmonic waves of frequencies ν_1 and ν_2 to yield two new frequencies $\nu_1 + \nu_2$ and $\nu_1 - \nu_2$. Since $\mathbf{g}$, given by the difference of $\mathbf{g}_1$ and $\mathbf{g}_2$, is a relatively small vector, the corresponding beam may enter the objective aperture when neither $\mathbf{g}_1$ or $\mathbf{g}_2$ or $\mathbf{g}'$ beams can do so, that is $\mathbf{g}\lambda < \beta_{obj}$, where λ is the wavelength and β_{obj} is the solid half-angle, defined by the objective aperture (see page 230).

Equation 82 may be represented graphically as in Figure 63, after Heidenreich,[13] taking η radians as the angle between $\mathbf{g}_1$ and $-\mathbf{g}_2$. From the geometrical

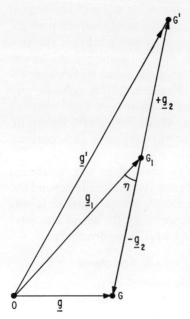

Figure 63   Reciprocal lattice representation of double diffraction as combination of reciprocal lattice vectors g_1 and g_2. After Heidenreich.[13]

properties of triangles we get:

$$|\mathbf{g}|^2 = |\mathbf{g}_1|^2 + |\mathbf{g}_2|^2 - 2\,|\mathbf{g}_1|\,|\mathbf{g}_2|\cos\eta \qquad (84)$$

For the simple *parallel moiré* case, that is, two overlapping crystals of the same orientation but slightly different spacing, we have $\eta = 0$ and $\cos\eta = 1$, so that $|\mathbf{g}|^2 = (\mathbf{g}_1 - \mathbf{g}_2)^2$ and the parallel moiré spacing $D = 1/|\mathbf{g}|$ is

$$D = \frac{d_1 d_2}{d_2 - d_1} \qquad (85)$$

where $d_1 = 1/|\mathbf{g}_1|$ and $d_2 = 1/|\mathbf{g}_2|$.

For the simple *rotational moiré* case, that is, two overlapping crystals of the same spacing and orientation, but one rotated slightly with respect to the other, we have $\eta \neq 0$ and $|\mathbf{g}_1| = |\mathbf{g}_2|$, so that $|\mathbf{g}|^2 = 2|\mathbf{g}_1|^2 - 2|\mathbf{g}_1|^2 \cos\eta$. Hence the rotational moiré spacing D is given by:

$$D = d_1/(2\sin\eta/2) \approx d_1/\eta \qquad (86)$$

In both the parallel and rotational moiré cases a set of straight fringes is obtained. In the former, the fringes are parallel to the superposed sets, that is, perpendicular to g, whereas in the rotational moiré they are normal to the mean direction of the superposed sets, that is, parallel to g. In the parallel moiré, dislocations occurring in either crystal or at the interface can give rise to extra half-fringes in the moiré pattern.

A third simple case, less commonly encountered because of the critical requirements, is the *tilt moiré*, produced when two identical crystals are related by a small rotation about an axis lying in the interface.[13] In Figure 63 this corresponds to G_1 being out of the plane of the drawing.

A combination of a difference in spacing and in orientation gives moiré fringes of spacing:

$$D = \frac{d_1 d_2}{[d_1{}^2 + d_2{}^2 - 2d_1 d_2 \cos\eta]^{1/2}} \qquad (87)$$

which for small values of η becomes

$$D = \frac{d_1 d_2}{[(d_1 - d_2)^2 + d_1 d_2 \eta^2]^{1/2}} \qquad (88)$$

We have so far considered simple two-beam indirect lattice resolution cases, in which the interface lies parallel to the foil surface, so that a single set of parallel fringes of uniform intensity results. When the interface is inclined to the foil surface, more complicated moiré patterns are observed with the same directions and spacings as above, but with the intensity modulated by thickness fringes. If more than one diffracted beam enters the image, more complex patterns result, as in the case of the direct resolution of lattices. Dislocations may be imaged in moiré patterns (see Amelinckx[100] for a detailed discussion).

In the case of materials such as clay or layer minerals of large spacing, moiré fringes may form according to Equation 83, that is, g', g_1, and g_2 all pass through the objective apertures, giving a lattice resolution image with mixed periods.

Moiré fringes are very commonly observed, for example, in polycrystalline foils, where grains overlap; in superposed epitaxial films (e.g., oxide patches on foils); and in lamellar crystals, in which the rotation of adjoining layers can occur, because of tilt or twist boundaries, or precipitates. They have been used[153] to determine the mismatch at individual coherent or semicoherent cobalt precipitates in copper. Moiré fringes are sometimes confused with dislocation arrays; however, a distinction is usually possible.[12,100]

For a full interpretation of the contrast detail and intensities, the dynamical theory is needed (see Refs. 12, 13, 100, 134, and 154). Hashimoto et al.[134] concluded that the intensity and spacing of the moiré fringes are affected by crystal thickness and deviation from the exact reflecting condition.

VI. MAJOR FIELDS OF APPLICATION

A complete account of the applications of conventional electron microscopy would require several volumes. The reader in search of detail is referred to the annual electron microscopy conferences of the Electron Microscopy Society of America, and the proceedings of the European and International conferences on the electron microscope, held every two years in alternating sequence. References 2, 12, 100, and 155 also contain extensive bibliographies. We shall briefly discuss some of the major fields of application in materials science and mention some other fields.

A. Defects

Dislocations may be classified as one-dimensional effects, those extended in one dimension. Information may be obtained, using diffraction contrast, on the number per unit volume, arrangement, and Burgers vector, and also whether the dislocations are perfect or extended. Extended dislocations, consisting of partial dislocations connected by a stacking fault, fall into the category of two-dimensional defects, which also include dislocation loops, and stacking faults which may be bounded by dislocations, grain boundaries, or crystal surfaces. The extent, number, and arrangement of faults, and whether they are intrinsic or extrinsic, may be determined by diffraction contrast experiments.

Three-dimensional defects, including twins, tetrahedral faults, and clusters, either of vacancies or interstitials, may be studied and the nature of the cluster determined by contrast experiments.

Isolated point defects cannot be studied at present. In principle, if monolayer samples were prepared and the resolution were adequate, it might prove possible to image atoms, and thus detect vacant sites and, conceivably, interstitial atoms. It is unlikely, however, that individual point defects will ever be imaged in projected images of multilayer samples. Estimates of the concentration of vacancies existing at high temperature may be made from quenching and aging experiments, in which the density and size of either clusters, or dislocation loops, that is, collapsed vacancy disks, are measured. This technique has been extensively applied to pure metals, and to alloys, in which vacancy-solute interactions may occur.

Radiation damage, which may result in the simultaneous presence of both interstitial and vacancy clusters, has also been intensively studied by the thin film technique, using diffraction contrast experiments to distinguish the two types of defect.

The thin film technique has been extensively applied to the study of the deformation of metals and alloys, whether by tension, creep, fatigue, or other modes. Much of the work has involved detailed studies of densities, Burgers vectors, arrangement, and interactions of dislocations in stretched single crystals. Some dynamic experiments have been carried out on thin films in the microscope but their behavior generally differed from that of bulk samples.

The lattice resolution technique has already permitted observations[15] of groups of closely spaced dislocations that would not be separately distinguishable by diffraction contrast. The technique thus should greatly extend our knowledge in this direction in the future.

The effects of annealing on crystals containing various kinds of defects has been extensively studied by the thin film technique. Using a hot stage,[156] the kinetics of recovery phenomena such as the growth of dislocation loops and tetrahedra may be determined, and hence information on mechanisms obtained. Recrystallization may be studied, although here dynamic studies have tended to give results differing from bulk behavior. Possibly, the use of thick samples coupled with high voltage microscopy may overcome this difficulty.

Another important area of application is in the determination of stacking fault energy, using diffraction contrast. Several different methods exist.[12]

B. Solid State Transformations

Transmission electron microscopy using diffraction contrast has, to a large extent, replaced replication techniques for the study of precipitation and phase transformations. However, extraction replicas remain useful for particle identification purposes which can be difficult in situ. When the structure is coarse compared with the transparency thickness, the uniform thinning necessary to obtain samples for the thin film technique can be extremely troublesome. The replica technique is then still useful but is gradually being replaced by the scanning electron microscope which has the advantage of directness and the possibility of obtaining additional information, for example, by the use of the x-ray signal for analysis. References to much of the literature on precipitation using the thin film technique up to 1963 are to be found in the review by Kelly and Nicholson.[157]

Information is obtainable by the thin film technique on the size, shape, and size distribution of second phase particles. The presence of strain fields around particles, whether due to coherency strains or other causes, may be detected. The sign of the strain may be determined by diffraction contrast experiments. The nature of the interface, that is, whether coherent, semicoherent, or incoherent may be determined. The orientation relationship between particles and matrix may be defined. The lattice mismatch between individual coherent particles and the matrix may be determined by means of moiré fringe or strain field contrast analysis. Effects such as the production of dislocations during precipitation or phase transformation may be studied. The interaction of particles and dislocations during deformation may be observed. Dynamic experiments may be carried out inside the electron microscope in many of these areas, using special stages for heating, cooling, stretching, and so forth.

Other areas in which the thin film technique has proved valuable are in the study of martensitic transformation, which may involve microtwin and dislocation formation, and in phase transformations such as the fcc to hcp transformation during the cooling of cobalt, where stacking faults may play a role.

A great deal of valuable information on order-disorder transformations has been obtained by the thin film technique.[12,107,158]

C. Structure and Properties of Thin Films

The transmission technique, employing diffraction contrast, lends itself to the study of the nucleation and growth of thin films, epitaxially or otherwise; there is a considerable literature on this subject [see, for example, Anderson[53]]. With the aid of an electron transparent substrate and an evaporation source inside the microscope, the nucleation and growth of such films may be observed directly. Defects inside the films may be studied and information obtained on their origin. Moiré fringes may be obtained if an epitaxial film is examined on the substrate, giving information on the mismatch and defects present. The lattice resolution technique still in its infancy promises also to be a powerful tool for such studies.

Because of their two-dimensional nature and the proximity of the surfaces, the deformation-, recovery-, recrystallization-, and precipitation behavior of thin films often differs from that of bulk samples. The transmission technique of diffraction contrast has been employed for studies in all of these areas.

Magnetic domain structures in ferromagnetic materials have been extensively studied by the technique of Lorentz microscopy, whereby they are directly revealed as a result of small electron beam deflections, caused by the magnetic field.[12] Materials such as cobalt, iron, nickel, and permalloy have been studied in this fashion. The domain arrangement in films again tends to differ from that in bulk materials.

D. Fracture

Thin sections prepared electrolytically or chemically from the vicinity of fractures have been employed to study the mechanism of fracture in metals by the diffraction contrast technique. Such features as microvoid formation at inclusions can be studied. Fractography employing replicas is also a powerful tool here, although tending now to be supplanted by scanning electron microscopy. Dynamic experiments in the electron microscope again indicate that thin film behavior differs from that of bulk samples. Crack studies in bulk samples have been limited by the problem of preparing thin sections without enlarging the cracks. Ultramicrotomy appears to offer considerable potential for this purpose.

E. "Noncrystalline" Materials

Other fields of application of electron microscopy of great importance, but outside the present scope, include histology and cytology, bacteriology, and biological macromolecules [see Siegel[159] and the various electron microscope conference volumes]. Materials that may be studied include wood, natural and synthetic fibers, bone, plants, polymers, and pigments. The electron microscope is a valuable tool for identifying and characterizing particulate materials, whether organic inorganic, crystalline, or noncrystalline. This is clearly of vital importance in relation to the present international concern over problems of pollution.

VII. REFERENCES

1. W. James, *Electron Microscopy and Microanalysis of Metals,* J. A. Belk and A. L. Davies, Eds. Elsevier Publishing Co., New York, 1968, Chap. 1.
2. G. Thomas, *Transmission Electron Microscopy of Metals,* John Wiley & Sons, New York, 1961.
3. V. E. Cosslett, *Electron Optics,* Clarendon Press, Oxford, 1950.
4. V. K. Zworykin, G. A. Morton, E. G. Ramberg, J. Hillier, and A. W. Vance, *Electron Optics and the Electron Microscope,* John Wiley & Sons, New York, 1945.
5. C. E. Hall, *Introduction to Electron Microscopy,* McGraw-Hill Book Co., New York, 1953.
6. J. W. Menter, *Advan. Phys.,* **7,** 299 (1958).
7. M. E. Haine, *Advan. Electronics,* **6,** 295 (1954).
8. M. E. Haine and T. Mulvey, *J. Sci. Instrum,* **31,** 326 (1954).
9. M. E. Haine and V. E. Cosslett, *The Electron Microscope,* E. and F. N. Spon, London, 1961.
10. V. E. Cosslett, *Brit. J. Appl. Phys.,* **7,** 10 (1956).
11. F. A. Jenkins and H. E. White, *Fundamentals of Optics,* 1951, McGraw-Hill Book Co., New York, 1951.
12. P. B. Hirsch, A. Howie, R. B. Nicholson, D. W. Pashley, and M. J. Whelan, *Electron Microscopy of Thin Crystals,* Butterworths, Washington, 1965.
13. R. D. Heidenreich, *Fundamentals of Transmission Electron Microscopy,* Interscience Publishers, New York, 1964.
14. H. Boersch, *Z. Phys.,* **44,** 202 (1943).
15. V. A. Phillips and J. A. Hugo, *Acta Met.,* **18,** 123 (1970).
16. R. D. Heidenreich, W. M. Hess, and L. L. Ban, *J. Appl. Crystallogr.,* **1** (1), 1 (1968).
17. V. A. Phillips and J. A. Hugo, *Proc. Electron Microscopy Society of America,* 27th Annual Meeting, C. J. Arceneaux, Ed., Claitor's Publishing Division, Baton Rouge, La., 1969, p. 180.
18. J. F. Nankivell, *Brit. J. Appl. Phys.,* **13,** 126 (1962).
19. O. C. Wells, *Brit. J. Appl. Phys.,* **11,** 199 (1960).
20. A. W. Agar and R. W. Horne, *Techniques for Electron Microscopy,* 2nd ed., D. H. Kay, Ed., Blackwell Scientific Publications, Oxford, England, 1965, Chap. 2.
21. V. A. Phillips, *Experimental Methods of Materials Research,* H. Herman, Ed., Interscience Publishers, New York, 1967, p. 51.
22. D. H. Kay, Ed., *Techniques for Electron Microscopy,* 2nd ed., Blackwell Scientific Publications, Oxford, England, 1965.

23. B. V. Whiteson, A. Phillips, and V. Kerlins, *Techniques of Metals Research,* R. F. Bunsah, ed., 2, (1), Interscience Publishers, New York, 1968, Chap. 14.
24. G. Henry and J. Plateau, *Microfractography,* translated by B. Thomas, Editions Métaux, Paris, 1966.
25. A. Phillips et al., *Electron Fractography Handbook,* Tech. Report, ML-TDR-64-416, U.S. Air Force Materials Laboratory, 1965.
26. "Electron Fractography," ASTM, STP 436, American Society for Testing and Materials, 1968.
27. R. M. Fisher, *J. Appl. Phys.,* **24,** 113 (1953).
28. W. J. McG. Tegart, *Electrolytic and Chemical Polishing of Metals,* 2nd ed., Pergamon Press, London, 1959.
29. I. S. Brammar and M. A. P. Dewey, *Specimen Preparation for Electron Metallography,* American Elsevier Publishing Co., New York, 1966.
30. W. Bollman, *Proc. Stockholm Conf. Elect. Micr., Almquist and Wiksell,* Stockholm, Sweden, 1957, p. 316.
31. J. A. Hugo and V. A. Phillips, *Brit. J. Appl. Phys.,* **40,** 202 (1963).
32. A. M. Glauert and R. Phillips, *Techniques for Electron Microscopy,* 2nd ed., D. H. Kay, Ed., Blackwell Scientific Publications, Oxford, England, 1965, Chap. 8.
33. V. A. Phillips, *Praktische Metallographie,* 4, 637 (1967).
34. V. A. Phillips, *Imperfections in Crystals,* J. B. Newkirk and J. H. Wernick, Eds., Interscience Publishers, New York, 1962, p. 179.
35. V. Puspanen, *J. Appl. Phys.,* **19,** 876 (1948).
36. J. A. Hugo and V. A. Phillips, unpublished work.
37. G. R. Booker and R. Stickler, *Brit. J. Appl. Phys.,* **13,** 446 (1962).
38. V. A. Phillips and J. A. Hugo, *Proc. 26th Annual EMSA Meeting,* C. J. Arceneaux, Ed., Claitor's Publishing Division, Baton Rouge, La., 1968, p. 342.
39. J. Washburn, A. Kelly, and G. K. Williamson, *Phil. Mag.,* **5,** 192 (1960); also with G. W. Groves, ibid., p. 991.
40. P. B. Hirsch, A. Kelly, and J. W. Menter, *Proc. 3rd Int. Conf. Electron Microscopy,* Royal Microscopical Society, London, 1954, p. 231.
41. P. B. Hirsch, R. W. Horne, and M. J. Whelan, *Phil. Mag.,* **1,** 677 (1956).
42. S. R. Keown and F. B. Pickering, *J. Iron Steel Inst. (London),* **200,** 757 (1962).
43. G. Induni, *Helv. Phys. Acta,* **20,** 463 (1947).
44. R. Castaing, *Proc. 3rd Int. Conf. Electron Microscopy, 1954,* Royal Microscopical Society, London, 1956, p. 379.
45. M. Paulus and F. Reverchon, *J. Phys. Radium,* **22,** 103A (1961).
46. N. J. Tighe, *Ultrafine Grain Ceramics,* Syracuse University Press, New York, in press.
47. V. A. Phillips and J. A. Dash, *J. Appl. Phys.,* **33,** 568 (1962).
48. T. Evans and C. Phaal, *Phil. Mag.,* **7,** 843 (1962).
49. G. A. Bassett, *Techniques for Electron Microscopy,* 2nd ed., D. H. Kay, Ed., Blackwell Scientific Publications, Oxford, England, Chap. 13.
50. D. W. Pashley, *Techniques for Electron Microscopy,* 2nd ed., D. H. Kay, Ed., Blackwell Scientific Publications, Oxford, England, 1965, Chap. 3.
51. K. H. Behrndt, *Thin Films,* 1963 Seminar, published by American Society for Metals, Ohio, 1964, Chap. 1.
52. L. Holland, *The Vacuum Deposition of Thin Films,* John Wiley & Sons, New York, 1956.

278   Conventional Electron Microscopy

44
53. J. C. Anderson, Ed., *The Use of Thin Films in Physical Investigations*, 1965, NATO, Academic Press, New York, 1966.
54. D. W. Pashley, *Advan. Phys.*, **5**, 173 (1956).
55. V. A. Phillips, *Phil. Mag.*, **5**, 571 (1960).
56. D. W. Pashley and A. E. B. Presland, *Phil. Mag.*, **7**, 1407 (1962).
57. D. W. Pashley, *Proc. Roy. Soc., Ser. A*, **255**, 218 (1960).
58. J. W. Matthews, *Phil. Mag.*, **7**, 915 (1962).
59. A. B. Glossop and D. W. Pashley, *Proc. Roy. Soc., Ser. A*, **250**, 132 (1959).
60. D. W. Pashley and A. E. B. Presland, [Symposium on] Mechanical Properties of Intermet. Compounds, Philadelphia, 1959, Electrochemical Society, 1960, p. 211.
61. V. A. Phillips, *J. Appl. Phys.*, **33**, 712 (1962).
62. P. B. Price, *J. Appl. Phys.*, **32**, 1746 (1961).
63. P. B. Price, *Phil. Mag.*, **5**, 473, 873 (1960).
64. E. Suito and N. Uyeda, *Proc. 3rd Int. Conf. on Electron Microscopy 1954*, Royal Microscopical Society, London, 1956, p. 223.
65. L. Reimer, *Z. Metallk.*, **50**, 37 (1959).
66. L. Reimer, *Naturwissenschaften*, **46**, 68 (1959).
67. P. Duwez, R. H. Willens, and W. Klement, *J. Appl. Phys.*, **31**, 1136, 1500 (1960).
68. G. Thomas and R. H. Willens, *Acta Met.*, **12**, 191 (1964).
69. P. B. Hirsch, P. G. Partridge, and R. L. Segall, *Phil. Mag.*, **4**, 721 (1959).
70. K. Thomas and K. F. Hall, *Phil. Mag.*, **4**, 531 (1959).
71. P. B. Hirsch and U. Valdré, *Phil. Mag.*, **8**, 237 (1963).
72. V. A. Phillips, *Advances in Electron Metallography*, STP No. 396, American Society for Testing and Materials, Philadelphia, 1966, p. 21.
73. V. A. Phillips, *Acta Met.*, **11**, 1139 (1963).
74. P. R. Swann and J. Nutting, *J. Inst. Metals*, **88**, 478 (1960).
75. G. A. Bassett, *Phil. Mag.*, **3**, 1042 (1958).
76. H. Bethge, *Surface Sci.*, **3**, 33 (1965).
77. J. G. Allpress and J. V. Saunders, *Phil. Mag.*, **9**, 645 (1964).
78. G. R. Henning, *Appl. Phys. Lett.*, **4**, 52 (1964).
79. J. L. Robins, T. N. Rhodin, and R. L. Gerlach, *J. Appl. Phys.*, **37**, 3893 (1966).
80. D. J. Stirland, *The Use of Thin Films in Physical Investigations*, 1965, NATO, J. C. Anderson, Ed., Academic Press, New York, 1966, p. 163.
81. C. S. Barrett and T. B. Massalski, *Structure of Metals*, 3rd ed., McGraw-Hill Book Co., 1966.
82. A. Taylor, *X-Ray Metallography*, John Wiley & Sons, New York, 1961.
83. A. Guinier, *X-Ray Crystallographic Technology*, 1952, Hilger and Watts, London, 1952.
84. R. H. Alderson and J. S. Halliday, *Techniques for Electron Microscopy*, 2nd ed., D. H. Kay, Ed., Blackwell Scientific Publications Oxford, England, 1965, Chap. 15.
85. S. Kikuchi, *Japan J. Phys.*, **5**, 83 (1928).
86. A. Howie, *Techniques for Electron Microscopy*, 2nd ed., D. H. Kay, Ed., Blackwell Scientific Publications, Oxford, England, 1965, Chap. 14.
87. H. Wilman, *Proc. Phys. Soc.*, **233**, 416 (1948).
88. M. von Heimendahl, W. Bell, and G. Thomas, *J. Appl. Phys.*, **35**, 3614 (1964).
89. R. M. Otte, J. Dash, and H. F. Schaake, *Phys. Status Solidi*, **5**, 527 (1964).

90. E. Levine, W. L. Bell, and G. Thomas, *J. Appl. Phys.*, **37**, 2141 (1966).
91. P. R. Okamoto, E. Levine, and G. Thomas, *J. Appl. Phys.*, **38**, 289 (1967).
92. G. Thomas, *Trans. Met. Soc. AIME*, **233**, 1608 (1965).
93. A. R. Stokes and A. J. C. Wilson, *Proc. Cambridge Phil. Soc.*, **38**, 313 (1942).
94. E. Bauer, *Techniques of Metals Research*, Vol. 2, (2), R. F. Bunsah, Ed., Interscience Publishers, New York, 1969, Chap. 15.
95. R. C. Newman and D. W. Pashley, *Phil. Mag.*, **46**, 927 (1955).
96. R. W. G. Wyckoff, *Crystal Structures*, 1948–60 series and 1963 rev. ed., Interscience Publishers, New York.
97. W. B. Pearson, *A Handbook of Lattice Spacings and Structures of Metals*, Vol. 1, 1958; Vol. 2, 1st ed. 1967; Pergamon Press, London.
98. Varian Associates, Palo Alto, Calif.
99. A. G. Balzers, *Abtlg. Korpuskularstrahl-Geräte*, 8634, Hombrechticon, Switzerland, 1970.
100. S. Amelinckx, *The Direct Observation of Dislocations*, Academic Press, New York, 1964.
101. M. J. Makin, *Metallography*, **1**, (1), 109 (1968).
102. R. W. G. Wyckoff, *The World of the Electron Microscope*, Yale University Press, New Haven, Conn., 1958.
103. H. Hashimoto, A. Howie, and M. J. Whelan, *Proc. Roy. Soc., Ser. A*, **269**, 80 (1962).
104. V. A. Phillips, *Advances in Electron Metallography*, STP No. 396, 1966, American Society for Testing and Materials, Philadelphia, p. 21.
105. P. Delavignette and R. W. Vook, *Phys. Status Solidi*, **3**, 648 (1963).
106. P. B. Hirsch, A. Howie, and M. J. Whelan, *Phil. Trans. Roy. Soc., Ser. A*, **252**, 499 (1960).
107. M. J. Marcinkowski, *Report of Conference on Electron Microscopy and Strength of Crystals*, 1963, G. Thomas and J. Washburn, eds., Interscience Publishers, New York, p. 333.
108. W. J. Tunstall, P. B. Hirsch, and J. W. Steeds, *Phil. Mag.*, **9**, 99 (1964).
109. R. Gevers, *Phil. Mag.*, **7**, 59 (1962).
110. R. Gevers, *Phil. Mag.*, **7**, 651 (1962).
111. R. Gevers, *Phil. Mag.*, **8**, 769 (1963).
112. N. F. Mott and F. R. N. Nabarro, *Proc. Phys. Soc.*, **52**, 86 (1940).
113. V. A. Phillips and J. D. Livingston, *Phil. Mag.*, **7**, 969 (1962).
114. M. Ashby and L. M. Brown, *Phil. Mag.*, **8**, 1083, 1649 (1963).
115. V. A. Phillips and J. D. Livingston, unpublished.
116. L. M. Brown, *From Molecule to Cell*, Symposium on Electron Microscopy, Modena, 1963, P. Buffa, Ed., Tipografia S. Pio $\overline{X}$, Rome 1964, p. 99.
117. S. L. Sass, T. Mura, and J. B. Cohen, *Phil. Mag.*, **16**, 679 (1967).
118. G. C. Weatherly, *Phil. Mag.*, **17**, 647 (1968).
119. V. A. Phillips, *Acta Met.*, **14**, 1533 (1966).
120. L. E. Tanner, *Phil. Mag.*, **14**, 111 (1966).
121. H. Gleiter, *Phil. Mag.*, **18**, 847 (1968).
122. H. F. Merrick, Ph.D. Thesis, Cambridge University, Cambridge, England, (1963).
123. A. J. Ardell and R. B. Nicholson, *Acta Met.*, **14**, 1295 (1966).
124. A. J. Ardell and R. B. Nicholson, *J. Phys. Chem. Solids*, **27**, 1793 (1966).
125. A. J. Ardell, *Phil. Mag.*, **16**, 147 (1967).
126. A. Fourdeux, R. Gevers, and S. Amelinckx, *Phys. Status Solidi*, **24**, 195 (1967).

127. J. Van Landuyt, R. Gevers, and S. Amelinckx, *Phys. Status Solidi*, 10, 319 (1965).
128. H. Hashimoto, Int. Conf. on Electron Diffraction and Crystal Defects, Melbourne 1965, I 0-1.
129. P. Guyot, *Phys. Status Solidi*, 28, 349 (1968).
130. J. W. Menter, *Proc. Roy. Soc., Ser. A*, 236, 119 (1956).
131. J. W. Menter, *Phil. Mag. Suppl.*, 7, 299 (1958).
132. J. M. Cowley, *Acta Cryst.*, 12, 367 (1959).
133. J. M. Cowley and A. F. Moodie, *Proc. Phys. Soc.*, 120, 486 (1957).
134. H. Hashimoto, M. Mannami, and T. Naiki, *Phil. Trans. Roy. Soc. Ser A.*, 253, 459 (1961).
135. T. Komoda, *Optik*, 21, 93 (1964).
136. A. Howie and M. J. Whelan, *Proc. Roy. Soc., Ser. A.*, 263, 212 (1961).
137. G. A. Bassett, J. W. Menter, and D. W. Pashley, *Proc. Roy. Soc., Ser. A.*, 246, 345 (1958).
138. J. Hornstra, *J. Phys. Chem. Solids*, 5, 129 (1958).
139. V. A. Phillips and J. A. Hugo, unpublished work.
140. M. Mannami, *J. Phys. Soc. Japan*, 17, 1160 (1962).
141. J. R. Parsons and C. W. Hoelke, *J. Appl. Phys.*, 40, 866 (1969).
142. J. R. Parsons and C. W. Hoelke, 4th European Regional Conference on Electron Microscopy, Rome, 1, 133 (1968).
143. T. Komoda, E. Suito, N. Uyeda, and H. Watanabe, *Nature*, 181, 332 (1958).
144. T. Komoda, *Japan. J. Appl. Phys.*, 7, 27 (1968).
145. V. A. Phillips and J. A. Hugo, unpublished work.
146. J. A. Hugo, V. A. Phillips, and B. W. Roberts, *Nature*, 226, 144 (1970).
147. K. Yada, *Acta Cryst.*, 23, 704 (1967).
148. W. C. T. Dowell, *J. Phys. Soc. Japan*, 17, Suppl. B-II, 175 (1962).
149. W. C. T. Dowell, *Optik*, 20, 535 (1963).
150. D. W. Pashley (1959), see J. A. Chapman and M. J. Whelan, *J. Appl. Phys.*, 11, 31 (1960).
151. T. Komoda, *Japan. J. Appl. Phys.*, 5, 603 (1966).
152. T. Komoda, *Hitachi Rev.*, 15 (9), 345 (1966).
153. V. A. Phillips, *Acta Met.*, 14, 271 (1966).
154. R. Gevers, *Phil. Mag.*, 7, 1681 (1962).
155. G. Thomas and J. Washburn, Eds., *Report of Conference on Electron Microscopy and Strength of Crystals*, Interscience Publishers, New York, 1963.
156. G. Thomas, *High-Temperature High-Resolution Metallography*, H. I. Aaronson and G. S. Ansell, Eds., Gordon and Breach, New York, 1967, pp. 217–280.
157. A. Kelly and R. B. Nicholson, "Precipitation hardening," in *Progr. in Materials Science*, Vol. 10, (3), B. Chalmers, Ed., Pergamon Press, New York, 1963, pp. 149–391.
158. D. W. Pashley, *Modern Developments in Electron Microscopy*, B. M. Siegel, Ed., Academic Press, New York, 1964, Chap. 5.
159. B. M. Siegel, Ed., *Modern Developments in Electron Microscopy*, Academic Press, New York, 1964.

6

SPECIALIZED ELECTRON MICROSCOPY

I. High Voltage Electron Microscopy 282
   A. Introduction 282
   B. Instrumentation 283
   C. Electron Penetration 288
   D. Image Resolution 289
   E. Applications 290

II. Reflection Electron Microscopy 292

III. Energy-selecting and Energy-analyzing Microscopy 297
   A. Introduction 297
   B. Energy-Analyzing Microscope 299
   C. Energy-Selecting Microscope 310

IV. Electron Emission Microscopy 316
   A. Introduction 316
   B. Instrumentation 317
   C. Resolution 321
   D. Contrast 321
   E. Applications 322

V. Field Emission and Field-ion Microscopy 325
   A. Field Emission Microscopy 325
   B. Field Ion Microscopy 330
   C. Combination of "Atom Probe" and Field Ion Microscope 338

VI. Other Techniques 342
   A. "Proton Scattering Microscope" 342
   B. Mirror Electron Microscope 347
   C. High Resolution Scanning Transmission Electron Microscope 352

D. High Voltage Transmission Scanning Electron Microscope 358
1. Image Contrast 360
E. Scanning High Energy Electron Diffraction 362

VII. References 366

I. HIGH VOLTAGE ELECTRON MICROSCOPY

A. Introduction

Electron microscopes with accelerating voltages of up to 1 million volts or more have recently become commercially available, although as yet only a few units have been installed and there is limited operating experience. The subject has been reviewed by Dupouy,[1] who also gives a fairly extensive bibliography. Voltages have been limited to 100 keV for the past decade by design limitations, although in some instances 150 keV have been attained. Above this level, radical design changes are necessary, resulting in a great increase in the size, complexity, and cost of the instruments. Ultimately, some reduction in size may be possible through the introduction of superconducting lenses.

The principal advantages claimed for high voltage instruments are, first, increased penetration that permits the use of thicker samples which should be more representative of bulk samples and show bulk behavior in dynamic experiments; and second, improved resolution. These claims will be discussed in more detail later. It is unlikely that improved resolution will be achieved in first generation high voltage instruments; it appears that further development will be necessary before the resolution of current (third) generation 100 keV instruments is matched. Furthermore, the use of thick samples is not compatible with optimum resolution.

Since the practical sample thickness in conventional ($\sim$100 keV) microscopes is probably limited not so much by the penetration as by the incoherent scattering of electrons, which leads to a low signal-to-noise ratio, it is entirely possible that, with the addition of energy selection and some form of image intensification, probably involving digitalization and computer processing of the image, a general purpose high resolution 100 to 250-keV electron microscope will evolve. The million or multimillion volt microscope would then assume the role of an expensive highly specialized tool for such purposes as in situ radiation (electron) damage studies, contrast experiments, and other physical studies of the interaction of electrons with matter at relativistic velocities. It is particularly noteworthy that resolutions < 2 Å have already been achieved with 100-keV instruments, using the phase contrast technique for the resolution of periodic structures. Atomic resolution may well prove possible, since this is still a new and relatively undeveloped field.

B. Instrumentation

The development of the high voltage microscope has been discussed by Dupouy,[1] Cosslett,[2-4] and Dupouy and Perrier.[5] In designing an instrument, it is desirable to keep the focal length of the magnetic lenses short, in order to reduce aberrations and to reduce the column height. The minimum focal length is already limited in 100-keV instruments by saturation of the iron casing and lens pole-pieces. In the absence of superconducting lenses, which would permit higher field strengths, it is necessary to scale up the whole design to handle more energetic electrons, keeping the magnetic induction along the axis the same but increasing the distance over which it is effective by increasing the pole-piece gap. The focal length is thus increased from a typical value of about 2 mm for 100-keV electrons to some 5 mm for 1-MeV electrons. The lens diameter is correspondingly increased and the lenses typically weigh several hundred pounds or as much as half a ton if radiation protection is built in. In the former case remote operation must be adopted or beam currents kept very low ($\sim$0.1 μA at 1 MeV) to keep the x-ray exposure of the operator below the tolerance level. In either case, the column height is increased, so that remote operation of the less accessible controls is necessary. Depending on the lens weights, either fork lift truck handling or a crane is necessary for dismantling the column. Even changing a filament may thus become a major operation unless automated.

Unlike the 100 keV design where a single-stage electron gun is adequate, the high voltage microscope requires a multistage gun design, greatly increasing the length and diameter because of insulation problems. While individual stages can be designed to run up to 150 keV, more conservative designs employ up to 30 stages for 1 MeV. A further major design choice lies between putting the generator and accelerator into a pressurized gas vessel or using air insulation. The latter improves accessibility but requires a great deal of space. Furthermore, regular polishing of the large aluminum domes and stress rings is necessary to reduce microdischarges through the relatively poor air insulation which otherwise introduce voltage fluctuations. In the pressurized systems the generator and accelerator may be put in one big tank, or in two separate tanks with a pressurized connecting cable and tracks to permit moving them apart. The weight of two tanks may amount to about 15 tons, bringing the total weight including the microscope to 20 tons or more, which must be carried on a concrete block with provision for insulating the microscope from mechanical vibrations. All possible combinations of the above major options may be seen in current commercial microscopes.

A high voltage stability of a few parts per million is desirable. The generators used are normally scaled-up versions of those employed for 100-keV microscopes, consisting of a Cockcroft-Walton voltage doubling and rectifying stack with fluctuations limited by feedback control. The design may be simplified by limiting the voltage range, say, from 400 keV to 1 MeV, although there are clearly advantages in extending the range down to conventional voltages. It appears that there is some advantage in providing a continuously variable voltage, so that a value may be selected that is optimum for a particular sample material and thickness, and the voltage dependence of phenomena found. This practice might well be extended to conventional 100 keV instruments, in which provision is made only for voltage variation in steps of 20 keV or more.

The JEOLCO (USA) JEM-1000D is an example of a commercial 1-MeV microscope, in which the generator and 14-stage accelerator are enclosed in a single pressurized gas vessel above the column (Figure 1). Accelerating voltages of 500, 750, and 1000 keV are provided, with direct magnifications ranging from 1000 to 150,000 times. The column consists of a double condenser, plus objective, diffraction, intermediate, and projector lenses, giving a guaranteed resolution of 7 Å. A plate camera is provided.

In contrast to the above design, the 1-MeV instrument[6,7] at the United States Steel Corporation employs a Cockcroft-Walton accelerator, constructed by the Swiss firm, Emile Haefely Company, which is 17 ft high. This instrument is contained in a stainless steel room that is grounded. The accelerator is fed from a transformer rated at 6 kHz and 85 keV. The microscope, designed by the Radio Corporation of America, contains a number of interesting features. The column that weighs 6 tons and is 9 ft in height, contains 6 magnetic lenses and 3 sets of magnetic deflecting coils. A plate camera, image intensifier, and video-tape recorder are provided. The vacuum system is ultrahigh vacuum. Because of its size, the instrument is contained in a separate, specially designed building.

The GEC-AEI (Electronics) British 1-MeV EM7 instrument is shown in Figures 2 and 3. The design is based on a 750-keV instrument, built at the Cavendish Laboratory in England. Two separate tanks are employed, pressurized at 3 atm with SF_6 gas, for the generator and a 24-stage accelerator, respectively, and located on a platform above the column. These are made by Emile Haefely and Company of Basel, Switzer-

Figure 1 The JEM-1000D 1-MeV electron microscope. [Courtesy of JEOLCO (USA).]

land. The design details of the accelerator have been described by Reinhold.[8] The generator tank is about 9½ ft high and 7 ft in diameter and weighs about 7 tons. The separate tanks screen the accelerator from generator ripple and permit operation at 100-keV intervals from 100 keV to 1 MeV with a claimed d-c stability of $\pm 1 \times 10^{-5}$ over a 3-min period and ripple of $\pm 2.5 \times 10^{-6}$ peak to peak.

Six filaments are accommodated in a rotating turret, each of which can be removed through a gas lock, giving minimum down time. An electron beam current up to 100 μA is available from the conventional

Figure 2   Mock-up of the 1-MeV GEC-AEI EM7 electron microscope. (Courtesy of AEI Scientific Apparatus and Picker Nuclear.)

triode gun. Also, X-Y deflection coils are provided at the top of the column for steering the beam from the accelerator into the entrance aperture of the microscope. The six lenses provide a normal magnification range from 1000 to 160,000X, plus a low range from 100 to 1000X and extra high magnifications up to 1,600,000X by special switching. The column weighs about 5½ tons, bringing the total weight to about 20 tons, which is supported on a concrete block.

A double set of alignment coils below the second condenser permits beam shift, or a tilt of up to 1½° at 1 MeV about the object level to permit high resolution dark field operation. A goniometer stage permits 360° rotation of the sample and a double-tilt motion of ±30° on two axes at right angles, with anticontamination provision. Octopole correctors are provided for astigmatism correction on the condenser, the objective, and the first of the three projector lenses. A point resolution of 10 Å is guaranteed. The theoretical resolution of the objective lens is said to be 1.1 Å at 1 MeV.

Figure 3 Control desk of the 1-MeV GEI-AEI EM7 electron microscope. (Courtesy of AEI Scientific Apparatus and Picker Nuclear.)

The most powerful electron microscope yet constructed is a 3-MeV instrument, completed in 1969 at the Laboratoire d'Optique Electronique at Toulouse, France, which has a theoretical resolution of < 1 Å. The instrument was built by Dupouy et al.,[9] who completed a 1.5-MeV microscope in 1960.[5] The instrument employs 6 lenses and is extensively shielded with lead blocks to protect the operator against x-radiation. The column is 90 cm in diameter by 390 cm high and weighs about 20 tons. The high voltage generator and electron accelerating tube are housed in separate gas-filled tanks in a room above the column. The Compagnie

Générale pour les Systems et Projets Avancés, which built the generating equipment, plans to market the instrument commercially.

C. Electron Penetration

Cosslett[2] points out that, in amorphous materials and probably also in polycrystalline materials, the effective electron penetration is controlled by the cross sections for elastic and inelastic scattering. The latter controls the energy loss in the sample and thus the chromatic aberration produced; the former controls the intensity of the transmitted beams. The transmitted intensity I, expressed as a fraction of the incident intensity I_0 is given by the exponential relation:

$$\frac{I}{I_0} = e^{-S\rho t} \tag{1}$$

where S is the cross section per unit mass thickness for scattering outside a given angle; ρ and t are the sample density and thickness, respectively.

Equation 1 indicates that, for a given fractional transmission, the penetrable thickness is inversely proportional to S which to a first approximation[10] is proportional to $(c/v)^2$, where c is the velocity of light and v the electron velocity, and thus the penetration should vary as $(v/c)^2$. As the accelerating voltage is increased, v approaches c asymptotically, so that there is a diminishing return in penetration. The gain between 100 keV and 1 MeV on this criterion is only about a factor of 3 for all elements.[2]

Fortunately, for a single crystal, or a polycrystalline sample whose grain size exceeds the foil thickness, anomalously high transmission can occur near an orientation which satisfies the Bragg diffraction condition, provided that many beams rather than just two-beams are present. This is the case if the voltage is sufficiently high, that is, about 200 keV for gold and 1.5 MeV for aluminum, the critical voltage increasing with a decrease in atomic number. Thus the penetration in gold is predicted to be 10 to 12 times greater at 1 MeV than at 100 keV.[11] The prediction rests on certain assumptions that require experimental verification, in particular that the main energy loss process is thermal diffuse scattering, due to lattice vibrations.

The useful gain in penetration appears to be bounded between limits imposed by absorption and chromatic aberration because of inelastic scattering which limit the contrast and resolution.[12]

D. Image Resolution

Although an increase in voltage corresponds to a decrease in the wavelength λ of the electrons (see Table 1 in Chapter 5 on page 148) and hence to a decrease in the diffraction aberration Δr_d which is proportional to λ, it also leads to an increase in the focal length of the objective and a corresponding increase in the spherical aberration coefficient C_s. However, we saw from Equation 10, in Chapter 5 on page 153, that the resolving power Δr_{min} is proportional to $\lambda^{3/4} C_s^{1/4}$, so that the decrease in λ more than offsets the increase in C_s. On this basis Dupouy[1] argues, using C_s values of 3.2 mm for a 100-keV instrument and 6.5 mm for a 1-MeV instrument, that the theoretical resolving power would be 2.7 Å at 100 keV and 1.1 Å at 1 MeV, a gain of about 2.5 times at the higher voltage. While this is encouraging to the development of higher voltage instruments, a variety of factors enter into the practical resolution achieved, and it is by no means clear what the optimum voltage may be. The current trend in 100-keV high resolution instruments is toward shorter focal length objectives with reduced spherical aberration. A C_s value of 1.6 mm is attained in at least one of the newer instruments.

Dupouy[1] points out that, at 1 MeV, the theoretical gain in resolution should enable some atoms to be resolved. Since phase contrast appears necessary for this purpose, and techniques are already known for reducing the phase shift due to spherical aberration (and chromatic aberration) nearly to zero at 100 keV (or other voltages), the advantage of high voltage is by no means apparent, the more so since one would need to employ a monolayer sample.

In considering practical image resolution in the high voltage microscope, as opposed to the theoretical objective resolving power discussed above, a key consideration is the energy loss (ΔE) suffered by the beam through inelastic interactions with the specimen. The chromatic error defined as Δr_c, the radius of the disk of confusion referred to object space, is given by Equation 7 on page 151:

$$\Delta r_c = C_c \alpha \frac{\Delta E}{E} \tag{2}$$

where C_c is the chromatic aberration constant of the lens and α the semiangular aperture of the objective.

For 100 keV, $C_c = 1.3$ mm and $\alpha = 6.95 \times 10^{-3}$ radian, hence $\Delta r_c = 0.9$ Å per volt energy loss. For a specimen of given thickness, Δr_c decreases by more than a factor of 10 as the beam energy E increases from 100 keV to 1 MeV, since α (given by Equation 5, Chapter 5 on page 149) decreases somewhat, and ΔE, which is the energy loss per unit mass thickness (ρt) of the sample, also falls according to the Bethe energy relation.

As an example, Cosslett[2] points out that, for the objective in the Cavendish high voltage microscope, the computed value of Δr_c for a 1 μ thick, randomly oriented foil falls from 6300 to 210 Å for lead, and from 2350 to 75 Å for aluminum, on changing from 100 keV to 1 MeV. This large decrease in chromatic error explains the observed great increase in clarity in images of a thick sample, when the voltage is increased. Alternatively, the same chromatic error results if the specimen thickness is increased by a factor of 30 as the voltage is increased from 100 keV to 1 MeV. Unfortunately, the chromatic error at 1 MeV, produced by a 1 μ thick specimen of lead or aluminum, is still very large compared with the theoretical instrument resolution. It follows that, for a thick sample, one needs to use a high voltage microscope in order even to see a dislocation; but, to get high resolution at high voltage, a very thin ($<$ 100 Å) sample is necessary which will also be readily transparent at 100 keV. In view of other considerations, it is not obvious what voltage is optimum from the high resolution standpoint.

The discussion above of chromatic error produced in the sample points out the desirability of incorporating an energy-selecting device that would permit the image to be formed using only electrons of a narrow selected range of energy. This would also permit the study of energy loss processes and provide information on, for example, microcompositional variations. This would be advantageous, both for high and for conventional voltage microscopy. In combination with image intensification and computer processing of a digitalized image to eliminate noise, it should be readily possible to examine 1 μ thick samples to 100 keV, securing the advantages of bulk behavior, two-beam contrast conditions, easier interpretation, and economy in microscope cost. The claimed advantages of high voltage may therefore prove to be of a transitory nature, except for studies related to the physics of the interaction of materials with relativistic electrons.

E. Applications

While a study of thicker samples truly representative of the bulk is a primary motivation for the construction of high voltage microscopes, the useful voltage for metal samples appears to be limited by the threshold energy, above which there is a rapidly increasing probability for an electron to displace an atom from the crystal lattice. The threshold is 250 keV for aluminum, 495 keV for copper, and 725 keV for silver.[3] The use of higher voltages results in point defect formation and clustering, that is, radiation damage. Clearly, it should be possible to study electron damage and its annealing in situ in the high voltage microscope.

Preliminary results indicate that radiation damage may also occur in samples such as glass and polymers, resulting in the loss of crystallinity in polymers.[2,3,13] In organic substances it appears that ionization may take place, together with bond rupture and cross-linking. Since the loss of energy in the specimen, due to electron excitation, is roughly inversely proportional to the electron voltage,[2] the high voltage microscope should thus find application in polymer research.

Damage in biological material caused by electron excitation and probably also that caused by ionization,[2] is reduced as the voltage is raised. However, the probability of the complete removal of a carbon atom and the production of x-rays will increase with an increase in voltage. Although in situ damage studies should be possible, it is doubtful whether many organisms can survive for very long. The early claim[1,14] that bacteria, maintained in a microchamber at atmosphere pressure in air saturated with water vapor, survived examination at 540 keV to 1 MeV is still regarded by other workers with some skepticism.

In the light of evidence[15-17] that dislocations may run out or rearrange at room temperature in metals of moderate melting point such as aluminum, as a consequence of thinning to a thickness transparent to 100-keV electrons, there is a considerable advantage in using thicker samples. However, the thicker the sample, the lower the maximum dislocation density that can be usefully studied because of the problem of detail overlap in the projected image. For studies at high dislocation densities, it will still be necessary to employ thin samples and adopt a different approach such as pinning the dislocations with solute[18] to avoid movement.

There is evidence[19] from studies in a 500-keV microscope that magnetic domain structures in thick films may be quite different from those in thin films. A possible explanation of this is that only in thick films can an appreciable component of magnetization occur perpendicular to the plane of the film.

It is hoped that bulk phenomena such as recrystallization, deformation, phase transformations (precipitation, martensitic transformations, etc.) may be studied inside the high voltage microscope, using hot and cold stages, stretching stages, and so forth. Experiments of this kind on a 500-keV microscope have been reported by Fujita et al.[20]

Clearly, when the features of interest are larger than the transparency thickness at 100 keV, resort may usefully be made to a higher voltage instrument. Examples of such features are dislocation networks, large particles, components of composite materials, and whole cells or microorganisms. Figures 4 and 5 illustrate the potentiality. Cement and glass have been studied.[121] The technique has recently proved valuable, in conjunction with ion bombardment thinning, for studying lunar rock samples.[22]

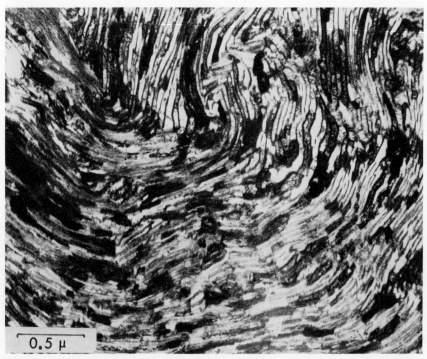

0.5 µ

Figure 4 Transmission micrograph of cross section of drawn oriented pearlite, taken at 800 keV. (Courtesy of R. M. Fisher, United States Steel Corporation Research Center.)

High voltage microscopes will certainly fulfill a permanent and valuable role in studying the physics of the interaction of energetic electrons with matter. Areas of interest include electron damage already mentioned, energy loss processes, and contrast theory. Possibly, new contrast mechanisms capable of giving new kinds of information will be found.

II. REFLECTION ELECTRON MICROSCOPY

The term "reflection electron microscopy" is essentially a misnomer, since usually the electrons inelastically scattered by atoms at or just below

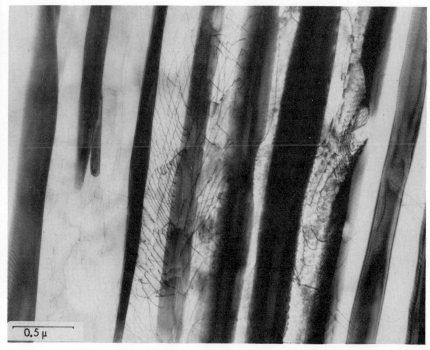

0.5 μ

Figure 5 Transmission micrograph of coarse pearlite, taken at 800 keV, showing interfacial dislocations. (Courtesy of R. M. Fisher, United States Steel Corporation Research Center.)

the surface of the specimen constitute the signal, rather than the diffracted electrons. The system used (Figure 6) has some similarity with oblique illumination in light microscopy. Previous work has been discussed by Halliday.[23] Although provision has been made in a number of commercial models of conventional electron microscope for the reflection technique, it has not become commonly used and will probably be displaced by scanning electron microscopy. Therefore, it is only described briefly.

The electron gun is tilted so that the electron beam is incident on the bulk sample at an angle θ_1 (angle of illumination) to the surface, while the scattered electron signal is taken off along the microscope axis at an angle θ_2 (angle of viewing) to the sample surface (Figure 6). Signal intensity is maximized at the expense of excessive image foreshortening if glancing incidence is employed and $\theta_1 = \theta_2$. Unfortunately, only a small region of the specimen is then in focus. Some compromise is necessary. In the original experiments by Ruska[24] the gun tilt, which

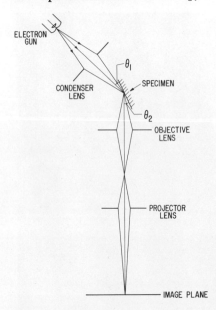

Figure 6 Schematic representation of a reflection electron microscope.

is the angle between the incident beam and the microscope axis, was 90° and a resolution of 20 to 30 μ was obtained. Von Borries and Janzen[25] reduced the tilt to 8° and obtained a resolution of a few hundred angstroms. At $\theta_2 = 26\frac{1}{2}°$, the image is foreshortened by a factor of 2.

As is apparent from the geometry, the fraction of the incident intensity ϕ, scattered into the image, decreases with an increase in the gun tilt, and the objective aperture must be increased in order to get sufficient intensity in the image. Page[26] showed that, if the magnification and objective aperture size are adjusted as the gun tilt is varied, so that a constant (minimum useful) image intensity is maintained, then the resolution δ is

$$\delta \propto \phi^{-\frac{1}{4}} \left(\frac{\Delta V}{V} \right)^{\frac{1}{2}} \tag{3}$$

where $\Delta V/V$ is the fractional energy spread of the scattered electrons (ΔV is of the order of 200 V at high scattering angles).

At small gun tilts, the radius δ_c of the disk of confusion referred back to the object plane is[23]

$$\delta_c = \alpha C_c \left(\frac{\Delta V}{V} \right) \tag{4}$$

where α is the semiaperture angle of the objective lens (typically 10^{-2}

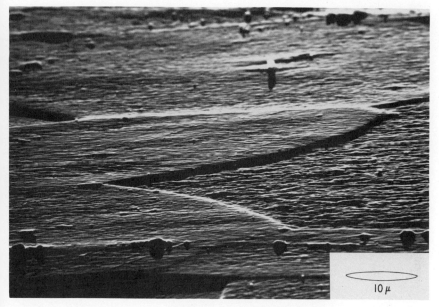

Figure 7 Reflection electron micrograph of the surface of zirconium. Angle of illumination $\theta_1 = 3°$; angle of viewing $\theta_2 = 6°$. (Courtesy of D. Chescoe, GEC-AEI Electronics, Harlow, England).

to 10^{-3} radian) and C_c is the chromatic aberration constant which is nearly equal to the focal length for a weak lens.

A fairly long focal length objective is necessary, as is apparent from the geometry, if a specimen of say 0.5 to 1 cm diameter is to be accommodated. Kushnir, Biberman, and Leukin[27] showed that, for a 6° gun tilt, $\Delta V/V$ is about 1.5×10^{-3}. Inserting typical values of $C_c = 1$ cm and $\alpha = 2 \times 10^{-3}$ radians into Equation 4, we obtain a value of 300 Å for the resolution δ_c in fair agreement with experiment.

It is apparent from Equation 4 that the resolution could be improved by reducing α, C_c, and $\Delta V/V$. Thus, using $C_c = 0.4$ cm, Dupouy and Fert[28] obtained approximately 200 Å resolution. By using a single crystal of etched germanium and employing mainly the Bragg reflected electrons, for which the energy spread is less, to form the image, Halliday and Newman[29] were able to obtain an exceptionally good resolution of approximately 80 Å. This approach requires a rather perfect crystal and would not, of course, be possible with a polycrystal. However, it indicates that better resolution could, in general, be obtained using an energy selecting device to reduce the energy spread of the inelastically scattered electrons, plus an image intensifier to enhance the weak signal.

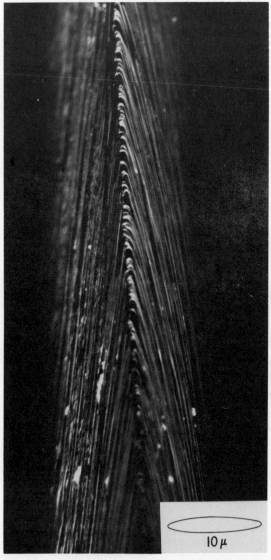

Figure 8 Reflection electron micrograph of a razor blade edge. $\theta_1 = 2°$; $\theta_2 = 8°$. (Courtesy of D. Chescoe, GEC-AEI Electronics, Harlow, England.)

The depth of field depends on the takeoff angle θ_2 and is typically about 20 μ. One can obtain high contrast from shallow surface tilts in a manner similar to oblique illumination in the optical microscope. Because of the shadowing effects it is difficult to examine usefully very rough samples. The high contrast obtained from topographical features is illustrated in Figures 7 and 8. Stereoscopic pairs of micrographs may be obtained.[1]

As discussed by Halliday,[23] it is desirable to employ a specimen manipulator, providing four independent specimen movements. A considerable temperature rise may occur in the sample as a result of the need to use a high intensity incident beam, and because it is mainly absorbed in the bulk sample. Contamination of the sample also tends to be a problem, causing changes in topography and increased energy spread of the scattered electrons.

Chromatic differences in magnification across the field of view, and chromatic differences in image rotation leading to distortion, result from the energy spread of the imaging electrons. These may seriously restrict the usable field when $\theta_1 + \theta_2$ is large, unless compensated by exciting the intermediate and projector lenses in opposition to the objective lens.[23]

III. ENERGY-SELECTING AND ENERGY-ANALYZING MICROSCOPY

A. Introduction

When high energy electrons are transmitted through thin solid films, they undergo scattering. Electrons scattered elastically, which are the ones normally considered in diffraction contrast theory, remain coherent with the incident beam and undergo no energy loss. However, some of the electrons are scattered inelastically, become incoherent, and suffer an energy loss. The fraction scattered inelastically increases with specimen thickness.

The two principal loss mechanisms responsible for losses up to about 50 eV are (a) interband transitions of single electrons, and (b) collective oscillations of the electron plasma. In the latter case a metal is regarded as analogous to a highly ionized gas which is composed of positive ions and virtually free electrons, with zero total change, referred to as a plasma. A metal can be regarded as an array of positive ions arranged on a crystal lattice, with valence or conduction electrons that are fairly free to move throughout the lattice. Although the average charge density is zero, the local electron density continually varies, since the electrons are always in thermal motion. A chance deficiency results in an excess positive charge attracting neighboring electrons which acquire momentum and overshoot. The excess negative charge repels electrons that overshoot, setting up plasma oscillations, that is, longitudinal oscillations in the electron gas analogous to sound waves. The fact that the loss spectrum of a binary alloy consists of losses intermediate between those of the elements present has been taken as evidence for the existence of plasmons.

Experimental work on the loss spectra of metals and alloys has been briefly surveyed by Klemperer and Shepherd;[30] additional references are given by Spalding and Metherell,[31] who studied plasmon losses in different phase regions observed across the phase diagram in Al-Mg alloys. The plasmon loss mechanism predominates in the lighter elements such as aluminum and magnesium, so that the principal loss peak occurs between about 7 to 15 eV, with a width at half maximum intensity of approximately 1 to 3 eV. For certain elements such as the transition metals, the interband transition mechanism predominates, so that the principal characteristic loss peak occurs between about 10 to 30 eV, with a half-width of approximately 10 to 15 eV.

Since the characteristic losses are a function of composition and vary with different phases, they can be related to the microstructure of thin foils. Efforts to combine electron microscopy and energy selection or energy analysis have been made recently by a number of individuals with several objectives in view. First, since the loss electrons emerging from a sample carry information on composition, one goal has been to develop a new technique of microanalysis of potentially higher spatial resolution than the electron microprobe, so that compositional variations related to defects such as grain boundaries and dislocations may be studied. Second, better resolution might be obtained with thick transmission microscopy samples if the loss electrons that result in chromatic error were removed. Or, conversely, for a given resolution, thicker samples could be examined if only electrons of a narrow energy range were used to form the image, although image intensification or computer processing would probably be necessary. Third, bright field images obtained in the electron microscope contain contrast effects associated with the loss electrons,[33-38] and in order to recognize these it is desirable to separate and study them. So far, these effects have been largely ignored in diffraction contrast theory and interpretation of images, with consequences that are difficult at present to assess properly.

Two different instrumental approaches are possible, although the two may be combined in a single instrument.[39] In the first, which may be called an energy-analyzing microscope, the complete energy spectrum is obtained from only a small part (strip) of the image or diffraction pattern, formed by a conventional electron microscope. In the second, or energy-selecting microscope, a specimen image is formed, using only electrons from a preselected band of the energy spectrum. Examples of these two approaches will now be described.

B. Energy-Analyzing Microscope

Metherell et al.[40-45] modified a 100-keV Siemens Elmiskop 1 electron microscope by adding a Möllenstedt type[46] electrostatic energy (velocity)

Figure 9 General view of energy-analyzing microscope, produced by modification of the Elmiskop 1 electron microscope. From Metherell,[47] after Cundy, Metherell, and Whelan.[44]

analyzer below the final image screen. This was a major modification, necessitating removal of the microscope control table, so that the column was supported instead on the analyzer vacuum chamber L (Figures 9 and 10). In Figure 10, J is an additional projection tube with viewing

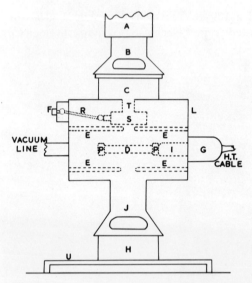

Figure 10 Schematic sectional diagram of lower portion of energy-analyzing microscope shown in Figure 1. From Metherell,[47] after Cundy et al.[44]

screen and window, and H is a plate camera. The 35 mm camera A and the normal projector tube and viewing port B of the Siemens microscope were retained. A trolley was located inside the normal camera chamber C which served to translate and rotate the entrance slit T of the analyzer into a position suitable for velocity analysis.

The analyzer (Figures 10 and 11) consisted[40,44] of a pair of cylindrical electrodes D at beam potential, mounted horizontally on a spare gun insulator between two horizontal earthed plates E with large entrance and exit slots. The normal gun movements were retained and an additional control provided to rotate the insulator, carrying the electrode pair D. Movable fine entrance slot S was about 5 μ or more wide by up to 18 mm long; its length and width were adjusted by controls F (Figure 10). Additional controls permitted slit S to be adjusted into parallelism with the electrodes and then moved off axis across their gap to locate the position of maximum dispersion. Apparently, in constructing and putting into

Figure 11   Internal details of the analyzer used by Cundy et al.[44] (Courtesy of The Institute of Physics and The Physical Society.)

operation this kind of analyzer, the proper alignment of the several slits is a matter of considerable difficulty. The analyzer and microscope column were supported on the table U.

A conventional electron microscope image was recorded in the standard 35-mm film camera A (Figure 10) which was then withdrawn, allowing the electron beam to pass into the analyzer. Provision was made for adjusting the voltage supply to the cylindrical electrodes in steps of a few volts, with means of voltage calibration. At 100 keV a mean dispersion of 0.10 mm/V was achieved in the camera H with an energy resolution of about 1 to 2 eV and exposure times of 2 sec to 2 min.

The principle of operation of the analyzer is illustrated in Figure 12 from Metherell.[47] The electron beam, apertured by the off-axis slit, passes through regions of high chromatic aberration which lie close to the two cylindrical electrodes b, maintained at the beam potential (see also Figures 10 and 11). The curvature of the loss lines in the energy spectra, shown

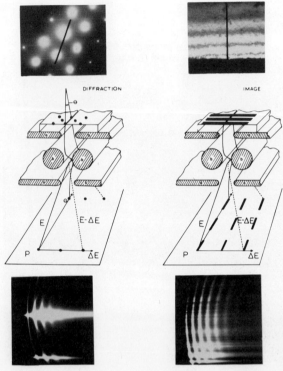

Figure 12   Principle of energy-analyzing microscope, using a Möllenstedt velocity analyzer. The operation of the analyzer at 80 keV is shown for the cases of diffraction (left) and image formation (right) on the wedge edge of an aluminum crystal. The black lines on the diffraction pattern and image (crossing the thickness contours) indicate the entrance slit position. From Metherell.[47]

at the bottom of Figure 12, is caused mainly by the end caps which are fitted to the center electrodes of the analyzer (Figures 10 and 11). These destroy the perfect cylindrical symmetry required for linearity.[47] There is also a significant effect caused by the projector lens.[47]

The advantage of this technique was that energy analysis from very small regions could be combined with electron microscopy at magnifications up to 20,000X. Image details down to about 100 Å could be recognized in the energy spectrum. The technique has been applied to the microanalysis of precipitated phases,[31,48-52] the study of solute gradients in alloys,[52-56] and to the study of the preservation of image contrast by inelastic scattering.[57-59] Some examples from each of these three classes of application will now be discussed.

Cundy and Grundy[49] applied the technique to the microanalysis of

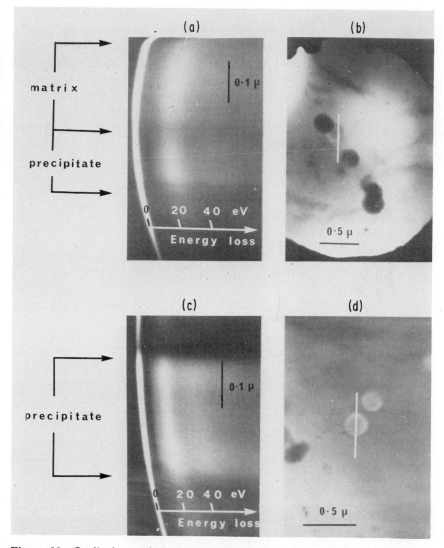

Figure 13 Qualitative analysis of internally oxidized Ni-0.05%Si alloy, using an energy-analyzing microscope.[44] (*a*) Energy spectrum transmitted through the line shown on the electron micrograph (*b*) using 80 keV; (*c*) and (*d*) a similar pair, using 100-keV beam voltage. From Cundy and Grundy.[49]

an internally oxidized Ni-0.05%Si alloy. In Figure 13a and c, taken with 80 and 100-keV incident electrons, respectively, the bright line on the left corresponds to electrons that lost no energy and, as may be seen by reference to the densitometer traces in Figure 14a and b, the energy loss increases as one moves to the right. The curvature of the lines in Figure 13a and c results from asymmetry in the analyzer field. Since transmission samples were used, the precipitate spectra included some matrix, and the difference between the traces in Figures 14a and b represents the precipitate. Energy losses are independent of the accelerating

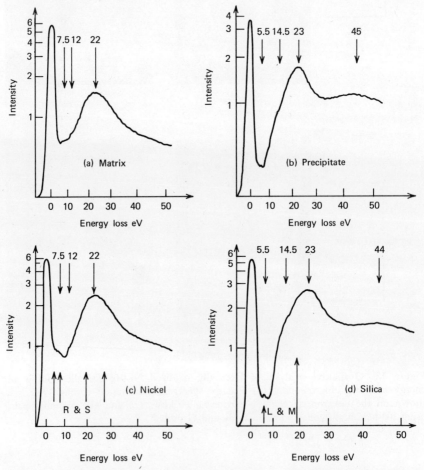

Figure 14 Densitometer traces of energy loss spectra. (a) From matrix region in Figure 13a; (b) from precipitate region in Figure 13c; (c) from pure nickel standard; and (d) from amorphous silica standard. From Cundy and Grundy.[49]

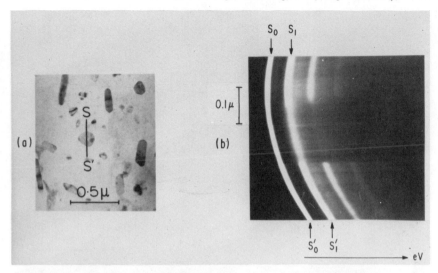

Figure 15   Application of energy-analyzing microscope to θ precipitates in Al-4 wt% Cu alloy. (*a*) Transmission micrograph, showing entrance slit position *SS'* across θ particle; (*b*) shows corresponding energy loss spectrum. From Cundy et al.[50]

voltage. A comparison with the reference spectra obtained from pure nickel (Figure 14*c*) and amorphous silica (Figure 14*d*) permitted identification of the precipitate as amorphous silica.

Cundy, Metherell, and Whelan[50] determined the energy loss spectra across a θ precipitate in an Al-4%Cu alloy, apparently using 80-keV electrons. The different character of the loss across the precipitate (Figure 15*a*) is clearly seen in Figure 15*b*, where S_0S_0' and S_1S_1' correspond to the zero loss and plasma loss electrons, respectively. They concluded that, for successful microanalysis, it was necessary that a precipitate extend from top to bottom of the foil, unless the characteristic loss spectra from the matrix and from the precipitate are well defined and well separated. In order to study the concentration gradient near a boundary, it should be carefully aligned parallel to the incident beam to avoid overlap effects.

Spalding, Villagrana, and Chadwick[55] applied the energy-analyzing microscope to the study of the copper distribution in various lamellar Al-CuAl$_2$ eutectics. Figure 16 shows the typical transmission microstructure of the coarse eutectic that they studied; the interlamellar spacing is about 2 μ. The energy loss spectrum across a single such interface between the Al and CuAl$_2$ phases, aligned parallel to the beam is seen in Figure 17. The plasmon loss peak of about 15 eV in the Al-rich phase

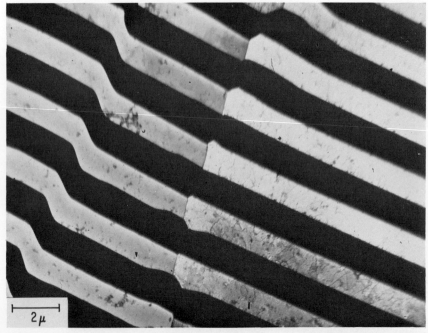

Figure 16   Typical transmission electron micrograph of Al-CuAl₂ eutectic, studied by Spalding et al.[55] The $CuAl_2$ θ phase is in dark contrast and the Al-rich phase in light contrast. A mismatch boundary between parallel lamellae is apparent. Micrograph taken by F. D. Lemkey. (Courtesy of G. A. Chadwick.)

is broadened and increased slightly in amount in the $CuAl_2$ phase. The strong loss peaks around 30 and 45 eV in the Al-rich phase are barely apparent in the $CuAl_2$ phase (Figure 17).

Microdensitometer traces at 100 Å intervals from, and parallel to, an interphase boundary, similar to that in Figure 17 but for a crystal aged at room temperature, permitted Figure 18 to be constructed which shows the plasmon loss variation as a function of distance from the boundary. The plasmon loss was constant at 15.25 ± 0.03 eV across most of the Al-rich phase (Region I) and decreased gradually on approaching the boundary (Region II), changing fairly abruptly at the boundary from 14.97 ± 0.03 eV to 16.34 ± 0.09 eV throughout the $CuAl_2$ θ phase (Region IV). The intermediate value of 15.58 ± 0.09 eV, found in the approximately 100-Å wide boundary Region III, was thought to be most probably caused by surface plasma oscillation at the interface.

The composition profile shown in Figure 19 was derived from the plasmon loss curve in Regions I and II of Figure 18, using a cali-

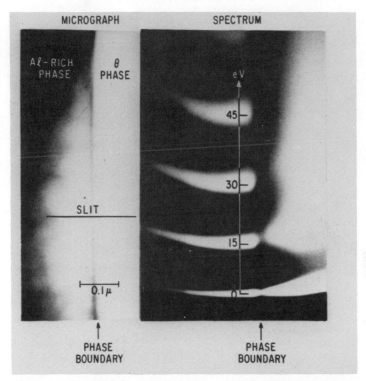

Figure 17   Energy loss spectrum across an interface between aluminum and θ phases in eutectic of Figure 16. After Spalding et al.[55] (Courtesy of G. A. Chadwick.)

bration curve showing plasmon loss versus weight percent copper measured on a series of Al-Cu solid solution alloys. The copper concentration increased from 0/1.5% to about 4% near the interphase boundary and was considered to be in solution, since no precipitates were seen in the microscope image. In contrast to this result obtained on the air-cooled and aged crystal, a splatt-quenched crystal showed a constant 0/1.5% concentration right to the interphase boundary.

The results of Spalding et al.[55] demonstrate the quantitative capability of microanalysis measurements by this technique. They were also able to deduce that strain-field effects were absent near the interface, since a constant plasmon loss was observed across the Al-rich phase in a quenched crystal. The concentration gradient observed near the interface in the slowly grown and aged crystal is not fully understood.

Spalding and Metherell[31] determined the plasmon losses for the α, β,

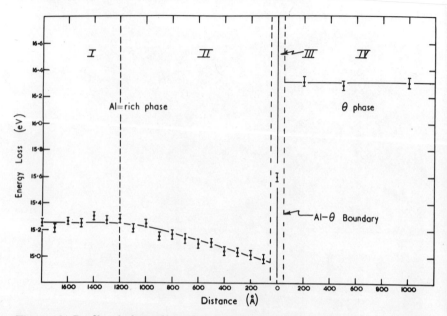

Figure 18 Profile of plasma loss relative to the interphase boundary in an Al-CuAl₂ eutectic, air-cooled after solidification and aged several months at room temperature. From Spalding et al.[55] (Courtesy of G. A. Chadwick.)

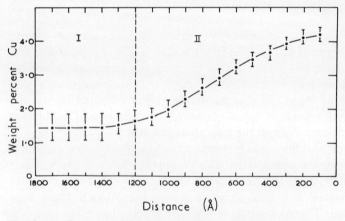

Figure 19 Copper concentration profile in the Al-rich phase, derived from Figure 18, showing copper concentration increases on approaching the interphase boundary. After Spalding et al.[55] (Courtesy of G. A. Chadwick.)

308

γ, and δ phases in the binary Al-Mg alloy system, again combining transmission electron microscopy, selected area electron diffraction, and velocity analysis. Their results are illustrated in Figures 20 and 21. The ratio of the plasmon loss peak halfwidth W_1 to the zero loss peak halfwidth W_0 (Figure 20) was found independent of the magnesium content. The

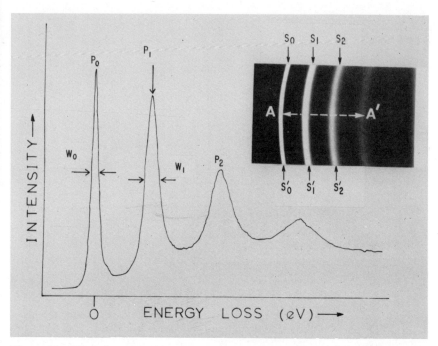

Figure 20 Typical energy spectrum (inset) on pure aluminum, using an 80-keV beam potential. The energy profile was obtained by a microdensitometer trace along AA'; P_0 and P_1 correspond to the zero and 15.2-eV plasmon loss lines, S_0S_0' and S_1S_1' respectively. From Spalding and Metherell.[31]

plasmon loss differed for the four phases found in the system (Figure 21) and varied linearly with composition in the α-, γ-, and δ phases. Using the information in Figure 21, it was possible in an Al-95 at. % Mg alloy to identify precipitates approximately 100 Å in size as γ phase by energy loss determination.

Cundy et al.[44] were able to demonstrate, using combined electron microscopy and energy analysis (Figure 22), that essentially the same contrast effects would be obtained with inelastically scattered electrons as with the elastically scattered electrons normally considered responsible for diffraction contrast. It will be noted in (*b*) that the distribution of zero loss, plasmon loss (15.2 eV), and approximately 30-eV loss electrons across the bend contour is much the same. Furthermore, quantitative intensity measurements on the loss electrons are possible as a function of deviation from the Bragg reflecting positions. The results of Metherell[47]

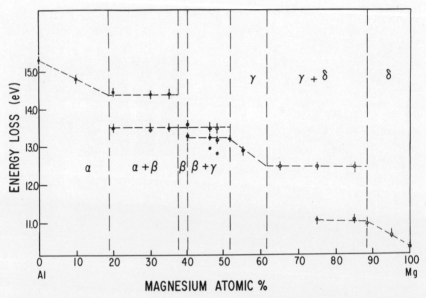

Figure 21 Plasmon loss versus composition for Al-Mg alloy series across the phase diagram. The broken vertical lines indicate the phase boundaries after Hansen.[60] After Spalding and Metherell.[31]

in Figure 12 show analogous results for thickness fringes in aluminum. This case is discussed in detail by Cundy et al.[57]

C. Energy-Selecting Microscope

Energy-selecting microscopes were invented independently by Castaing and Henry[61] and by Watanabe and Uyeda.[62] The former used as analyzer a double magnetic prism and an electrostatic mirror, placed after the first intermediate lens. Electrons of a particular energy were selected by an aperture, and the image was formed by a second intermediate and a projector lens. By removing the aperture, a conventional electron microscopic image could be formed.

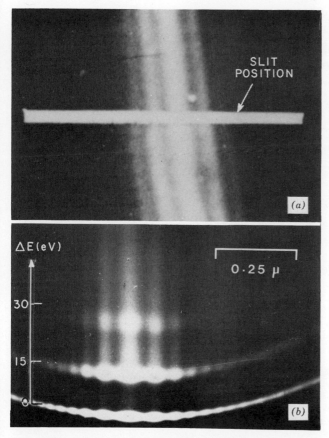

Figure 22 Single crystal of aluminum (*a*) transmission electron micrograph of (111) bend extinction contour at 80-keV beam potential. The white line shows the entrance slit position employed in obtaining the corresponding energy spectrum (*b*). From Cundy et al.[44] (Courtesy of The Institute of Physics and The Physical Society.)

Watanabe and Uyeda,[62] whose instrument is shown schematically in Figure 23, employed a Möllenstedt analyzer[46] which was placed after the intermediate lens. Here the first image of the specimen was formed by the objective lens in the plane of the first (entrance) slit S_1 to the analyzer. The energy spectrum from the analyzer was formed in the plane of the second slit S_2, placed parallel to S_1. The lateral position of S_2 was adjusted to select electrons with a certain energy, which then passed through a projector lens to form an image of the strip of the first image passed by S_2, on the screen. Alternately, S_2 could be opened up to provide a complete spectrum. Externally, the instrument was very similar in appearance to a conventional electron microscope.

In order to obtain a complete image, the first and final images were scanned by applying synchronized voltages to the two sets of deflector plates D_1 and D_2. The width of slit S_2 determined the energy spread, which was about 1 eV at an accelerating voltage of 50 keV.

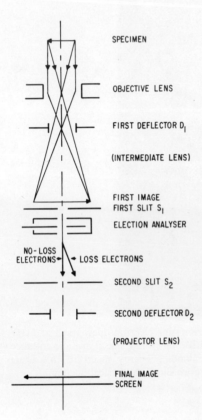

SPECIMEN

OBJECTIVE LENS

FIRST DEFLECTOR D_1

(INTERMEDIATE LENS)

FIRST IMAGE
FIRST SLIT S_1

ELECTION ANALYSER

NO-LOSS ELECTRONS → ← LOSS ELECTRONS

SECOND SLIT S_2

SECOND DEFLECTOR D_2

(PROJECTOR LENS)

FINAL IMAGE
SCREEN

Figure 23 Schematic ray diagram of energy-selecting microscope. After Watanabe.[62-64]

In an alternative mode of operation, a selected-energy, selected-area diffraction pattern was obtained by using an intermediate lens with a field limiting aperture (Figure 23) to project the diffraction pattern formed in the back focal plane of the objective lens onto the plane of slit S_1. Scanning was used, as above, to obtain a complete selected-energy diffraction pattern.

The instrument was usable as a conventional 100-keV electron microscope at up to 200,000X. The electrostatic analyzer was only operable at 75 keV or lower. Some examples of the results obtained by Watanabe[63-65] are shown in Figures 24 to 27. All the specimens examined,[64] listed in Figure 24, had characteristic energy loss peaks situated at about 10 to 25 eV. However, it will be noted that the lower energy characteristic loss peaks are often sharper and may also be large enough

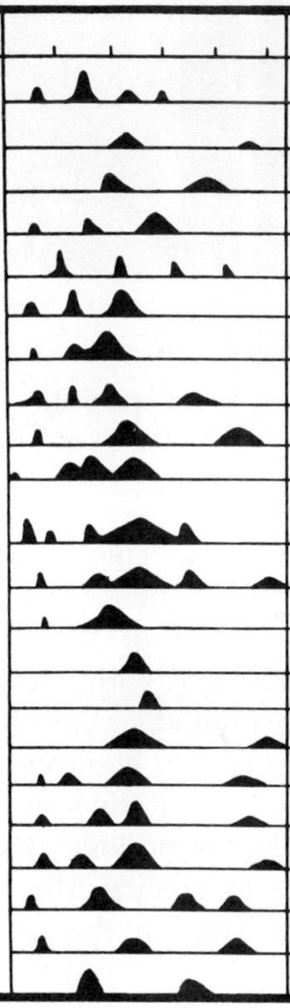

		Present Author
Al		6.5 14.8 23 29.5 45
Al₂O₃		22 5 46
Be		19 38 56
BeO		5.7 16.5 28 57
Mg		10.3 20.6 32 43
MgO		4.5? 5.5 11.4 25
Sn		6.3 13 19.5
SnO₂		5.5 12.5 19.5 35 63
Si		5.5 22.5 45
SiO₂		12.5 16.2 24.5
Ag		3.4 8 17 5 25 34
Au		6.5 17 5 25 34 50 62 ?
Cu		7 19.5
Co		22.9 63.3
Cr		26 54
Fe		23.2 49 62
Ni		6.5 12 22.5 45
Sb		6.5 18 24.5 46
Ti		6.5 13 24 50
Tl		4.7 17.4 34.5 43
W		7 22 44 54
Ge		16.4 33.8

Figure 24   Table of energy loss peaks. From Watanabe.[64] (Courtesy of Hermann & Cie, Paris.)

to use for microanalysis of an alloy sample. Unfortunately, transition and noble metals and most compounds have peaks too broad to permit them to be used for identification. The large sharp peaks present in aluminum at 15 eV, magnesium at 10 eV, and beryllium at 19 eV are, on the other hand, very suitable. In order to obtain useful analytical information, it is clear that careful choice of the alloy studied is necessary.

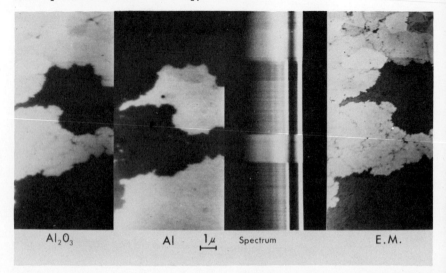

Al_2O_3 Al 1μ Spectrum   E.M.

Figure 25 Transmission selected-energy micrographs of a partially oxidized aluminum film at 75 keV. From Watanabe.[64] (Courtesy of Hermann & Cie, Paris.)

Watanabe[64] was able to distinguish aluminum and aluminum oxide in a selected-energy image of an oxidized aluminum film (Figure 25). He claimed that, by using characteristic loss electrons of 10 to 20 eV, two-dimensional distribution of elements was observable with a resolution better than 0.1 μ. Dislocations in molybdenum disulfide were observed in both the no-loss and 23-eV loss images (Figure 26), confirming that the inelastically scattered electrons, which are those suffering an energy loss, can produce diffraction contrast just as do the no-loss, elastically scattered electrons. Watanabe[63,65] was not able to confirm the finding of Kamiya and Uyeda[66] that the contrast of dislocations in the loss image differed from that in the ordinary bright field image.

Electron diffraction patterns[63] of an electropolished aluminum crystal (Figure 27) illustrate the remarkable reduction of the diffuse background around the central spot, obtained by eliminating the loss electrons. The pattern obtained with 15-eV loss electrons is somewhat more diffuse than the no-loss pattern. The 30-eV loss pattern is even more diffuse. In later experiments, Watanabe[67] established that the amorphous oxide layer, formed on aluminum during electrothinning, was an important source of small angle no-loss scattering. These electrons were presumably eliminated by the slits in Watanabe's equipment, as well as the loss electrons, accounting for some unknown fraction of the remarkable reduction in the diffuse background in Figure 27.

Figure 26 Transmission micrographs, showing dislocation networks in MoS₂ film at 40 keV. (*a*) Ordinary bright field image; (*b*) no-loss image; (*c*) 23-ev loss image. From Watanabe.⁶³

Watanabe⁶⁴ found that extinction contours were obtained in character-istic-loss images as in no-loss images. Diffuse streaks, connecting the spots in a diffraction pattern of germanium, were only observed in the no-loss pattern, showing that they were due to no-loss electrons. In the case of amorphous material, such as carbon particles, the 23-eV loss image looked similar to the ordinary dark field image, indicating that the con-trast in the latter was principally caused by inelastically scattered elec-trons. Fresnel fringes were absent in the loss images formed by incoherent waves, as expected from theory.

The brightness of an image, formed solely by electrons of a selected energy loss characteristic of a particular element, is proportional to the concentration of the element. Thus microanalysis is, in principle, possible with high image resolution. In practice the element resolution possible

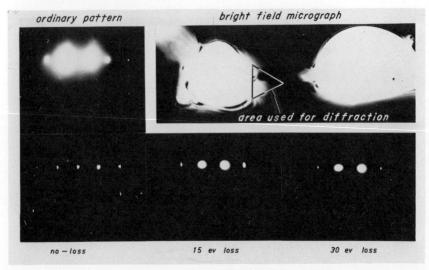

Figure 27   Electron diffraction patterns of Al single crystal at 40 keV. From Watanabe.[63]

depends upon the nature of the alloy, and the availability of a (sharp) characteristic loss element peak distinguishable from the peaks of other elements present.

Beside its possible application for microanalysis, it is apparent that the technique is a powerful tool for obtaining a better understanding of the origin of contrast in the conventional electron microscope. Because of their different energy, loss electrons are brought to focus in a different plane from no-loss electrons for a given objective lens setting. Since all of these electrons are superimposed in normal images, the resolution is improved in principle in no-loss and selected-energy loss images. The gain is greater with thick samples, since the proportion of loss electrons increases with the thickness. The gain in resolution obtained[63] with selected-energy loss images was less than expected as a result of the weak signal that necessitated long exposures. This might be obviated by the use of image intensification.

IV. ELECTRON EMISSION MICROSCOPY

A. Introduction

Many microscopes have been built, and at least one is commercially available, in which electrons emitted directly from the sample are imaged by electromagnetic or electrostatic lenses directly onto a fluorescent screen. Electron emission microscopes, as they are called, differ from the scanning electron microscope principally in the absence of scanning and in the direct imaging feature. The field emission microscope (page 325) is also a type of electron emission microscope but relies on the use of a very high electrostatic field to pull electrons out of the specimen surface, necessitating special specimen and microscope geometry. In the electron emission microscope, electron emission from the surface of the sample, which is usually a flat plate, is produced by ultraviolet excitation, ion bombardment, electron bombardment, or heating with or without an activator coating on the sample. These varied techniques and their applications have been critically reviewed by Wegmann,[68] who includes an extensive number of references. A bibliography on thermionic emission microscopy is given in the reviews by Eichen[69] and Möllenstedt and Lenz.[70]

The contrast in the image results directly from absolute differences in emission across the specimen surface, unlike the scanning electron microscope, in which differential contrast effects can be produced by electronic processing of a signal and a variety of signals is available. Electron emission varies with orientation, crystal structure, the presence of stress gradients, and composition. Thus silicon and chromium solutes decrease, but carbon and copper increase emission in iron.

B. Instrumentation

As pointed out by Wegmann,[68] there is a close parallel between an optical microscope, in which the specimen is illuminated by reflected oblique light and viewed by an objective and eyepiece, and the emission electron microscope (Figure 28). In the latter, electrons excited from the sample in one of a number of ways constitute the signal, instead of reflected light; and two (or three) lenses, which may be either magnetic or electrostatic, are used to form a magnified image on a fluorescent screen or photographic emulsion. The specimen is maintained at a high negative potential relative to the grounded anode, so that electrons can be accelerated away from the sample surface. Ion gun excitation is shown in Figure 28, as well as a heater to raise the sample temperature.

The commercial Balzers KE-3 model Metioscope,[68] made in Liechtenstein, is similar to that in Figure 28 but is a much more versatile and elaborate instrument. It will serve to illustrate most of the possible modes of operation found in a variety of previous instruments (Figures 29 and 30). The specimen, typically a flat metallographically prepared sample, introduced via an airlock, can be heated up to 1200°C in the vacuum

($\sim 10^{-6}$ torr) while under observation. A thermocouple is used to measure its temperature. The special stage permits a 3-mm diameter field to be examined and provides ±190° rotation about the optical axis. A 30 to 50 keV accelerating voltage can be applied, and three magnetic lenses provide a magnification range of 100 to 12,000X. The objective lens has approximately a 10-mm focal length and is provided with a stigmator.

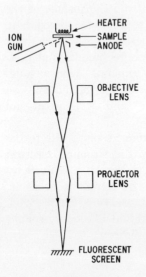

ION GUN
HEATER
SAMPLE
ANODE

OBJECTIVE LENS

PROJECTOR LENS

FLUORESCENT SCREEN

Figure 28 Schematic illustrating image formation in the electron emission microscope.

The plate camera can be replaced by a 36-mm film camera for single exposures or ciné recording.

The specimen can be moved under vacuum into a preparation chamber equipped with ion etching or polishing; a double evaporation source for coating; and a quenching device, consisting of a conducting rod which can be brought into contact with the sample. The specimen can be rotated, while exposed to the 10 to 30-keV ion beam at 0 to 90° inclination.

The instrument can be used in the following modes.

1. Photoemission employing electrons released by ultraviolet (UV) irradiation, provided by one to four UV lamps with quartz optics, whose light is mirrored onto the specimen from the anode. Point resolutions of approximately 150 Å are obtainable. Switching from one lamp to an opposite lamp will reverse the orientation contrast. Variation in the lamp centering permits variation of the principle UV wavelength. Sample rotation about the optical axis during viewing permits contrast variation.

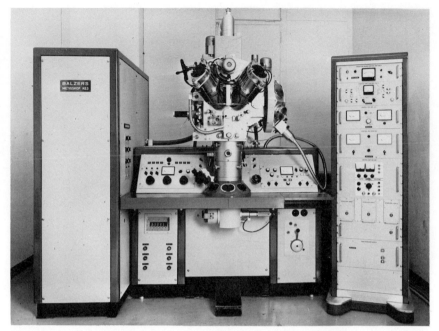

Figure 29 Balzers emission electron microscope Metioscope KE3. (Photograph courtesy of L. Wegmann, Balzers Aktiengesellschaft, Liechtenstein.)

The UV system can be hydraulically withdrawn to protect it from contamination, while the other modes are in use.

2. Secondary electron emission under ion bombardment using a neutral particle source. Gas atoms or molecules accelerated as ions at 5 to 20 keV and finally neutralized, impinge on the sample at a glancing angle up to 15°. The ions tend to produce relief contrast, rather than just surface contrast, and are useful at the temperatures of transition to thermal emission.

3. Thermionic electron emission. Since high temperatures are commonly necessary to obtain sufficient thermal emission, an alternate stage provided with electron beam heating is used to raise the sample temperature to 1000 to 2000°C.

In the intermediate temperature ranges superposition of two kinds of emission may be possible. Between them, the three modes of operation permit the imaging of a suitable sample at temperatures up to 2000°C, facilitating dynamic reaction studies.

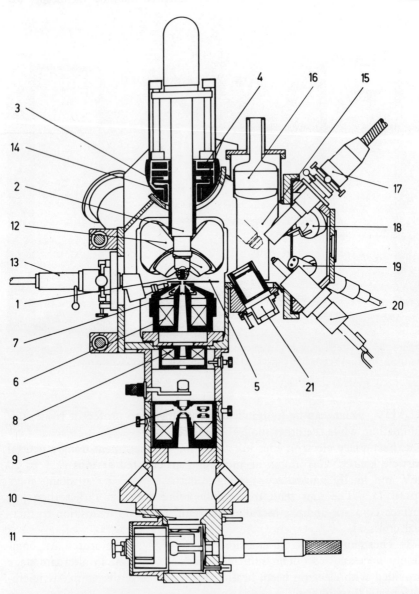

Figure 30 Cross section of Balzers KE3 Metioscope: 1, specimen; 2, specimen holder with heating device; 3, specimen rotating stage; 4, specimen displacement stage; 5, anode interchanging device; 6, objective lens with a stigmator; 7, aperture diaphragms; 8, intermediate lens; 9, projector lens; 10, screen; 11, plate camera; 12, UV illumination systems; 13, neutral particle gun; 14, tilting specimen support; 15, preparation chamber; 16, cooling trap; 17, ion gun; 18, surface temperature probe; 19, evaporation source; 20, specimen cooling device; 21, airlock. From Zaminer et al.[71]

C. Resolution

As pointed out by Brandon,[72] whatever the mode of excitation, the resolution is limited by the fact that, at the surface to be imaged, the energy spread of the electrons is appreciable in proportion to their energy, so that they have a significant component of lateral velocity. This is in contrast to other electron optical techniques, in which the energy spread is small relative to the total energy at the object plane. Increase in the accelerating field improves the resolution up to about 200 Å, at which electrical breakdown occurs between the specimen and the anode.

D. Contrast

The image contrast obtained varies with the nature of the sample and the mode of electron excitation.

Photoemission varies with orientation, the nature of a constituent, and the wavelength of the ultraviolet light. Emission decreases as the work function of the surface is increased, so that the cleanliness of the surface is a factor. The *work function* is the energy needed to raise an electron from the top of the conduction band in a metal to a position just outside the sample surface. The work function varies with the metal and the structure of the surface. It follows that contrast changes may be produced by light oxidation, resulting from heating in the presence of a little oxygen, analogous to heat tinting in the optical microscope.

The intensity of secondary electron emission produced by ion bombardment depends mainly on the ion flux and on the composition of the surface, and is relatively insensitive to the ion energy and the crystallographic orientation of the sample surface. If the surface is uneven, and the ions are incident at a low angle, stray contrast differences arise, since the ion flux varies over the surface. The contrast is analogous to that of a shadowed replica in the conventional electron microscope or to that of an obliquely illuminated optical microscope sample. It is desirable to work at low ion energies, in order to reduce sputtering which produces time-dependent changes in sample topography and image detail that may be hard to interpret. These effects have tended to discourage the use of ion bombardment excitation.

Contrast in the case of thermionic emission, as with photoemission, is again related to differences in the work function over the sample surface, that is, it is an intrinsic property of the material. The temperature dependence of emission may be expressed by the relation:

$$I = aT^2 \exp\left(\frac{-b}{T}\right) \tag{5}$$

where I is the current density at the specimen surface, a a constant for a particular material, b a constant related to the work function, and T the absolute temperature.

Extensive use has been made of an adsorbed layer of an activator to reduce the work function of the specimen surface, in order to obtain adequate thermionic emission, particularly at temperatures of 450 to 1300°C which are of interest in the case of steels (see further detail under Applications). Strong contrast effects then result from changes in the absorption characteristics produced by variations in sample composition and orientation. Care is necessary in the interpretation of images obtained using an activator, since complexities of contrast and artifacts may result. The subject is rather fully discussed elsewhere.[69,73]

E. Applications

Typical fields of application[69-71,73-77] of the electron emission microscope include (dynamic) studies of grain growth,[71,73,76,77] recrystallization, phase transformations,[69,73] oxidation,[71,74,75] reduction,[71] and diffusion reactions.[78] The instrument has two principle advantages over the high temperature optical microscope: (a) better resolution, and (b) the ability to reveal structures offering poor optical contrast. Since contrast results from intrinsic properties of the material, changes are revealed instantaneously, particularly in the photoemission and thermionic emission modes. Although bulk samples are used, the technique is open to the criticism that the observations are made on a surface under high vacuum, whose behavior may differ from that of the sample interior. Surface preparation is also clearly of importance.

The introduction of ultraviolet excitation is a relatively recent innovation, and most emission microscopy has been carried out using thermionic emission. The latter is only applicable to samples that will withstand the temperatures necessary to produce usable emission, without undergoing undesired structural changes or reacting with residual gases in the vacuum system. Phase transformations in steels have been the most popular subject of study. Since thermionic emission has often been inadequate at the temperatures of interest ($< \sim 1300$°C), extensive use has been made of an adsorbed layer of an activator to reduce the work function of the specimen surface. Below about 450°C, depending on the sample material, sufficient thermionic emission cannot be produced, even with the help of an activator, to produce a usable image on the fluorescent screen. Cesium is the most effective activator and is therefore used in the range 450 to 600°C. Because of its high vapor pressure, frequent

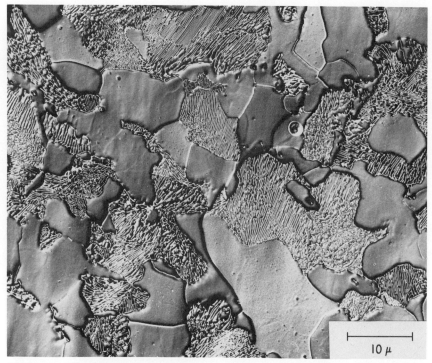

10 μ

Figure 31 Photoemission micrograph of a hypoeutectoid steel, consisting of ferrite and pearlite, electrolytically polished and chemically etched. Taken at room temperature using ultraviolet excitation. (Micrograph courtesy of L. Wegmann, Balzers Aktiengesellschaft, Liechtenstein.)

renewal is necessary. Other activators used in the range 600 to 1300°C include barium, strontium, thallium, and rubidium. The activator should not alloy with the sample.

Examples of the results obtainable with the electron emission microscope are shown in Figures 31 to 37. The great progress made in photoemission micrography is illustrated in Figures 31 to 35. Excellent phase discrimination is obtained in steel (Figures 31 and 32), hard metal (Figure 33), and even in an uncoated insulating rock (Figure 34). A resolution of 100 to 200 Å is demonstrated in the troostite phase of a steel (Figure 32), considerably superior to the light microscope and comparable with current commercial scanning electron microscopes. The superiority of photoemission over kinetic (neutral ion) emission microscopy, with regard to phase differentiation, is shown in Figure 35, taken on a Nimonic

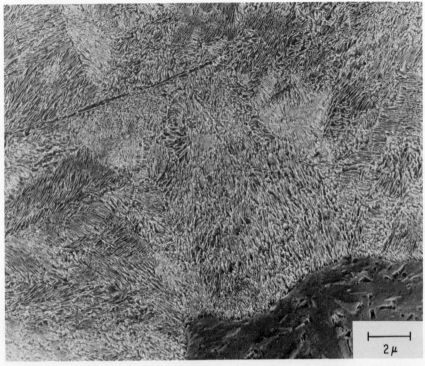

Figure 32 Photoemission micrograph of heat-treated 1.1% carbon steel, showing fine grained pearlite (Troostite) and martensite. Taken at room temperature, using ultraviolet excitation. (Micrograph courtesy of L. Wegmann, Balzers Aktiengesellschaft, Liechtenstein.)

alloy. Other examples of the application of photoemission microscopy are given by Wegmann.[79] This is an exciting new technique.

Thermionic emission is particularly useful for in situ studies at high temperature. Its capabilities are illustrated in Figures 36 and 37, the former showing austenitic grain contrast on the activated surface of a steel at approximately 1000°C, the latter showing tungsten carbide needles on a tungsten surface at approximately 1750°C, produced by admitting a little hydrocarbon when the temperature was 1400°C.[71] Zaminer, Graber, and Wegmann[71] were able to watch the spreading of a monolayer chemisorbed layer of oxygen over a tungsten surface at approximately 1400°C, using the thermionic emission technique and cinematography. By reducing the oxygen pressure, they were able to watch the desorption process. The sorption layer changes the work function which is also dependent on grain orientation.

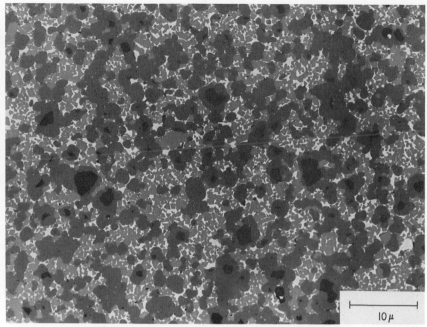

Figure 33 Photoemission micrograph of mechanically polished hard metal alloy, showing titanium carbide (black), tantulum-tungsten carbide (dark gray), and tungsten carbide (light gray) in a cobalt matrix (white). Taken at room temperature using ultraviolet excitation. (Micrograph courtesy of L. Wegmann, Balzers Aktiengesellschaft, Liechtenstein.)

V. FIELD EMISSION AND FIELD ION MICROSCOPY

Field emission and field ion microscopes have unique capabilities for the direct examination of the surfaces of suitable solids. With the field emission technique, information can be obtained on the distribution of adsorbed gas layers and on local variations in the work function of the surface; with the field ion microscope, which is of principal interest here, the atomic arrangement of crystal surfaces can be determined. Although the interpretation of the data is a matter of some difficulty, considerable progress has been made, as will become evident.

A. Field Emission Microscopy

Development of the field emission microscope dates from its invention by Muller in 1936.[80] The information obtained in field emission micros-

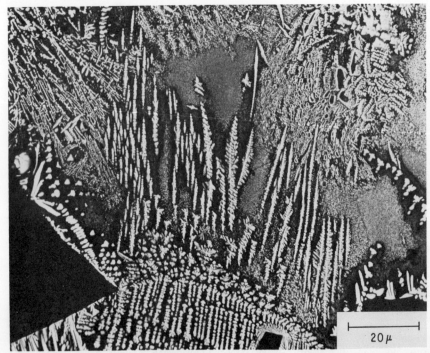

Figure 34 Photoemission micrograph of mechanically polished (uncoated) fine grained basalt rock. The bright dendrites are augite and the dark crystals are plagioclase. A fine grained devitrification mass is seen in the matrix. Taken at room temperature, using ultraviolet excitation. (Micrograph courtesy of L. Wegmann, Balzers Aktiengesellschaft, Liechtenstein.)

copy relates principally to the electronic state of the surface layer of atoms, since the image is formed by extraction of conduction electrons from the hemispherical tip of a specimen wire by means of a high negative electric field. Atoms cannot be imaged directly, since the resolution is typically about 30 Å.

The basic geometry is illustrated in Figure 38 after Brandon.[81] The wire sample, electropolished into a needlelike point with a radius r of approximately 2000 Å or less, is mounted in a bulb of radius R. The bulb is baked out and gettered to below 10^{-9} torr. The specimen is then "flashed" to a high temperature to remove surface contamination and to round off the tip. For some specimen materials with oxides of poor volatility, other cleaning procedures may be necessary. A high negative potential, typically of 5 to 10 kV is now applied to the specimen. Because of the small tip radius, high fields ($\sim$0.3 V/Å) are produced near the

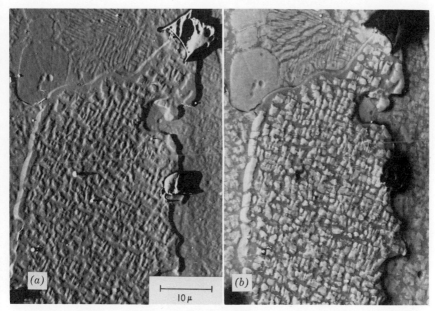

Figure 35 Emission micrographs of mechanically polished nickel-base Nimonic superalloy: (*a*) kinetic, that is, neutral particle, excitation at room temperature; (*b*) ultraviolet excitation at room temperature. Note that in (*a*) the γ′ particles are differentiated from the γ matrix by relief effects, whereas in (*b*) they appear in lighter contrast. (Micrographs courtesy of L. Wegmann, Balzers Aktiengesellschaft, Liechtenstein.)

tip, resulting in electron emission. Electrons tunneling out of the tip are accelerated toward a phosphor screen (anode), deposited on the inside of the bulb, thus producing an image. It is convenient to coat the phosphor with an evaporated tin oxide conductive coating, to which an anode connection can be made. The magnification given by cR/r, where c is a compression factor which depends on the tip profile, is of the order of 10^5. The tangential velocity of the electrons in the free electron gas limits the resolution to approximately 20 Å. Thus individual adsorbed atoms cannot be detected, but only larger aggregates.

The hemispherical emitter surface is actually faceted and made up of a variety of crystal planes. In a bcc metal the closer packed planes such as {110}, {211}, and {100} have a higher work function, and therefore appear in darker contrast than stepped regions which emit more readily. The orientation of the emitter and the identity of the planes can be deduced from the symmetry of the pattern, using standard orthographic projections. Planes with a high work function, that is, planes with high

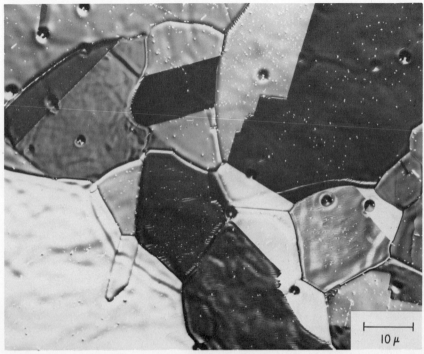

Figure 36 Thermionic emission micrograph of 0.4% carbon austenitic steel activated with evaporated barium layer. Taken at about 1000°C. Some grains show terracing, due to thermal etching. The bright spots are probably Ba0. (Micrograph courtesy of L. Wegmann, Balzers Aktiengesellschaft, Liechtenstein.)

atomic density such as the close-packed {110} plane in the bcc structure emit relatively few electrons and appear dark, relative to those with low work function such as the {111} planes. The angular separation of planes in the image provides an additional check. It is usually necessary to know the crystal structure.

Such materials as molybdenum, niobium, tungsten, zirconium, titanium, chromium, iron, and nickel are suitable for examination by field ion emission. A large number of metals have been grown in whisker form from the vapor for examination. It is also possible to vapor-deposit epitaxial layers of up to 500-Å thickness on the hemispherical tip of a tungsten emitter for examination.

If the field applied to the tip is increased by an order of magnitude and made positive, field evaporation occurs and atoms are stripped from the surface as positive ions without the need for thermal activation. This

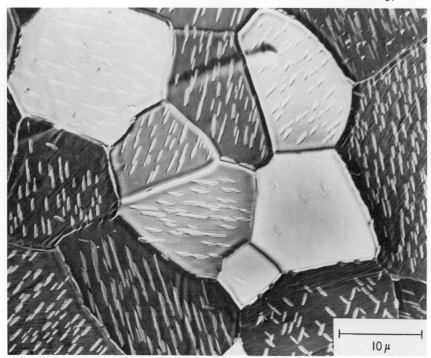

10 μ

Figure 37   Thermionic emission micrograph, showing epitaxial growth of tungsten carbide needles on tungsten by reaction of the hot metal surface with admitted hydrocarbons. Taken at about 1750°C. (Micrograph courtesy of L. Wegmann, Balzers Aktiengesellschaft, Liechtenstein.)

provides an alternate cleaning or atomic smoothing technique. Unfortunately, few metals can withstand the enormous hydrostatic stresses generated. Pulsed field techniques have been successfully employed with relatively poor vacua of about 10^{-9} torr, when it is desired to study clean surfaces. Since recontamination occurs fairly rapidly, it is more common to employ ultrahigh vacua.

The field ion technique has been employed for measurements of the orientation dependence of the activation energy for surface diffusion by observing the migration of material deposited on one side of the tip across the clean surface as a function of temperature. The activation energy for evaporation can be studied as a function of orientation. Processes such as surface segregation, absorption, and corrosion can also be studied.

The reader interested in further detail may refer to reviews by Brandon,[81] Ehrlich,[82] Gomer,[83] Good and Müller,[84] and Gretz.[85]

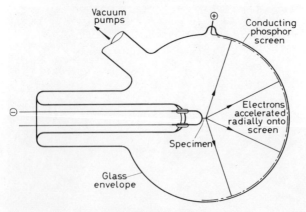

Figure 38 Schematic of field emission microscope. From Brandon.[81] (By permission of Butterworths Publishers.)

B. Field Ion Microscopy

The first field ion micrographs were published by Müller in 1951.[86] For full details of the techniques involved see Müller,[87-90] Brandon,[81,91] Brenner,[92] Ehrlich,[82] Bowkett and Smith,[93] and Ralph and Southon.[94] The technique is similar to field emission, except that the image is formed by ionized gas atoms, rather than by electrons. The resolution of the field emission technique is limited by the tangential velocity component of the emitted electrons and by a diffraction effect associated with the wavelike characteristics of electrons. In the field ion microscope, the emitter is cooled, so that the tangential velocity component is decreased. Diffraction becomes negligible, since the de Broglie wavelength connected with the large ion mass is so small. The resolution is then limited, not by these factors, but rather by the field distribution on the specimen surface and by the fact that the ions originate in free space in a region slightly above the surface.

The tip is conveniently cooled, after Brenner,[92] using a commercially available miniature hydrogen liquifier[95] (Cryotip). Details of the principles, construction, and operation of miniature hydrogen liquifiers may be found elsewhere.[95,96] Brenner's microscope is shown in Figure 39. Using hydrogen and nitrogen supplied from standard gas cylinders, 2 to 5 cc of liquid hydrogen is generated and maintained in a sealed glass or thin-walled, stainless steel portion of the microscope, thus reducing the safety hazards. The use of a standard "O" ring seal permits the removal from, or insertion of, the liquifier into the microscope. The temperature of the specimen holder assembly may be varied between 17 and 80°K by controlling the inlet and exhaust pressures of the gases.

When the field at the tip is increased to approximately 4.5 V/Å by applying voltages of the order of 10 to 20 keV, helium atoms at the tip surface become ionized and are accelerated toward the fluorescent

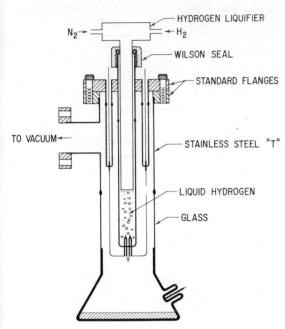

Figure 39   Field ion microscope. From Brenner.[92]

screen which is maintained negative with respect to the tip. The ions used for imaging are not truly emitted from the surface, as the electrons are in the field emission microscope, but are produced by the tunneling of a valence electron from a neutral gas atom 4 to 5 Å above the surface into the tip. Gas atoms, usually helium, from the few microns pressure atmosphere maintained in the tube, approach the tip by kinetic motion, aided by attraction due to polarization in the inhomogeneous field near the tip emitter. They arrive at the surface with an average kinetic energy E_a given by

$$E_a = kT_g + \frac{1}{2}\alpha F^2 \qquad (6)$$

where k is the Boltzmann's constant, T_g the gas temperature, α the polarizability, and F the field strength at the point of impact. After arrival and accommodation to the lower tip temperature T_t, the molecules diffuse over the surface in a hopping motion, being prevented from escaping by the dipole attraction. On encountering a region of enhanced field strength over a protruding atom or a lattice step, they become ionized by tunnelling and are accelerated towards the screen by the high field. The overall field strength (imaging field) F_i, as distinct from the locally enhanced field, is

$$F_i = 3.7 \, V_I^{3/2} \tag{7}$$

where V_I is the ionization energy.

Typical values of F_i are 450, 370, and 230 MV/cm for helium, neon, and hydrogen respectively, that is, about 4 V/Å for helium and neon. It is of interest to note that, if the work function ϕ of the tip material is substituted for F_i, Equation 7 also gives the approximate field strength for the field electron emission, for example, 35 MV/cm for tungsten with $\phi = 4.5$ eV. Since the field at the tip is given approximately by the relation

$$F_i = \frac{V}{5 \, R} \tag{8}$$

where V is the voltage between the tip and the screen, and R is the radius of the tip, voltages in the region of 5 to 50 kV are needed for tip radii between 100 and 3000 Å.

The magnification is approximately

$$M = \frac{r}{R} \tag{9}$$

where r is the distance from tip to screen (typically a few cm), and M is typically about 10^6. Image intensity is increased by raising the gas pressure but this has to be limited to less than 10^{-2} torr to avoid undue scattering and loss of resolution by collision of the ions with other gas atoms on their way to the screen. The total image current is then of the order of 10^{-9} A. The background pressure of contaminant gases is kept below 10^{-6} torr to minimize corrosion of the tip, although most of these contaminant atoms are ionized well away from the specimen and do not reach the surface of the tip.

Since ionization occurs most frequently at high concentrations of the individual surface atoms, where both the field and helium supply are the greatest, these atoms appear as bright spots in the image (Figure best images. Forbes[97] recently made a revolutionary proposal that bright spots depend on the probability of finding an imaging gas atom in the right place to be ionized.

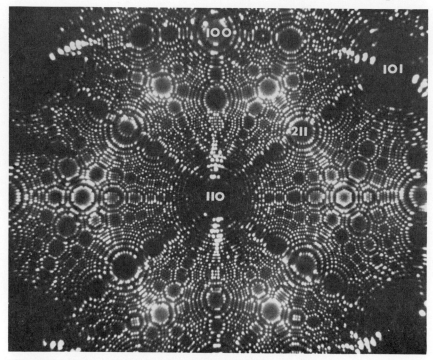

Figure 40 Field ion micrograph of (110) oriented tungsten tip in standard configuration. Surface formed by field evaporation; He image at 21°K; tip radius approximately 550 Å. (Private communication from Müller.)

Experimental difficulties arise because of the low level of image brightness. Thus Brenner,[92] even when employing liquid hydrogen cooling, fast lenses, and high speed recording film (ASA 3000) needed exposure times of approximately 1 min with helium and approximately 15 min when using neon as the imaging gas at 10 kV. This has led to the use of image intensifiers, either combined with the microscope tube[98] or externally.[92] An arrangement using an RCA cascading-image, converting intensifier of the latter type is shown schematically in Figure 41, after Brenner.[92] The microscope image is focused by a fast lens onto the first photocathode, producing photoelectrons that are accelerated and focused by electric and magnetic fields onto a phosphor-coated anode. The final image on the intensifier screen, having undergone three-stage intensification, is viewed through a microscope or recorded with a camera. Because of optical coupling losses, only about 10% of the overall luminous flux gain of some 5×10^4 is realized. This could

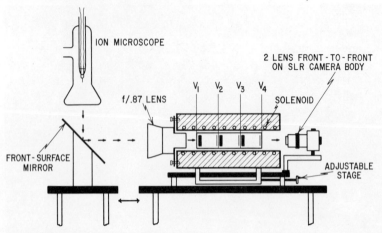

Figure 41 Use of the cascading-image, converting intensifier in conjunction with the ion microscope. From Brenner.[92]

be improved by coupling through fiber glass optics. An example of the reduction in exposure time obtainable is shown in Figure 42. Recently, excellent results have been obtained by Brenner,[99] using a much simpler device consisting of a channel plate which converts each ion to many secondary electrons that are then post accelerated to a phosphor screen, forming a relatively intense image.

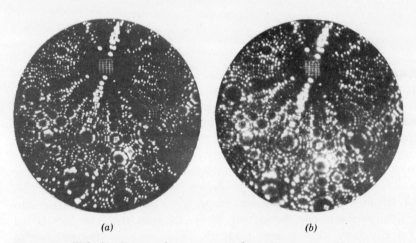

(a)   (b)

Figure 42 Field ion images of a tungsten surface, showing the decrease of exposure time using the intensifier arrangement of Figure 41. (a) Direct photography, no intensification, 6-min exposure time; (b) high gain intensification, 0.2-sec exposure. From Brenner.[92]

A sample can be prepared without thermal disorder of the surface by field evaporation alone, without thermal activation. Materials that can be examined are limited by the fact that field evaporation may occur below the threshold voltage required to produce ionization of the image gas. Thus it is difficult to examine metals such as copper, nickel, and iron using helium gas which has a threshold field of about 4.5 V/Å, as can be seen from Table 1 after Brandon.[81] One solution is to use

Table 1   Calculated Evaporation Field F_n (V/Å) for
Ion Charge $n = 1$, 2, or 3

Element	Atomic Number	F_1	F_2	F_3	Expected Ion
Au	79	4.5	4.3	~6.1	Au^{2+}
Be	4	5.4	4.6	76.0	Be^{2+}
C	6	14.3	8.2	15.8	C^{2+}
Co	27	4.3	3.7	~6.4	Co^{2+}
Cu	29	3.1	4.3	8.1	Cu^+
Fe	26	4.5	3.6	55.4	Fe^{2+}
Ir	77	7.9	~4.0	~5.4	Ir^{2+}
Nb	41	9.7	3.4	~4.4	Nb^{2+}
Ni	28	3.5	3.5	~6.8	Ni^{2+}
Pd	46	3.8	4.1	~6.5	Pd^{2+}
Pt	78	6.3	4.1	~5.4	Pt^{2+}
Re	75	8.2	4.3	~4.0	Re^{2+}
Ta	73	9.3	4.6	~3.8	Ta^{3+}
W	74	10.2	5.7	~4.3	W^{2+}

After Brandon.[81]

an image gas with a lower ionization potential such as neon; but a lower gas pressure must then be used to avoid adsorption on the tip, resulting in a weaker ion current and image. Resort may be made to an image intensifier.

Difficulties in interpretation of field ion images often arise because preferential evaporation of one atom leads to field enhancement at neighboring atoms. This atomic pitting may occur near lattice defects, at impurity atoms, or because of polarization effects at the surface. Field ion images from concentrated solid solution alloys tend to be far more irregular than would be expected if one alloy constituent alone were preferentially evaporated, as a result of changes in other terms in the field evaporation equation.[81]

In order to examine the interior of a specimen, as in studies of defects

Figure 43 Field ion image of ion-damaged platinum hemisphere of an 1800-Å average radius, ionized at 21°K. Single atoms are resolved at some of about 1000 crystal facets. Arrows indicate local lattice damage, due to the impact of heavy ions of 20 keV energy. From Müller.[134]

and the measurement of binding energies, field evaporation is employed to remove the surface atoms layer by layer. The electrostatic field required to do this is much greater than the best imaging field. This suggests the use of a pulsed field technique, in which either single pulses or a repetitive pulsed voltage, just sufficient to cause field evaporation, is superimposed on the best imaging voltage. Brenner[92] employed 10^3 μsec pulses/sec and successfully retained a sharp image, while precisely controlling the rate of field evaporation.

Applications of field ion microscopy include the detection of vacancies and substitutional or interstitial impurities;[87,89,100,101] dislocations;[87,89,100–104] stacking faults;[100,105] damage due either to cold work,[87] fatigue,[87] or irradiation;[87,106–109] grain boundaries;[89,106,110–112] subboundaries;[106] order-disorder;[89,100,106,113–115] and surface diffusion and other studies [see review by Ehrlich[116]]. Other applications are segregation

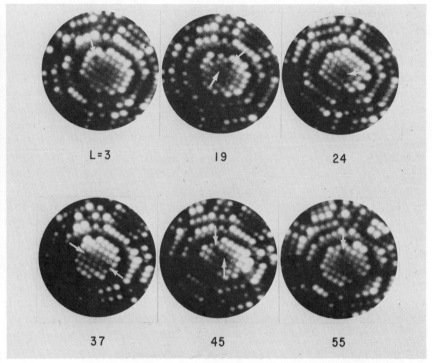

L=3   19   24

37   45   55

Figure 44 Field ion image of (120) plane of annealed platinum, field-evaporated at 21°K to remove 57 successive layers. Layers, 3, 19, 24, 37, 45, and 55, which contained vacancies, are shown. The vacancies were believed created during field evaporation. From Speicher et al.[127] (By permission of North Holland Publishing Company.)

to boundaries;[94,110,117,118] precipitates;[100,119,120] oxide nucleation on tungsten[91] and iridium;[121] faceting;[122] and the study of electrical contact phenomena.[123-124] See also Müller,[125] Hren and Ranganathan,[101] and Müller and Psong.[90] Figure 43 from Müller[134] shows the damage produced in platinum by heavy ions. Work by Speicher et al.[127] indicates that vacancies can be created during field evaporation (Figure 44).

The difficulties in interpretation of field ion micrographs will be appreciated when it is realized that many of the contrast effects are not fully understood and that one is looking for small changes in an area containing many thousands of atoms. A common technique in looking for changes is to superimpose the images from successive photographs, using green and red filters to form a colored image that shows the atoms common to both images in yellow and those present in only one image in red or green.

Ball models of the atom arrangements exposed by intersection of the spherical tip surface with the single crystal tip of a specified orientation have been used as a guide to the interpretation of images.[116] The problem is compounded when attempts are made to visualize the effects of defects such as dislocations, dislocation loops, and stacking faults on the surface arrangement of atoms. Considerable success has been achieved by Brandon and his co-workers[128-130] in the computer simulation of such images.

Although our discussion has been concerned mainly with the atom imaging of the surfaces of pure metals, it will be apparent from the applications already mentioned that substantial progress has been made in studying alloys, at least those of a more refractory kind. Thus studies have been made of ordering in Ni_4Mo,[115] and $PtCo$,[114] of precipitation of vanadium carbide in iron[119] and of tungsten silicide in tungsten,[120] and of interphase boundaries in tungsten-rhenium alloys.[94,118] The discrimination of one kind of atom from another in a solid solution by image appearance may not, in general, be possible; however, in the case of iridium it is claimed[117,121] that oxygen atoms give brighter image spots than iridium atoms.

C. Combination of "Atom Probe" and Field Ion Microscope

If the individual atoms in a field ion image could be identified, the ultimate in microanalysis might be realized. Considerable progress has recently been made in this direction. The approach pioneered by Müller et al.,[131-135] and also used by Brenner and McKinney[136] is to strip the atoms one by one, using a pulsing technique, and to analyze the resulting ion in a time of flight mass spectrometer. Although one cannot be sure which atom will be stripped next, it is possible to correlate a particular atom with a particular ion analyzed. A conventional field ion microscope is used with a 1-mm aperture[131,136] in the imaging screen, through which the atom whose image has been placed over the probe hole will pass when it is field-evaporated. This ion travels on to an ion detector, consisting of a 14-stage electron multiplier located at the end of a 1-meter long drift tube in Müller's equipment[131,132] (Figure 45). The probe hole is at the center of the screen and the atom is brought opposite it by tilting the liquid nitrogen- or liquid hydrogen-cooled microscope head by means of a micrometer-controlled gimbal system, providing ±45° motions.

Since the field stripping pulses and spectrometer signals are both on a time basis, field evaporation can be continued until application of a pulse strips the selected atom, and the appropriate detector output is selected. The use of a time base enables most of the noise problem to be overcome. The ratio of mass to charge m/n is[136]

$$\frac{m}{n} = KVt^2 \tag{10}$$

where K is a system constant, V the applied potential at the tip, and t the time of flight of the field-evaporated ion. In Brenner's first equipment[136] the time of flight measurements could be made in the presence

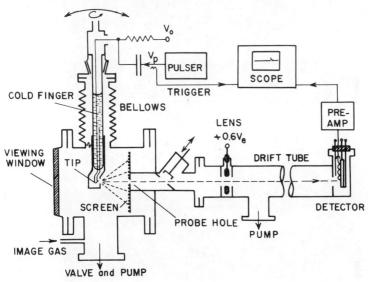

Figure 45 Sectional view of atom probe field ion microscope. From Müller.[137]

of the helium gas, whereas in the prototype equipment of Müller, Panitz, and McLane[131] it was necessary to pump out the gas to get an adequate mean free path. More recently, Müller states[134] that the helium imaging gas can be kept in the microscope during pulsing. The microscope currently used by Brenner[99] employs channel plate image intensification as an additional feature.

The results obtained are illustrated in Figures 46 and 47, after Müller et al.[132,138] Figure 46 shows the probe hole located over an impurity atom at a grain boundary in a doped NS-tungsten wire tip. The large inclusion I is thought to be a particle of dopant. Figure 47 shows the probe hole located on the sixth net plane step off the central 001 plane.

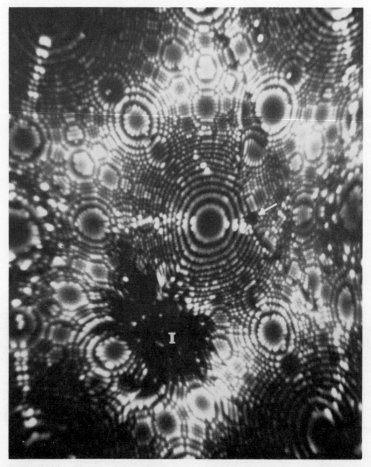

Figure 46 Atom probe, field ion microscope image of tungsten tip of about 420-Å radius, ionized at 78°K. The dark probe hole indicated by the arrow is located over an impurity atom at a grain boundary running roughly vertically. The black area *I* of approximately 100-Å diameter is interpreted as an inclusion. From Müller, McLane, and Panitz.[132]

A vacancy and several dislocations are clearly visible in the original micrograph.

The technique has a great potential for studying such problems as grain boundary segregation, long and short range ordering phenomena, and clustering, since the chemical nature of the species can be determined. Controversies relating to the interpretation of image features such as bright spots obtained by the adsorption of oxygen, nitrogen, and carbon

Figure 47   Atom probe, field ion microscope image of iridium tip ionized at 21°K. The arrow indicates the probe hole, illuminated by the filament of the vacuum gauge in the drift tube. Note the vacancy V and dislocations D. From Müller.[138]

monoxide on tungsten[131] should be settled. The state of ionization of field-evaporated ions can be determined; thus Brenner and McKinney[136] found that tungsten near the (110) plane evaporated predominantly as W^{3+}, whereas iridium evaporated predominantly as Ir^{2+}. Müller et al.[132] found W^{3+} and W^{4+} at 78°K with only W^{3+} from the [100] zone. Ir^{2+} and Ir^{3+} were equally abundant at 21°K, as well as 78°K. Molecular ions were also found[132] to represent oxides and nitrides, when oxygen and nitrogen were adsorbed on the specimen.

VI. OTHER TECHNIQUES

A. "Proton Scattering Microscope"

The "proton scattering microscope," which might better be called a proton blocking camera, holds promise as a complimentary technique to reflection high energy electron diffraction (RHEED) for obtaining simple crystallographic information on thin films. It differs from RHEED in that diffraction effects are not involved. An instrument is now commercially available.[139]

It is well established that a beam of fast ions can be steered along the open channels and planes of the lattice of a suitably oriented crystal. This phenomenon of *channeling* can occur for both light and heavy particles in the energy range from about 10 to 10^7 eV.[140-142] Studies[143,144] on the energy spectrum and spatial distribution of protons, elastically scattered from the surface layers of crystals, showed deficiencies in well-defined crystallographic directions, corresponding to close-packed rows and planes of atoms. These gave rise to characteristic patterns when an x-ray or nuclear film emulsion was placed in the path of the particles,[144-145] provided that the proton energies were greater than about 100 to 200 keV, so that reasonable penetration into the emulsion occurred.

Nelson[146,147] demonstrated that patterns visible to the naked eye could be observed with proton energies of 40 keV or less, using a phosphor such as zinc sulfide to coat the screen, thus making "proton microscopy" practical. The instrumental technique and application were further developed by Livesey and Butcher.[148-150] The analysis of the patterns for cubic crystals has been considered by Tulinov,[145] Chadderton,[151] and by Barrett, Mueller, and White.[152] The theory of the line intensities was discussed by Barrett.[153-154]

Chadderton[151] pointed out that, on a wave interpretation, the blocking patterns can be regarded as merely modified Kikuchi patterns, analogous to the electron case, produced by the elastic scattering of inelastically scattered proton waves. These proton Kikuchi bands move as though rigidly fixed to the crystal, so that the orientation change is accurately revealed by the direction and magnitude of the movement. The interpretation of the pattern is simple, since the lines are simply lines of intersection of crystallographic planes with the plane of the plate.[145]

A suitable equipment for recording proton blocking patterns is shown schematically in Figure 48; a commercially available[139] instrument is shown in Figure 49. The apparatus consists of an ion source and gun, a target (crystal) manipulator, mounted inside a stainless steel vacuum chamber operated at 10^{-4} to 10^{-5} torr, a transparent movable fluorescent screen and a viewing window through which photographs can be taken by means of a camera (Figure 48). Positive ions ($\sim30\%$ protons H^+, $\sim60\%$ H^{2+}, and $\sim10\%$ H^{3+} ions) are produced by a cold cathode d-c discharge in hydrogen, bled in through a needle valve. The ions are accelerated in a three-element ion gun, supplied with 3 to 30 keV.

The beam, typically of 20 keV and about 0.5-mm diameter, passes through a deflecting system, where the protons are separated out by

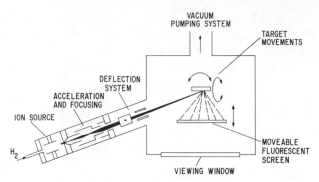

Figure 48 Schematic illustrating the "proton scattering microscope." After Livesey.[155]

a magnetic field, and the proton beam is deflected electrostatically onto a desired region of the crystal or alternatively scanned over the crystal. The crystal target can be rotated through $\pm60°$ in two perpendicular planes by external controls. The fluorescent screen is usually positioned about 15 mm from the crystal. Using the crystal manipulator, the crystal can be brought to within about 1° of an ideal orientation, by observing when the required axis, which is seen as a dark spot, coincides with cross hairs on the screen.

Examples of blocking patterns on (001) crystals of tungsten and copper from Livesey[155] are shown in Figures 50 and 51. A 0.5-mm proton beam diameter and 0.5-μA proton current were employed. Each dark spot represents the intersection of a row of atoms, projected to meet the fluorescent screen. For these cubic crystals, the pattern represents a gnomonic projection. The spot and line widths are proportional to the blocking angle ψ, and ψ is proportional to $d^{-3/4}$, where d is the interatomic distance.[155] Thus the largest darkest spots correspond to the closest packed direction which in tungsten is the $\langle 111 \rangle$ direction (Figure 50) and in copper is the $\langle 110 \rangle$ (Figure 51). The magnification of the blocking pattern

Figure 49 Proton scattering microscope, produced by Edwards High Vacuum International.[139]

can be varied by changing the target-to-screen distance. This distance can be calibrated using a crystal of a known material as the target.

The technique provides a rapid means of orienting crystals, using a goniometer holder and visual inspection of the pattern. Since it is a reflection technique, as normally used, and the penetration of 20 keV protons is typically less than 100 atom layers, epitaxial films may be examined on a substrate or the surface layer of a bulk sample examined. The mis-

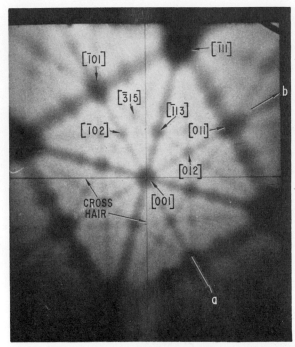

Figure 50 A 20-keV proton blocking pattern from a single (001) crystal of tungsten. From Livesey.[155]

orientation across a boundary may be determined with an accuracy of at least 1°, comparable with the back reflection x-ray technique or the use of electron diffraction cross-grating patterns. The Kikuchi line technique carried out in a transmission electron microscope is capable of much higher precision, partly because of the finer beams employed.

Radiation damage, due to the ion beam, may limit the usefulness of the proton microscopy technique for molecular crystals. In the case of rock salt, using 20-keV beam and sufficient beam current to produce a visible pattern on a fluorescent screen, the image was found[150] to become blurred within a minute or so because of damage. A contamination layer also tended to form on the sample during exposure to the ion beam in the 10^{-4} to 10^{-5} torr vacuum.[150] Polycrystalline samples may be studied, provided that the grain size is larger than the beam size; however, it would not appear easy to relate a particular grain with a particular blocking pattern, since no image of the grains is obtained. Using a calibrated hot stage, changes in crystal structure with temperature could be observed.

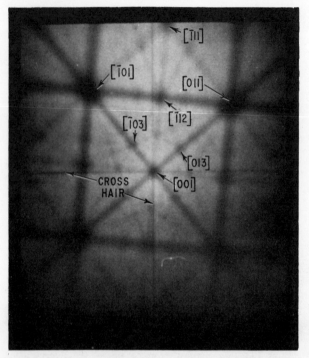

Figure 51 A 20-keV proton blocking pattern from a single (001) crystal of copper. From Livesey.[155]

The annealing characteristics of ion damage have been studied in an ion microscope provided with a hot stage, by first determining the blocking pattern of a silicon crystal with protons; then substituting ions such as 50 keV Ne⁺ to produce an amorphous layer; then switching back to protons and observing the return of crystallinity as a function of time.[146] For quantitative studies of blocking phenomena and for the determination of unknown crystal structures, densitometric studies of the films are necessary. The total charge necessary on the target to provide suitable films is greater the lower the atomic number of the target material, at least for the energy range 50 to 150 keV.[152] The use of small collimators to reduce the line width decreases the charge available, so that some compromise is necessary to obtain reasonable exposure times on nuclear emulsion of a few seconds.[152] The relationship between the proton energy distribution in the blocking pattern and the interaction at various depths in the photographic film is not fully established.[152] There is also difficulty in quantitatively distinguishing axial from planar intensity effects, for example, spots due to close-packed directions, from those due to intersecting lines from close-packed planes.[152] Certainly, there is much scope for further study.

The use of the term "proton scattering microscope" to describe equipment that does not produce an image of an object, but rather a proton blocking pattern, is particularly unfortunate, since a proton microscope is not only technically feasible but has, in fact, been built in prototype form by Magnan and DeBroglie.[156,157] [See the discussion by Hall[158].] The optical system consisted of an objective and two projector lenses of the electrostatic type. A resolution of 300 Å was achieved.[157] The DeBroglie wavelength for protons is much shorter than the electron wavelength for the same beam potential, so that, if the technological problems were solved, a resolving power superior to the conventional electron microscope might be achieved.[158]

B. Mirror Electron Microscope

In the mirror electron microscope, the (polished) sample surface is held at a small negative potential relative to the cathode from which the beam originated. The incident electron beam, typically at about 25-keV potential, is reflected at the equipotential surfaces just in front of the sample surface. Since the electrons travel very slowly while reversing their direction of motion, they are strongly affected by the surface microfield from the sample. The form of these fields depends on the topography, and on the magnetic and electrical properties of the surface. If the sample is, for example, a ferromagnetic or ferroelectric material or a semiconducting P-N junction with an applied bias, the magnetic or electrical field at the sample surface modulates the electron current density of the reflected beams, giving contrast effects in the mirror image. In this way magnetic domains or P-N junctions may be made visible. Surface potential differences of about 0.5 V may be detected. Surface topographical detail can also give rise to contrast effects. The specimen remains largely undisturbed by the electron beam.

The principle of the mirror electron microscope was discovered in Germany over 30 years ago, but most progress has occurred over the last 15 years.[159-172] The initial developments used electrostatic optics and were limited, even with two-stage optics, to magnifications of a few hundred. More recently magnetic lenses have beem employed[167,169,170] and magnifications of a few thousand realized[170] with a step height resolution of 300 Å or better and a lateral resolution of 0.5 μ or better. The lower limit of detectable surface potential is set by the thermal energy of the electrons which requires $\frac{1}{2}$ to 1 volt negative bias on the sample

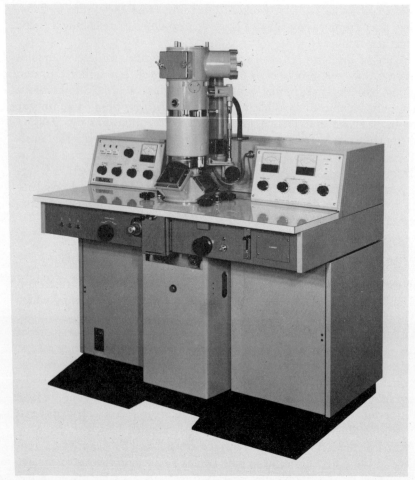

Figure 52   JEM-MI mirror electron microscope. From Someya and Watanabe.[171-172]

to prevent impingement. The closest distance of approach of an electron to the sample surface is limited by the practicable field in front of the specimen to 500 to 1000 Å.[170] The reflected electrons were accelerated onto a phosphor-coated screen in the arrangement of Barnett and Nixon[170] by the objective and two projector lenses. A double condenser was employed.

Somerja and Watanabe[171,172] developed a microscope employing the shadow projection method[170] (Figure 52) which is commercially available.[173] The microscope is capable of magnifications of 20 to approximately 1000 diameters, with a resolving power said to be better than 1000 Å. The electron optical system consists of magnetic condenser, intermediate, and projector lenses (Figure 53). The electron gun assembly

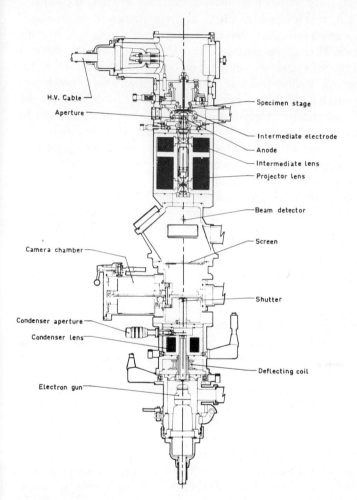

Figure 53 Sectional diagram of JEM-MI mirror electron microscope column. From Someya and Watanabe.[171–172]

is located below the console table. Accelerating voltages of 15 to 35 keV are provided, and the specimen can be biased to ±10 eV, relative to the beam voltage. A beam detector, consisting of a zircaloy crystal that emits light when impacted with electrons, is located on the optical axis above the screen, and facilitates beam alignment. The film camera is located below the screen and viewing chamber, the beam passing through a hole in the film, when it is in the recording position.

The specimen, (removable) intermediate electrode and anode constitute the electron mirror (Figure 53). There are two modes of operation,[174] using either a convergent or a divergent mirror field, the latter giving reversed contrast. If the intermediate electrode is not used, the distance between the specimen and the anode can be made as small as 3 mm without appreciable discharge at 35 keV; using divergent illumination, maximum magnification can be obtained. If the illumination is converged, both the field size and magnification are reduced; however, this can be avoided by inserting the intermediate electrode about 1 mm above the specimen. The geometrical optics of shadow projection imaging are such that sensitivity is gained at the expense of the spatial resolution which is now related to the size of the electron source.[170] This limitation may be overcome by the development of practicable high intensity point sources. Alternatively, point-to-point imaging might be used, although less sensitive, and there has been interest in such development.[175,176]

Mirror electron microscopy is a technique for the examination of surfaces and differs from the other techniques discussed of scanning, reflection, and emission microscopy, in that the sample surface is not appreciably affected by the bombardment of electron or ion beams nor does it emit electrons. Applications include studies of contact potentials, electrical conductivities, magnetic domain structures, and semiconducting and ferroelectric phenomena. By the use of a cryostat for the low temperature range, semiconductor surfaces have been observed at 77 to 1300°K.[168] Superconducting domains have been studied using a helium cryostat,[177] with about 1-μ resolution in bulk samples of lead, tantalum, and vanadium and evaporated lead films. In the case of Nb, Nb_3Sn, and Re, problems in producing a smooth surface were overcome by gluing a 15 μ thick mica sheet to the sample surface and coating with about 1000 Å of silver. This limited the domain resolution to 15 μ. Since dielectric films about 10 Å thick could be detected without destroying their structure, Spivak, Lukyanov, and Abalmazova[178] were able to study evaporation, polymerization, and the charge transfer of impurity films on bombardment by electrons.

The capabilities of the mirror electron microscope equipment (Figures 52 and 53) are illustrated in Figures 54 to 58. Topographical cleavage detail is clearly seen in a convergent mirror image of a cleaved rock salt crystal which was lightly coated with gold[171,172] (Figure 54). Contact potential images between a partial chromium layer on gold (Figure 55)

Figure 54 Convergent mirror image of cleavage surface of rock salt crystal, lightly coated with evaporated gold, showing cleavage steps. From Someya and Watanabe.[171,172]

and a partial gold layer on chromium (Figure 56) enable the sign of the potential to be determined; in the former the high intensity (white) line shows that the chromium film has a positive potential relative to the gold film, whereas in the latter the sign is reversed.

Mirror micrographs of the same area of the (100) oriented surface of a barium titanate crystal, coated with approximately 500 Å of evaporated gold, taken at 80°, 105° and 140°C by Someya, Azumi, and Kobayashi[179] (Figure 57a, b, c) show the important effect of temperature on the domain pattern. The correct choice of bias voltage is important in imaging a P-N junction in silicon. At zero bias, the junction is clearly seen (Figure 58a), but when a bias of -1.5 eV is applied to the sample (Figure 58b) breakdown occurs because of an electron avalanche phenomenon. Although rather a specialized technique, mirror electron microscopy appears to have some unusual and potentially valuable applications.

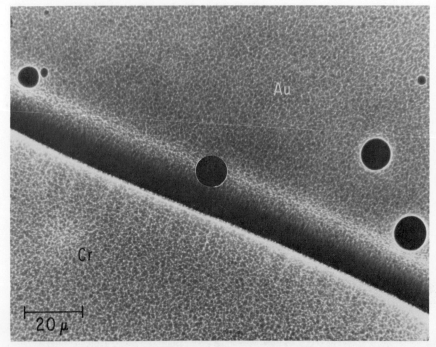

Figure 55   Contact potential mirror image between chromium and gold films. The chromium was evaporated onto the gold film. [Courtesy of JEOLCO (USA).[173]]

C. High Resolution Scanning Transmission Electron Microscope

Remarkable results have been obtained by Crewe and co-workers,[180-182] using a simple scanning transmission electron microscope employing a very fine electron beam and energy selection. A field-emitting tungsten tip,[183] operating in a vacuum of about 10^{-9} torr is used with one lens to provide a very fine electron beam with an effective diameter of about 30 Å. This type of source provides about a 1000-fold increase in brightness over a conventional hot filament, and, since it operates at room temperature, gives a measured energy spread as low as 0.192 eV,[184,185] compared with about 0.9 eV for a conventional filament.[186]

The gun that was used by Crewe, Wall, and Welter[182] to accelerate the electrons from the tip and refocus them into a fine probe is shown in Figure 59. An auxiliary lens is used to demagnify the image produced by the gun (Figure 60). The scanning arrangement is similar to one described by Oatley, Nixon, and Pease[187] and consists of two four-

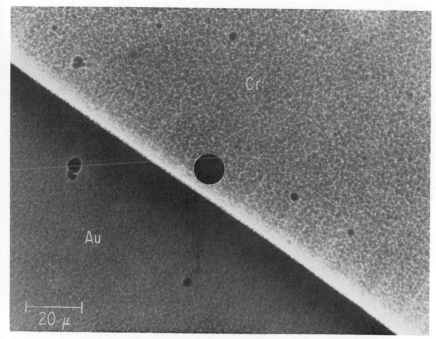

Figure 56 Contact potential mirror image between chromium and gold films. The gold was evaporated onto the chromium film. [Courtesy of JEOLCO (USA).[173]]

quadrant magnetic deflectors (Figure 60). This double-deflection arrangement permits the beam always to pass through the center of the auxiliary lens and gives a raster size on the specimen and hence a magnification independent of the axial position of the intermediate image (Figure 59). Astigmatism was corrected electrically with a double quadruple Rang device.[188]

Electrons transmitted through the sample pass through a variable aperture into a spherical electrostatic energy analyzer and thence through an adjustable slit into a photomultiplier. An image is developed in the usual way on a synchronously scanned cathode ray tube.

With the probe stationary on the sample, an energy loss spectrum can be determined. With the probe in the scanning mode, electrons of selected energy can be used to form a scanning image. The image contrast can be controlled either by adjusting the aperture before the analyzer, giving scattering contrast; or by adjusting the voltage on the analyzer, giving energy-loss contrast. In principle a diffraction pattern can be displayed by using parallel illumination on the sample.

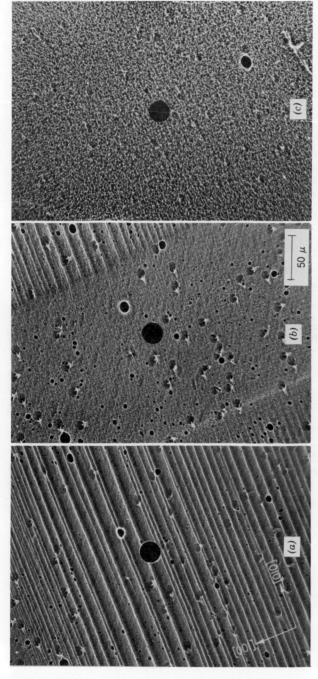

Figure 57 Mirror micrographs of the same area of a (100) surface of barium titanate crystal at (a) 80°C, (b) 105°C, and (c) 140°C. From Someya et al.[179]

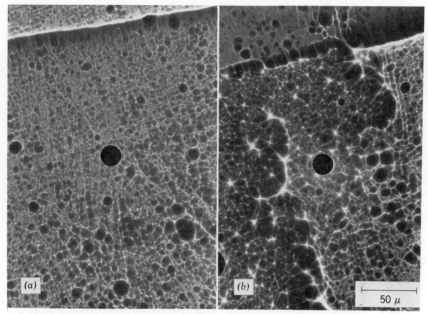

Figure 58 Mirror micrograph of *P-N* junction in silicon. (*a*) Zero bias, (*b*) —1.5 eV bias. [Courtesy of JEOLCO (USA).[173]]

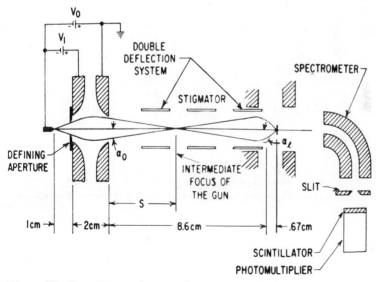

Figure 59 Ray diagram for scanning transmission electron microscope, where V_1 and V_0 are voltages between the tip and the first and second anodes, respectively; α_0 is the beam half-angle at the gun exit; α_1 is the beam half-angle at the auxiliary lens exit. From Crewe et al.[182]

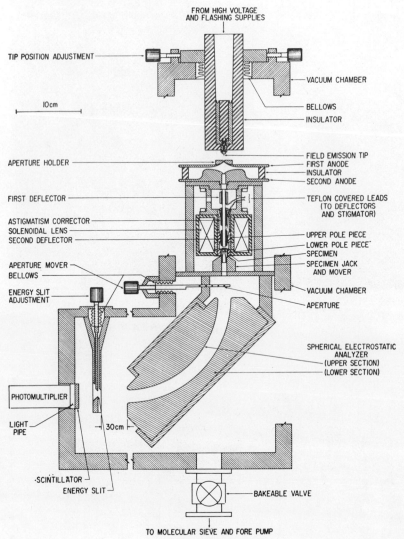

FROM HIGH VOLTAGE
AND FLASHING SUPPLIES

TIP POSITION ADJUSTMENT

VACUUM CHAMBER

10cm

BELLOWS
INSULATOR

FIELD EMISSION TIP
APERTURE HOLDER
FIRST ANODE
INSULATOR
SECOND ANODE

FIRST DEFLECTOR

TEFLON COVERED LEADS
(TO DEFLECTORS
AND STIGMATOR)

ASTIGMATISM CORRECTOR
SOLENOIDAL LENS
SECOND DEFLECTOR

UPPER POLE PIECE
LOWER POLE PIECE
SPECIMEN

APERTURE MOVER
BELLOWS

SPECIMEN JACK
AND MOVER

ENERGY SLIT
ADJUSTMENT

VACUUM CHAMBER
APERTURE

SPHERICAL ELECTROSTATIC
ANALYZER
(UPPER SECTION)
(LOWER SECTION)

PHOTOMULTIPLIER

LIGHT
PIPE

30cm

SCINTILLATOR
ENERGY SLIT

BAKEABLE VALVE

TO MOLECULAR SIEVE AND FORE PUMP

Figure 60   Cross section of scanning transmission electron microscope. From Crewe et al.[182]

A resolution < 100 Å was said to be consistently obtained with the microscope shown in Figure 60 with a best resolution of 30 Å. Recently, the design was slightly modified[189] by shortening the focal length of the single lens from 1 to 0.6 mm to produce a second beam crossover at the exit of the lens, permitting the insertion of a contrast aperture below the lens. With this arrangement, phase contrast imaging is possible

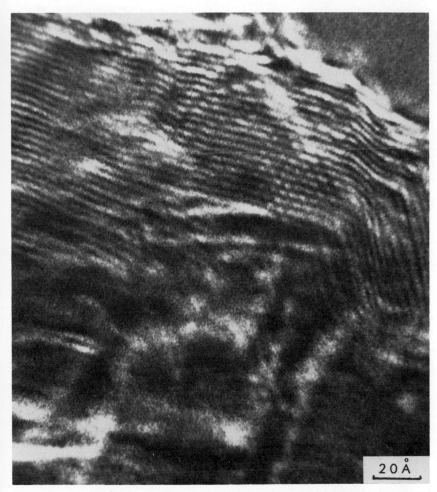

Figure 61   Transmission scanning micrograph of partially graphitized carbon black, showing 3.4-Å period of basal planes. From Crewe and Wall.[189]

and 5 Å point resolution is claimed at 18 keV. The basal planes of the 3.4 Å period were resolved in graphite (Figure 61).[189] The resolution gap between scanning transmission and conventional transmission microscopes has thus been substantially decreased. Since the area scanned has to be made very small, in order to achieve the high magnification needed to image fine detail, it would appear that the conventional transmission microscope will continue to have the advantage of obtaining resolutions < 5 Å over relatively large areas.

D. High Voltage Transmission Scanning Electron Microscope

A simple but ingenious 600-keV transmission scanning electron microscope has been constructed by Cowley and Strojnik[190-195] which appears capable of resolutions < 10 Å. The high voltage supply[191,194] consists of a 12-stage Cockcroft-Walton high voltage generator, electron gun, and 20-stage accelerating column, enclosed in an oil tank with a claimed stability better than 1 part in 10⁵. The pointed filament electron beam

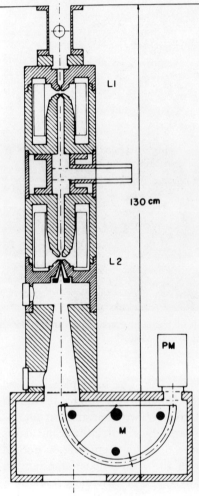

Figure 62 Cross section of the column of a high voltage transmission scanning electron microscope, showing the two electromagnetic lenses L1 and L2 and the magnetic velocity analyzer M. (From Cowley and Strojnik.[190,193,194]

source is demagnified by two, short focal length electromagnetic lenses L1 and L2 (Figure 62) to form a fine crossover at the specimen plane which is situated below the pole-piece of the objective lens L2. The beam can be scanned over the specimen by means of electrostatic deflector plates located within the objective pole-piece. The convergence angle of the beam is adjusted by a variable aperture shown in Figure 63a.

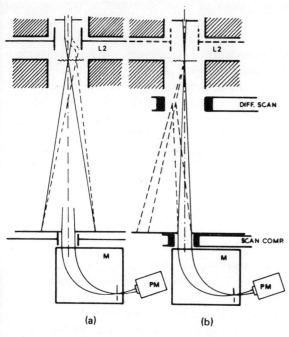

(a)   (b)

Figure 63 Showing principle scanning modes of 600-keV transmission scanning microscope. (a) Beam scanned over sample to produce image; (b) diffraction pattern scanned over aperture of energy analyzer. From Cowley and Strojnik.[190,193,194]

A fluorescent screen is located about 40 cm below the sample and images the transmitted beam as a bright spot surrounded by the diffraction pattern. When a bright field scanning image is desired, the beam is scanned over the sample and the transmitted beam allowed to pass through the aperture into the electromagnetic velocity analyzer M and thence to the scintillator photomultiplier detector PM. The signal of a selected energy is used to form an image on a synchronously scanned cathode ray tube in the usual way. Dark field images are obtained by guiding a part of the diffraction pattern into the analyzer, instead of the transmitted beam. The diffraction pattern can be imaged by scanning the diffraction pattern over the aperture of the analyzer (Figure 63b).

The instrument is of great simplicity, compared with current high voltage transmission microscopes. Only two lenses are necessary; these employ only a hundred turn winding of water-cooled copper tubing, carrying up to 100 A, which makes the lens size smaller (about 8 in. diam. by 10 in. length) and lighter than conventional lenses that employ tens of thousands of turns of wire carrying a few hundred milliamperes.[192,194]

The magnification can be increased to any desired degree by decreasing the specimen area scanned. With suitable electronics, a diffraction pattern and image can be displayed simultaneously on separate cathode ray displays. Unlike (high voltage) transmission electron microscopes, there are no lenses between the sample and the detector, so that no problems arise because of chromatic lens aberrations in conjunction with inelastic scattering in thick samples. In addition energy selection and analysis is provided. With a fine 100-Å-diameter stationary beam, selected area diffraction can be carried out on very tiny areas. By extension of the electronics used with the electron optics indicated in Figure 63b, which has been successfully tried out on a conventional 100-keV microscope, intensities can be measured electronically as an alternative to photographic recording.[194] Convergent beam electron diffraction patterns, and Kossel or Kikuchi line patterns, obtained from < 100 Å-diameter crystal regions, can be recorded by a film camera (not shown in Figure 62) which is located immediately above the viewing screen at the entrance to the energy analyzer (Figure 62).[193]

Although relatively few results have yet been obtained, this prototype instrument offers exciting possibilities and a unique combination of desirable features.

1. Image Contrast

Cowley[194,195] reports that the high resolution images, as in the case of the high resolution scanning transmission electron microscope of Crewe (see page 352), show diffraction effects similar to those seen in conventional transmission electron micrographs, including Fresnel fringes, phase contrast effects, and lattice fringes. That this should be so, is not obvious from geometrical optics but, as Cowley pointed out,[195] is to be expected on the principle of reciprocity. Pogany and Turner[196] discussed reciprocity and showed that it could be applied successfully to n-beam electron diffraction and electron microscopy. The principle is illustrated after Cowley[195] in Figure 64. In a conventional electron microscope (Figure 64a), electrons from a distant point B' strike the sample, the scattered electrons then being focused to an image point A'. In the transmission scanning electron microscope (Figure 64b), the situation is reversed, electrons from a point in the source A being focused on the specimen and the intensity observed at a distant point detector B. On the principle of reciprocity, the intensity at A', due to a source point at B', will be identical with that at B, due to the source point A. The argument can be extended to incoherent sources of finite size.

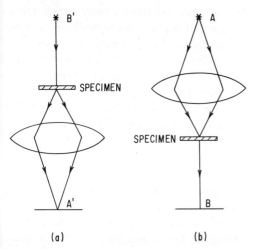

(a)   (b)

Figure 64 Schematic showing basic elements of electron-optical systems for (*a*) conventional transmission electron microscope; (*b*) transmission scanning electron microscope. After Cowley.[195]

Cowley[195] points out that scanning the incident beam over the sample is equivalent to scanning a conventional transmission image over a detector in order to record it. The enlarging of the source A in the scanning instrument is equivalent to the enlarging of the detector A' in a conventional instrument, causing a loss of resolution. Also, increasing the detector area B in the scanning microscope is equivalent to increasing the source size in a conventional microscope; thus the same image contrast is expected in the scanning instrument if the entrance aperture to its detector matches the illumination angle (typically, 10^{-3} radian) in the conventional microscope. Similarly, variation of the lens aperture in the scanning microscope is equivalent in its effect on image contrast to variation in objective aperture size in the conventional transmission microscope.

Cowley[195] concludes that the whole imaging theory developed for the transmission electron microscope (see Chapter 5) will, in general, apply to the interpretation of transmission scanning electron micrographs.

E. Scanning High Energy Electron Diffraction

Grigson[197-199] built the first high energy scanning reflection equipment to employ magnetic scanning of the diffraction pattern across an aperture. Earlier equipment such as that of Horstmann and Meyer[200] employed slower mechanical scanning. Grigson measured the intensity of electrons passing through the aperture by means of a phosphor scintillator, a light pipe, and low noise photomultiplier tube. The intensity was then displayed on an oscilloscope or x-y recorder versus the scanning coil current (distance). Magnetic radial scans could be carried out in 50 msec compared with 300 sec or more with mechanical scanning, and diffraction patterns could readily be scanned in two angular dimensions. Unlike electrostatic scanning, magnetic scanning does not affect focusing, and the angular scale is independent of the beam voltage.[199] An energy filter was added by Grigson, Denbigh, and Nixon[201] which was improved by Denbigh and Grigson,[202] making it possible to reveal diffraction rings with large d spacings previously obscured by loss electrons.

In 1963 a second generation instrument was built by Grigson,[203] specifically for transmission electron diffraction studies on evaporated films, and was used for this purpose by Grigson, Dove, and Stilwell.[204] A further direct recording reflection equipment was built in 1965; see Tompsett and Grigson[205,206] and Tillett and Grigson.[207] This instrument was used by Tompsett, Heritage, and Grigson[208] to study the growth of aluminum, silver, gold, nickel, 80/20 nickel/iron, iron, and germanium films.

The contrast in reflection electron diffraction is greatly improved by filtering out the inelastically scattered electrons.[205,206] The equipment of Tompsett and Grigson was used for a number of further studies.[207-209] The earliest stage of discontinuous film growth by low angle transmission electron diffraction was observed by Tompsett, Heritage, and Grigson[208] and Heritage and Tompsett.[209]

In 1969, Tompsett, Sedgewick, and St. Noble[210] built a horizontal scanning electron diffraction unit, incorporating energy filtering. A similar design of equipment, suitable for both reflection and transmission electron diffraction, is available commercially.[211] The equipment used by Tompsett and his co-workers (Figure 65) was an all metal and glass, bakeable system capable of ultrahigh vacuum operation, consisting of a triode gun, single magnetic lens, specimen stage, magnetic scanning coils, phosphor screen with central hole, energy filter, and electron detector. The filter was of the mesh type, biased relative to the gun cathode, which

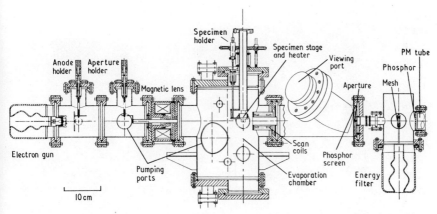

Figure 65 Sectional diagram of scanning high energy electron diffraction (SHEED) instrument. From Tompsett et al.[210] (Courtesy of The Institute of Physics and The Physical Society.)

could be operated up to 50 keV. An energy resolution of 10^{-4}, that is, 5 eV at 50 keV was obtained. Electrons passing through the mesh impinged on a P11 phosphor-coated glass window, viewed external to the vacuum by a photomultiplier. A fast response x-y recorder was used to record the signal (Figure 66).

The large, stainless steel, high vacuum specimen chamber (Figure 65)

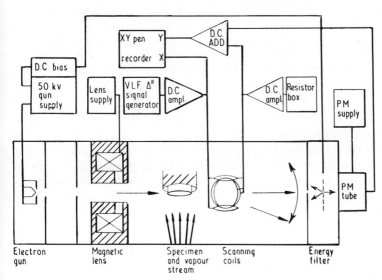

Figure 66 Schematic of scanning high energy electron diffraction instrument. From Tompsett et al.[210] (Courtesy of The Institute of Physics and The Physical Society)

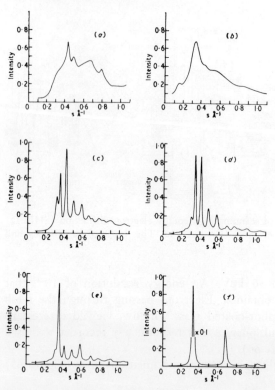

Figure 67 Intensity profiles versus reciprocal d spacing (s), obtained in reflection in SHEED instrument, showing the stepwise growth of lead oxide deposited at 1000 Å/sec in situ on aluminum-coated glass substrate. (a) Substrate only; (b) through (f) after 1,2,3,6, and 90-sec evaporation. From Tompsett et al.[210] (Courtesy of The Institute of Physics and The Physical Society.)

was provided with several ports for pumping and the attachment of vacuum gauges and devices for in situ experiments. Provision was made for heating the specimen stage. A shuttered evaporation source, refillable through an airlock, permitted in situ studies of the growth of evaporated films on a heated substrate. Figure 67 is an example of the results obtained;[210] Figures 67c, d, and e show a mixture of the orthorhombic and tetragonal phases. The effectiveness of this high vacuum electron diffractometer is apparent. For further results on lead monoxide films see Ref. 212.

The same type of equipment may be used with two-dimensional scanning to produce three-dimensional displays similar to those of Tompsett and Grigson[206] (Figure 68). One set of orthogonal coils is used to sweep

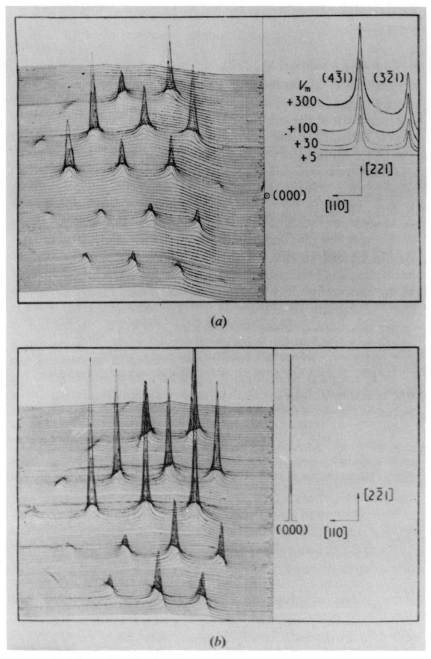

Figure 68 Three-dimensional reflection electron diffractograms of etched single crystal of β-brass, using 15 keV incident beam: (a) using unfiltered signal; (b) using loss-electron filtered signal. Insets show energy distribution across some of the peaks. From Tompsett and Grigson.[206] (Courtesy of The Institute of Physics and The Physical Society.)

the diffraction pattern to and fro along the x axis; the other shifts the pattern orthogonally by a small angular increment once per cycle. The x axis of the plot corresponds to the diffraction angle or reciprocal d-spacing. The y axis corresponds to the sum of the intensity plus the second scattering angle. The display is a set of "vertical" sections of reciprocal space and has complete geometrical correspondence with the original diffraction pattern. Measurements from the top of a peak to the base line of the trace give true intensities. The positions of the projected peaks give direct measurements of the spot spacings that are lattice vectors in reciprocal space. Normal diffraction patterns may be observed by means of a viewing port and fluorescent screen, seen in Figure 65, and photographically recorded.

Scanning high energy electron diffraction has been reviewed by Grigson.[213]

VII. REFERENCES

1. G. Dupouy, Advances in Optical and Electron Microscopy, Vol. 2, R. Barer and V. E. Cosslett, Eds., Academic Press, New York, 1968, pp. 167–250.
2. V. E. Cosslett, Sci. Progr. Oxford, 55, 15 (1967).
3. V. E. Cosslett, Science Journal, 4 (12) 38 (1968).
4. V. E. Cosslett, Physics Today, 21, 23 (1968).
5. G. Dupouy and F. Perrier, J. Microscopie, 1, (3–4) 167 (1962).
6. R. M. Fisher, Metal Progr., 93, February 66 (1968).
7. R. M. Fisher and J. S. Lally in Electron Microscopy 1968, Vol. 1, 4th European Reg. Conf., Rome, D. S. Bocciarelli, Ed., Tipografia Poliglotta Vaticana, Rome, p. 15.
8. G. Reinhold, 1969 National Particle Accelerator Conference, Washington, D.C.
9. G. Dupouy, F. Perrier, R. Fabre, L. Durrieu, and R. Cathelinaud, C. R. Acad. Sci., Paris, Ser B, 269, 867 (1969).
10. V. E. Cosslett, J. Roy. Microscop. Soc., 81, 1 (1962).
11. A. Howie, Phil. Mag., 14, 223 (1966).
12. R. M. Fisher, L. E. Thomas, and J. S. Lally, in Proc. 27th Annual Meeting Electron Microscopy Society of America, 1969, St. Paul, Minn., C. J. Arceneaux, Ed., Claitor's Publishing Division, Baton Rouge, La., 1969, p. 110.
13. L. E. Thomas, C. J. Humphreys, W. R. Duff, and D. T. Grubb, Radiation Effects, Gordon and Breach, New York, 1970.
14. G. Dupouy, F. Perrier, and L. Durrieu, C. R. Acad. Sci., Paris, 251, 2836 (1960).
15. R. K. Ham, Phil. Mag., 7, 1177 (1962).
16. J. C. Grosskrentz and G. G. Shaw, Phil. Mag., 10, 961 (1964).
17. R. K. Ham and M. G. Wright, Phil. Mag., 10, 937 (1964).
18. J. R. Low and A. M. Turkalo, Acta Met., 10, 215 (1962).
19. R. P. Ferrier and I. B. Puchalska, Phys. Status Solidi, 28, 335 (1968).
20. H. Fujita, Y. Kawasaki, E. Furubayashi, S. Kajiwara, and T. Taoka, Japan J. Appl. Phys., 6, 214 (1967).

21. K. F. Hale and M. H. Brown, *Micron*, 1, (4) 434 (1970).
22. S. V. Radcliffe, A. H. Heuer, R. M. Fisher, J. M. Christie, and D. T. Griggs, *Science*, 167, 638 (1970).
23. J. S. Halliday, *Techniques for Electron Microscopy*, 2nd ed., D. H. Kay, Ed., 1965, Blackwell Scientific Publications, Oxford, England, pp. 525–545.
24. E. Ruska, Z. *Phys.*, 83, 492 (1933).
25. B. von Borries and S. Z. Janzen, *Ver. Deut., Ing.*, 85, 207 (1941).
26. D. H. Page, Brit. *J. Appl. Phys.*, 9, 60 (1958).
27. Y. M. Kushnir, L. M. Biberman, and N. P. Levkin, *Bull. Acad. Sci. URSS, Fer. Phys.*, 15, 306 (1951).
28. G. Dupouy and C. Fert, *Proc. 3rd Int. Conf. on Electron Microscopy*, 1954, Royal Microscopical Society, London, 1956, p. 384.
29. J. S. Halliday and R. C. Newman, *Brit. J. Appl. Phys.*, 11, 158 (1960).
30. O. Klemperer and J. P. G. Shepherd, *Brit. J. Appl. Phys.*, 14, 89 (1963).
31. D. R. Spalding and A. J. F. Metherell, *Phil. Mag.*, 18, 41, (1968).
32. G. Möllenstedt and O. Rang, Z. *Angew. Phys.*, 3, 187 (1951).
33. H. Boersch, Z. *Physik*, 134, 156 (1953).
34. O. Rang, *Optik*, 11, 327 (1954).
35. T. Ichinokawa and R. Uyeda, *Proc. Japan Acad.*, 30, 857 (1954).
36. R. Uyeda, *J. Phys. Soc. Japan*, 10, 256 (1955).
37. Y. Kamiya and R. Uyeda, *J. Phys. Soc. Japan*, 16, 1361 (1961).
38. H. Watanabe and R. Uyeda, *J. Phys. Soc. Japan*, 17, 569 (1962).
39. A. J. F. Metherell, "Energy Analyzing and Energy Selecting Electron Microscopes," 1970, in *Advances in Optical and Electron Microscopy*, Vol. 4, R. Barer and V. E. Cosslett, Eds., in press.
40. A. J. F. Metherell, 1965, Ph.D. Thesis, University of Cambridge, England.
41. A. J. F. Metherell, S. L. Cundy, and M. J. Whelan, *Int. Conf. on Electron Diffraction and the Nature of Defects in Crystals*, 1965, Melbourne, published and distributed for the Australian Academy of Science by Pergamon Press, New York, 1966, IN-3.
42. A. J. F. Metherell and M. J. Whelan, *Brit. J. Appl. Phys.*, 16, 1083 (1965).
43. A. J. F. Metherell and M. J. Whelan, *J. Appl. Phys.*, 37, 1737 (1966).
44. S. L. Cundy, A. J. F. Metherell, and M. J. Whelan, *J. Sci. Instrum.*, 43, 712 (1965).
45. A. J. F. Metherell, *Optik*, 25, 250 (1967).
46. G. Möllenstedt, *Optik*, 5, 499 (1949).
47. A. J. F. Metherell, private communication.
48. S. L. Cundy, A. J. F. Metherell, and M. J. Whelan, *6th Int. Congress on Electron Microscopy*, 1966, Kyoto, Vol. 1, R. Uyeda, Ed., Marwzen Co., Tokyo, p. 87.
49. S. L. Cundy and P. J. Grundy, *Phil. Mag.*, 14, 1233 (1966).
50. S. L. Cundy, A. J. F. Metherell, and M. J. Whelan, *Phil. Mag.*, 17, 141 (1968).
51. S. L. Cundy, Ph.D. Thesis, 1968, University of Cambridge, England.
52. D. R. Spalding, Ph.D. Thesis, University of Cambridge, England, 1970.
53. S. L. Cundy, A. J. F. Metherell, M. J. Whelan, P. N. T. Unwin, and R. B. Nicholson, *Proc. Roy. Soc., Ser. A*, 307, 267 (1968).
54. D. R. Spalding, R. E. Villagrana, and G. A. Chadwick, *4th European Reg. Conf. Elec. Microscopy*, 1968, Rome, Vol. 1, D. S. Bocciarelli, Ed., Tipografia Poliglotta Vaticana, p. 347.
55. D. R. Spalding, R. E. Villagrana, and G. A. Chadwick, *Phil. Mag.*, 20, 471 (1969).

56. D. R. Spalding, J. W. Edington, and R. E. Villagrana, *Phil. Mag.,* **20,** 1203 (1969).
57. S. L. Cundy, A. J. F. Metherell, and M. J. Whelan, *Phil. Mag.,* **15,** 623 (1967).
58. R. F. Cook and A. Howie, *Phil. Mag.,* **20,** 641 (1969).
59. S. L. Cundy, A. Howie, and U. Valdré, *Phil. Mag.,* **20,** 147 (1969).
60. M. H. Hansen, *Constitution of Binary Alloys,* 2nd ed., McGraw-Hill Book Co., New York, 1958.
61. R. Castaing and L. Henry, (a) *C. R. Acad. Sci., Paris,* **255,** 76 (1962); (b) *J. Microscopie,* **2,** 5 (1963); (c) *J. Microscopie,* **3,** 133 (1964).
62. H. Watanabe and R. Uyeda, *J. Phys. Soc. Japan,* **17,** 569 (1962).
63. H. Watanabe, *Japan. J. Appl. Phys.,* **3,** (8), 480 (1964).
64. H. Watanabe, *X-Ray Optics and Microanalysis,* 1965, Orsay, R. Castaing, P. Deschamps, and J. Philibert, Eds., Hermann & Cie, Paris, 1966.
65. H. Watanabe, *Hitachi Rev.,* **14,** (6) 20 (1965).
66. Y. Kamiya and R. Uyeda, *J. Phys. Soc. Japan,* **16,** 1361 (1961).
67. H. Watanabe, *Japan. J. Appl. Phys.,* **6,** 808 (1967).
68. L. Wegmann, *Praktische Metallographie,* **5,** (5), 241 (1968).
69. E. Eichen, (a) *Techniques of Metals Research,* Vol. 2, Part 1, R. F. Bunshah, Ed., Interscience Publishers, New York, 1968, pp. 177–219; (b) *High-Temperature High-Resolution Metallography,* H. I. Aaronson and G. S. Ansell, Eds., Gordon and Breach, New York, 1967, pp. 167–216.
70. G. Möllenstedt and F. Lenz, *Advances in Electronics and Electron Physics,* Vol. 18, Academic Press, New York, 1963, pp. 251–329.
71. C. Zaminer, R. Graber, and L. Wegmann, *J. Vac. Sci. Technol.,* **6,** (1), 269 (1969).
72. D. G. Brandon, *Modern Techniques in Metallography,* Van Nostrand Reinhold Co., New York, 1966.
73. W. L. Grube and S. R. Rouze, *High-Temperature High-Resolution Metallography,* H. I. Aaronson and G. S. Ansell, Eds., Gordon and Breach, New York, 1967, pp. 313–346.
74. H. Düker, *Z. Metallk.,* **51,** (6), 314 (1960).
75. H. Düker, *Z. Metallk.,* **51,** (7), 377 (1960).
76. S. R. Rouze and W. L. Grube, *5th Int. Congress for Electron Microscopy,* 1962, Vol. 1, S. S. Breese Jr., Ed., Academic Press, New York, 1962, CC-6.
77. J. Nutting and S. R. Rouze, *5th Int. Congress for Electron Microscopy,* 1962, Vol. 1, S. S. Breese Jr., Ed., Academic Press, New York, 1962, CC-7.
78. M. Auwarter, H. K. Pulker, and C. Zaminer, *Z. Physik,* **224,** 298 (1969).
79. L. Wegmann, *Z. Angew. Phys.,* **27,** (3), 199 (1969).
80. E. W. Müller, *Z. Physik,* **37,** 838 (1936).
81. D. G. Brandon, *Modern Techniques in Metallography,* 1st ed., Van Nostrand Reinhold Co., New York, 1966, pp. 177–222.
82. G. Ehrlich, *Advances in Catalysis,* Vol. 14, Academic Press, New York, 1963, pp. 255–427.
83. R. Gomer, *Field Emission and Field Ionization,* Harvard University Press, Cambridge, Mass., 1961.
84. R. H. Good and E. W. Müller, *Handbuch der Physik,* Vol. 21, S. Flügge, Ed., Springer-Verlag, Berlin, 1956, pp. 176–231.
85. R. D. Gretz, in *High-Temperature High-Resolution Metallography,* Vol. 38, H. I. Aaronson and G. S. Ansell, Eds., Metallurgical Society Conferences, Gordon and Breach, New York, 1967, pp. 63–166.
86. E. W. Müller, *Z. Physik,* **131,** 136 (1951).

87. E. W. Müller, *Advances in Electronics and Electron Physics*, Vol. 13, Academic Press, New York, 1960, pp. 83–179.
88. E. W. Müller, *Direct Observation of Imperfections in Crystals*, J. B. Newkirk and J. H. Wernick, Eds., Interscience Publishers, New York, 1962, pp. 77–99.
89. E. W. Müller, Proc. Int. Conf. on Crystal Lattice Defects, Tokyo, 1962, *J. Phys. Soc. Japan, Suppl. II*, 18, 1–16 (1963).
90. E. W. Müller and Psong, *Field Ion Microscopy—Principles and Applications*, American Elsevier, New York, 1969.
91. D. G. Brandon, in *High-Temperature High-Resolution Metallography*, Vol. 38, H. I. Aaronson and G. S. Ansell, Eds., Metallurgical Society Conferences, Gordon and Breach, New York, 1967, pp. 1–62.
92. S. S. Brenner, *High-Temperature High-Resolution Metallography*, H. I. Aaronson and G. S. Ansell, Eds., Gordon and Breach, New York, 1967, p. 281.
93. K. M. Bowkett and D. A. Smith, *Field-Ion Microscopy*, American Elsevier Publishing Co., New York, 1970.
94. B. Ralph and M. J. Southon, *Science Journal*, 2, 50 (1966).
95. Cyro-Tip Instruction and Operating Manual, Air Products and Chemical, Allentown, Pa.
96. B. Yates and F. E. Hoare, *Cryogenics*, 2, 84 (1961).
97. B. G. Forbes, *Nature*, 230, 165 (1971).
98. D. G. Brandon, S. Ranganathan, and D. S. Whitmell, *Brit. J. Appl. Phys.*, 15, 55 (1964).
99. S. S. Brenner, private communication, 1970.
100. D. G. Brandon, *Brit. J. Appl. Phys.*, 14, 474 (1963).
101. J. J. Hren and S. Ranganathan, Eds., *Field-Ion Microscopy*, based on Invited Lectures, March 1966, at University of Florida, Gainesville, Plenum Press, New York, 1968.
102. D. A. Smith, R. Morgan, and B. Ralph, *Phil. Mag.*, 18, 869 (1968).
103. M. A. Fortes and B. Ralph, *Phil. Mag.*, 18, 787 (1968).
104. M. A. Fortes, D. A. Smith, and B. Ralph, *Phil. Mag.*, 17, 169 (1968).
105. D. A. Smith, M. A. Fortes, A. Kelly, and B. Ralph, *Phil. Mag.*, 17, 1065 (1968).
106. D. G. Brandon, M. Wald, M. J. Southon, and B. Ralph, *Acta Met.*, 12, 324–331 (1964).
107. M. A. Fortes and B. Ralph, *Phil. Mag.*, 14, 189 (1966).
108. K. M. Bowkett, L. T. Chadderton, H. Norden, and B. Ralph, *Phil. Mag.*, 15, 415 (1967).
109. B. Ralph, J. A. Hudson, and R. S. Nelson, *Phil. Mag.*, 18, 839 (1968).
110. B. Ralph and D. G. Brandon, *Phil. Mag.*, 8, 919 (1963).
111. S. S. Brenner, *Metal Surfaces*, 1962, W. D. Robertson and N. A. Gjostein, Eds., American Society for Metals, Cleveland, 1963, pp. 305–330.
112. B. Morgan and B. Ralph, *Acta Met.*, 15, 341 (1967).
113. D. G. Brandon, B. Ralph, S. Ranganathan, and M. S. Wald, *Acta Met.*, 12, 813 (1964).
114. H. N. Southworth and B. Ralph, *Phil. Mag.*, 14, 383 (1966).
115. B. Ralph, B. G. LeFevre, and H. Grenga, *Phil. Mag.*, 18, 1127 (1968).
116. G. Ehrlich, in *Surface Phenomena of Metals*, Monograph No. 28, Society of Chemical Industries, 1968, pp. 13–38.
117. M. A. Fortes and B. Ralph, *Acta Met.*, 15, 707 (1967).
118. B. Ralph and D. G. Brandon, *Phil. Mag.*, 8, 919 (1963).
119. D. M. Schwartz, A. T. Davenport, and B. Ralph, *Phil. Mag.*, 18, 431 (1968).
120. R. Morgan, R. G. Faulkner, and B. Ralph, *J. Iron Steel Inst. (London)*, 204, 943 (1966).

121. M. A. Fortes and B. Ralph, *Proc. Roy. Soc.*, *Ser. A*, **307**, 431 (1968).
122. S. S. Brenner, *Surface Sci.*, **2**, 496 (1964).
123. O. Nishkikawa and E. W. Müller, *Proc. Holm Seminar on Electric Contact Phenomena*, Illinois Institute of Technology, Chicago, Ill., November 1968, p. 193.
124. E. W. Müller and O. Nishikawa, "Adhesion or Cold Welding of Materials in Space Environment," Special Technical Publication No. 431, American Society for Testing and Materials, Philadelphia, 1968, p. 67.
125. E. W. Müller, in *Vacancies and Interstitials in Metals*, North Holland Publishing Co., Amsterdam, 1969, pp. 557–572.
126. E. W. Müller, *Science*, **149**, 591 (1965).
127. C. A. Speicher, W. T. Pimbley, M. J. Attardo, J. M. Galligan, and S. S. Brenner, *Phys. Lett.*, **23**, (3), 17 (1966).
128. D. G. Brandon and A. J. Perry, *Phil. Mag.*, **16**, 131 (1967).
129. D. G. Brandon and A. J. Perry, *Phil. Mag.*, **17**, 255 (1968).
130. D. G. Brandon and A. J. Perry, *Phil. Mag.*, **18**, 353 (1968).
131. E. W. Müller, J. A. Panitz, and S. B. McLane, *Rev. Sci. Instrum.*, **39**, (1), 83 (1968).
132. E. W. Müller, S. B. McLane, and J. A. Panitz, *4th European Regional Conf. on Electron Microscopy*, Rome, 1968, Vol. 1, D. S. Bocciarelli, Ed., Tipografia Poliglotta Vaticana, Rome, 1968, p. 135.
133. E. W. Müller, in *Molecular Processes on Solid Surfaces*, Battelle-Kronberg Colloquium, May 1968, E. Drauglis, R. D. Gretz, and R. I. Jaffee, Eds., McGraw-Hill Book Co., New York, p. 67.
134. E. W. Müller, "Field-Ion Microscopy and the Electronic Structure of Metal Surfaces," *Quart. Rev.*, Chemical Soc., London, **23** (2), 177–186 (1969).
135. E. W. Müller, S. B. McLane, and J. A. Panitz, *Surface Sci.*, **17**, (2), 430 (1969).
136. S. S. Brenner and J. T. McKinney, *Appl. Phys. Lett.*, **13**, (1), 29 (1968).
137. E. W. Müller, private communication.
138. E. W. Müller, *Naturwiss.*, **57**, 222 (1970).
139. Edwards High Vacuum International, Crawley, Sussex, England.
140. R. S. Nelson and M. W. Thomson, *Phil. Mag.*, **8**, 1677 (1963).
141. J. A. Davies, G. R. Piercy, F. Brown, and M. McCargo, *Phys. Rev. Lett.*, **10**, 399 (1963).
142. S. Datz, C. Erginsay, G. Liebfried, and H. O. Lutz, *Ann. Rev. Nucl. Sci.*, **17**, 129 (1967).
143. E. Bøgh and E. Uggerhøj, *Phys. Lett.*, **17**, 116 (1965).
144. A. F. Tulinov, V. S. Kulikauskas, and M. M. Malov, *Phys. Lett.*, **18**, 304 (1965).
145. A. F. Tulinov, *Sov. Phys. Usp.*, **8**, (6), 864 (1966).
146. R. S. Nelson, *Phil. Mag.*, **15**, 845 (1967).
147. R. S. Nelson, *New Scientist*, **34**, 138 (1967).
148. R. G. Livesey and G. Butcher, *Proc. 4th Int. Vacuum Congress*, 203 (1968).
149. R. G. Livesey and G. Butcher, *J. Sci. Instrum.*, (*J. Phys. E*) **1**, (2), 820 (1968).
150. R. G. Livesey, *Industrial Research*, 62 (March 1969).
151. L. T. Chadderton, *Physics Lett.*, **23**, 303 (1966).
152. C. S. Barrett, R. M. Mueller, and W. White, *J. Appl. Phys.*, **39**, (10), 4695 (1968).
153. C. S. Barrett, *Advances in X-Ray Analysis*, Vol. 12 (Proc. Denver Conf. on

Applications of X-Ray Analysis, 1968), Plenum Press, New York, 1969, p. 72.

154. C. S. Barrett, *Trans. Met. Soc. AIME*, **245**, 429 (1969).
155. R. G. Livesey, private communication, Central Research Laboratories, Edwards High Vacuum International.
156. C. Magnan, *Nucleonics*, **4**, 52 (1949).
157. P. Chanson and C. Magnan, *Compt. Rend.*, **233**, 1436 (1951).
158. C. E. Hall, *Introduction to Electron Microscopy*, McGraw-Hill Book Co., New York, 1953, p. 186.
159. G. Bartz, G. Weissenberg, and D. Wiskott, *Proc. London Conf. on Electron Microscopy*, Royal Microscopical Society, London, 1954, pp. 395–404.
160. I. Mayer, *J. Appl. Phys.*, **26**, 1228 (1955).
161. L. Mayer, *J. Appl. Phys.*, **28**, 975 (1957).
162. H. Bethge, J. Hellgardt, and J. Heydenreich, *Exp. Tech. Phys.*, **8**, 49 (1960).
163. G. Spivak, I. A. Pryamkova, D. V. Felison, A. N. Kabava, L. Lazareva, and A. I. Shilina, *Bull. Acad. Sci. USSR, Phys. Ser.*, **25**, 698 (1961).
164. L. Mayer, *J. Phys. Soc. Japan, Suppl. B-I*, Proc. Int. Conf. on Magnetism and Crystallography, **17**, 547 (1961–1962).
165. G. Bacquet and A. Santouil, *Compt. Rend.*, **255**, 1263 (1962).
166. G. V. Spivak and R. D. Ivanov, *Izv. Akad. Nauk SSSR, Ser. Fiz.*, **27**, (9), 1203 (1963).
167. M. E. Barnett and W. C. Nixon, *Proc. European Reg. Conf. on Electron Microscopy*, Vol. A, Czechoslovak Academy of Science, Prague, 1964, pp. 37–38.
168. E. Igras and T. Warminski, *Phys. Status Solidi*, **9**, (1), 79 (1965).
169. M. E. Barnett and W. C. Nixon, *6th Int. Congress for Electron Microscopy*, Kyoto, 1966, Maruzen, Tokyo, p. 231.
170. M. E. Barnett and W. C. Nixon, *J. Sci. Instrum.*, **44**, 893 (1967).
171. T. Someya and M. Watanabe, *4th European Reg. Conf. on Electron Microscopy*, Vol. 1 Rome, 1968, D. S. Bocciarelli, Ed., Tipografia Poliglotta Vaticana, p. 97.
172. T. Someya and M. Watanabe, *Proc. 26th Annual Meeting Electron Microscopy Society of America*, New Orleans, 1968, C. J. Arceneaux, Ed., Claitor's Publishing Division, Baton Rouge, La., 1968, pp. 300–301.
173. JEOLCO (USA), Medford, Mass.
174. M. E. Barnett and W. C. Nixon, *Optik*, **26**, 310 (1967).
175. A. B. Bok, J. Kramer, and J. B. LePoole, *Proc. 3rd European Reg. Conf. on Electron Microscopy*, Prague, 1964, Vol. A, M. Titbach, Ed., Czechoslovak Academy of Sciences, Prague, 445.
176. W. Schwartze, *Naturwissenschaften*, **15**, 448 (1965).
177. O. Bostanjoglo and G. Siegel, *Cryogenics*, **7**, (3), 157 (1967).
178. G. V. Spivak, A. E. Lukyanov, and M. G. Abalmazova, *Izv. Akad. Nauk SSSR, Ser. Fiz.*, **28**, (8), 1382 (1964).
179. T. Someya, T. Azumi, and J. Kobayashi, *J. Phys. Soc. Japan, Suppl.*, **28**, 374 (1970).
180. A. V. Crewe, *Science*, **154**, 729 (1966).
181. A. V. Crewe, *Proc. 6th Int. Congress for Electron Microscopy*, Kyoto, 1966, Vol. 1, R. Uyeda, Ed., Maruzen Co. Tokyo, 1966, p. 625.
182. A. V. Crewe, J. Wall, and L. M. Welter, *J. Appl. Phys.*, **39**, 5861 (1968).
183. A. V. Crewe and M. Isaacson, *Proc. 26th Annual Electron Microscopy Society*

372   Specialized Electron Microscopy

of America Meeting, 1968, New Orleans, C. J. Arceneaux, Ed., Claitor's Publishing Division, Baton Rouge, La., p. 359.
184. R. D. Young, *Phys. Rev.,* 113, 110 (1959).
185. R. D. Young and E. W. Müller, *Phys. Rev.,* 113, 115 (1959).
186. G. Möllenstadt and H. Dücker, *Z. Naturforsch.,* 89, 89 (1953).
187. C. W. Oatley, W. C. Nixon, and R. F. W. Pease, *Advan. Electron Phys.,* 21, 181 (1965).
188. O. Rang, *Optik,* 5, 518 (1949).
189. A. V. Crewe and J. Wall (a) *Proc. 27th Annual Electron Microscopy Society of America Meeting,* 1969, St. Paul, Minn., C. J. Arceneaux, Claitor's Publishing Division, Baton Rouge, La., 1969, p. 172; (b) *J. Mol. Biol.,* 48, 375 (1970).
190. J. M. Cowley and A. Strojnik, *4th European Reg. Conf. on Electron Microscopy,* Rome, 1968, Vol. 1, D. S. Bocciarelli, Ed., Tipografia Poliglotta Vaticana, p. 71.
191. A. Strojnik, *4th European Reg. Conf. on Electron Microscopy,* Rome, 1968, Vol. 1, D. S. Bocciarelli, Ed., Tipografia Poliglotta Vaticana, p. 31.
192. A. Strojnik, *Proc. 27th Annual Electron Microscopy Society of America Meeting,* St. Paul, Minn., 1969, C. J. Arceneaux, Ed., Claitor's Publishing Division, Baton Rouge, La., 1969, p. 104.
193. J. M. Cowley and A. Strojnik, *Proc. 27th Annual Electron Microscopy Society of America Meeting,* St. Paul, Minn., 1969, ed. C. J. Arceneaux, Ed., Claitor's Publishing Division, Baton Rouge, La., 1969, p. 106.
194. J. M. Cowley, *Aust. Physicist,* 6, (12), 183 (1969).
195. J. M. Cowley, *Appl. Phys. Lett.,* 15, (2), 58 (1969).
196. A. P. Pogany and P. S. Turner, *Acta Cryst.* A24, 103 (1968).
197. C. W. B. Grigson, *Nature,* 119, 647 (1961).
198. C. W. B. Grigson, Kyoto Conf. (1961) *J. Phys. Soc. Japan,* 7B, (II), 298 (1962).
199. C. W. B. Grigson, *Electron Control,* 12, 209 (1962).
200. M. Horstmann and G. Meyer, *Z. Physik,* 159, 563 (1960).
201. C. W. B. Grigson, P. N. Denbigh, and W. C. Nixon, 5th Int. Congress for Electron Microscopy, Philadelphia, Pa., 1962, Vol. 1, S. S. Breese, Jr., Ed., Academic Press, New York, Paper JJ7.
202. P. N. Denbigh and C. W. B. Grigson, *J. Sci. Instrum.,* 42, 305 (1965).
203. C. W. B. Grigson, *Rev. Sci. Instrum.,* 36, 1587 (1965).
204. C. W. B. Grigson, D. B. Dove, and G. R. Stilwell, *Nature,* 205, 1198 (1965).
205. M. F. Tompsett and C. W. B. Grigson, *Nature,* 206, 923 (1965).
206. M. F. Tompsett and C. W. B. Grigson, *J. Sci. Instrum.,* 43, 430 (1966).
207. P. I. Tillett and C. W. B. Grigson, *Nature,* 214, 77 (1967).
208. M. F. Tompsett, M. B. Heritage, and C. W. B. Grigson, *Nature,* 215, 498 (1967).
209. M. B. Heritage and M. F. Tompsett, *J. Appl. Phys.,* 41, 407 (1970).
210. M. F. Tompsett, D. E. Sedgewick, and J. St. Noble, *J. Sci. Instrum. (J. Phys. E),* 2, Ser. 2, 587 (1969).
211. Vacuum Generators, Sussex, England, marketed in the United States by Veeco Instruments, Plainview, N.Y.
212. M. F. Tompsett and J. St. Noble, *Thin Solid Films,* 5, 81 (1970).
213. C. W. B. Grigson, "Advances in Electronics and Electron Physics," Suppl. 4, Electron Beam and Laser Beam Technology, 1969, pp. 187–290.

7

X-RAY MICROSCOPY

I. Introduction 374

II. X-ray Microradiography 374
    A. Introduction 374
    B. Factors Affecting Contrast 376
        1. Absorption 376
        2. Choice of Radiation 380
    C. Factors Affecting Resolution 382
        1. Photographic 382
        2. Geometrical Blurring 384
        3. Fresnel Diffraction 386
    D. Contact Microradiography 387
    E. Point Projection Microradiography 388
    F. Quantitative Microradiography 390

III. Autoradiography 393
    A. Introduction 393
    B. Methods of Adding Activity 394
    C. Factors Affecting Resolution 396
        1. Particle Energy 396
        2. Geometrical Factors 398
        3. Photographic Grain Size 398
    D. Emulsions for Autoradiography 399

IV. X-ray Methods for Direct Imaging of Defects 401
    A. Introduction 401
    B. Back Reflection Berg-Barrett Method 402
    C. Lang Transmission Method 407
    D. Borrman Anomalous Transmission Method 409
    E. Instantaneous Video Display of X-ray Topographs 411

V. Scanning X-ray Distribution Microscopy 415

VI. References 416

I. INTRODUCTION

Since the refractive index of matter for x-rays is only a few parts in 10^5 less than unity, the focal length of any lens with a useful degree of magnification would have to be several kilometers, so that a refracting microscope of the type used for light is impractical. Extensive efforts have been made[1-3] to develop a microscope based on reflecting optics, but the technical problems such as the aberrations consequent upon grazing incidence and the smoothness of the mirrors necessary to get good resolution, have proved discouraging. Magnifications of only about 100X and resolution of only 1 to 2 μ over a very limited field have been achieved. Commonly employed methods of x-ray microscopy, therefore, differ substantially from those of light microscopy. The methods that will be discussed are x-ray microradiography (also known as x-ray absorption micrography), autoradiography, x-ray methods for direct imaging of defects (x-ray topography), and scanning x-ray distribution microscopy. Neutron radiography is excluded from discussion, since microscopic detail is not at present revealed. X-rays interact with atomic electrons, so that x-ray attenuation by matter is a monotonically increasing function of atomic number. Neutrons interact with atomic nuclei, so that neutron attenuation is a function of the nuclear properties and hence is randomly related to the atomic number. Since neutrons interact strongly with hydrogen nuclei, metallic hydrides and other hydrogeneous materials such as plastics, string, and hair may be detected, even when sandwiched between metals; also, hydrogen embrittlement can be studied. Likewise neutron-opaque boron can be detected. Certain heavy elements such as uranium and plutonium can be distinguished. Radioactive materials that would expose a photographic emulsion can be studied. The reader interested in neutrography is referred to Ref. 4 through 8.

II. X-RAY MICRORADIOGRAPHY

A. Introduction

Within five years of the discovery of x-rays in 1895 by Roentgen,[9] Heycock and Neville[10,11] demonstrated that additional detail of segregation structures in cast gold-sodium and gold-aluminum alloys could be revealed by making contact radiographs, using white radiation, and enlarging them optically. Little further development of this new technique of x-ray microradiography occurred until the late 1930s when fine grained emulsions became available, permitting much greater useful enlargements. While much can be done with little more than an x-ray source unit and an enlarger, a variety of x-ray microscopes are now commercially available that employ the point projection technique first developed in practical form in 1951 by Cosslett and Nixon.[12] The basic difference between these two systems is illustrated in Figure 1.

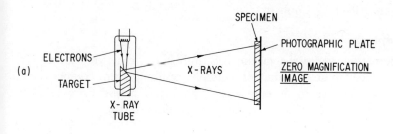

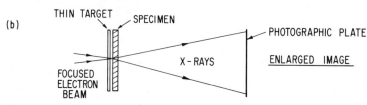

Figure 1 Schematic arrangements for (*a*) contact x-ray microradiography; (*b*) point projection x-ray microradiography.

The commonly used (low voltage) microradiography techniques, in which the image is produced by x-rays typically of 60 keV or less that are transmitted through the sample to a photographic emulsion, should be distinguished from the so-called "high-voltage microradiography" developed since 1940 by Trillat. In the latter,[13,14] filtered incident x-rays of short wavelength from a tube operating at 150 to 200 keV excite photoelectrons which produce a (reflection) image on an x-ray insensitive emulsion in close contact with the sample, giving contrast, due to compositional differences, of a thin surface layer of the sample or topographical detail. The negative detail is subsequently enlarged. In a transmission version of this technique, photoelectrons produced by the transmission of x-rays through a thin lead sheet, which is placed in contact with an electron transparent sample, impinge on an emulsion behind and in contact with the sample. Alternatively, the lead may be placed behind the sample and the film in front. Bubbles in a paint film on a sheet of iron have been successfully imaged by the latter technique, the iron substrate replacing the lead.[14] These transmission techniques are essentially electron-absorption microscopy, not x-ray absorption microscopy, and would seem limited in application, since the same contrast mechanisms are available in electron microscopy where much higher resolutions can be obtained.

B. Factors Affecting Contrast

1. Absorption

In microradiography the specimen is uniformly irradiated with x-rays, and differences in absorption, for example, resulting from compositional differences, give rise to differences in transmitted intensity, revealed as a variation in photographic blackening of the film or plate.

If a beam of parallel monochromatic x-rays passes through a thickness y of a particular element (Figure 2), its initial intensity I_0 is reduced

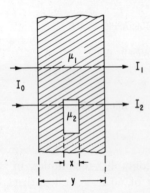

Figure 2   Absorption of x-radiation in a specimen of absorption coefficient μ_1, containing a region of different coefficient μ_2.

to a value I_1 according to an exponential relation (Lambert's law):

$$I_1 = I_0 \exp(-\mu_1 y) \tag{1}$$

where μ_1 is the linear absorption coefficient.

The linear absorption coefficient is constant for a particular material and wavelength λ but decreases with λ, so that a shorter wavelength is more penetrating. Lambert's law indicates that the intensity decreases from its initial value exponentially, the more rapidly the greater the linear absorption coefficient, and shows why a small increase in thickness *greatly* increases the exposure time when μ is large.

If the sample (Figure 2) contains a region such as an inclusion of thickness x and absorption coefficient μ_2, then

$$I_2 = I_0 \exp(-\mu_2 x) \exp(-\mu_1 (y - x))$$

hence

$$\frac{I_2}{I_1} = \exp (\mu_1 - \mu_2)x \qquad (2)$$

which is independent of the thickness y of the homogeneous portion of the specimen.

The response of an emulsion exposed to x-rays of constant incident intensity I for time t (exposure $= It$) is determined in terms of the optical density D, measured by the absorption of light in the image, and by analogy with Equation 1 is

$$D = \log_{10} \left(\frac{i_0}{i}\right) \qquad (3)$$

where i_0 is the incident light intensity, and i the transmitted light intensity. Although D is linearly proportional to the x-ray exposure for small densities up to approximately 1, the exposures in microradiography are commonly such that the range of blackening is on the linear part of the optical density versus logarithmic x-ray exposure curve of the emulsion. Thus the optical density D is related to the intensity I of the x-rays transmitted through the sample by the relation:

$$D = \gamma \log (It) + k \qquad (4)$$

where γ is the response of the emulsion to the x-rays, and k a constant.

The difference $(D_1 - D_2)$ of the density, due to the inclusion in Figure 2, is thus:

$$D_1 - D_2 = \gamma \log \frac{(I_1 t)}{(I_2 t)}$$

If we assume $\mu_2 > \mu_1$, so that

$$D_1 - D_2 = \gamma \log \exp (\mu_2 - \mu_1)x,$$

then

$$D_1 - D_2 = 0.4343\gamma\Delta\mu x \qquad (5)$$

Thus the difference $(D_1 - D_2)$ is independent of the sample thickness y and depends only on the thickness of the inclusion x, the difference $\Delta\mu$ in absorption coefficients, and the response of the emulsion. However, the visible effect (i.e., the contrast) in the microradiograph depends on $(D_1 - D_2)/D_1$, so that there will be a loss if the sample thickness is increased beyond the point at which the full thickness of the inclusion is included. The eye can readily detect differences as small as 5% in blackening. The use of too great a thickness will result in an unduly long exposure time. The emulsion response γ is not usually controllable, since the choice of emulsion is governed by the necessity for a fine grain size. The most important factor in Equation 5, governing the contrast, is then $\Delta\mu$, the differential absorption. Some values of $\Delta\mu$ for iron alloys are shown in Table 1 after Beattie.[15] Factors governing the absorption will now be considered.

Table 1   Difference in Absorption Coefficients ($\mu_A - \mu_{Fe}$) for Iron Alloys for K_α Radiation

		Target Material			
Absorbing Element A	Atomic No. Z	Cr $\lambda = 2.29$ Å	Fe $\lambda = 1.93$ Å	Co $\lambda = 1.79$ Å	Ni $\lambda = 1.66$ Å
Si	14	−457	−303	−254	−2944
Ti	22	1762	1097	880	−2030
V	23	−438	1967	1580	−1470
Cr	24	−257	2947	2330	−850
Mn	25	−161	−98	3080	−520
Mo	42	3582	2477	1980	−1110

After Beattie.[15]

The value of the linear absorption coefficient $\mu(\text{cm}^{-1})$ depends on the number of atoms in the path of the x-ray beam and hence on the state of aggregation. Division by the density ρ gives a quantity μ/ρ $(\text{cm}^2\text{g}^{-1})$, known as the mass absorption coefficient which is independent of the physical state. By using m to designate the product ρy (g cm^{-2}), known as the mass thickness (y = thickness in cm), Lambert's law can be rewritten:

$$I = I_0 \exp\left(\frac{-\mu}{\rho}\right) m \tag{6}$$

The experimentally determined mass absorption coefficient is actually made up of two quantities:

$$\frac{\mu}{\rho} = \frac{\tau}{\rho} + \frac{\sigma}{\rho} \tag{7}$$

where τ/ρ, the true mass absorption coefficient, represents energy absorbed and reemitted, either as secondary (fluorescent) radiation or as photoelectrons, and σ/ρ is the mass scattering coefficient. The latter can usually be neglected in the wavelength range used for microradiography. The scattering may be either incoherent with the change of wavelength (Compton scattering) or coherent, including Bragg diffraction.

The true absorption coefficient varies greatly with atomic number Z and wavelength λ, for it depends on the efficiency of the x-rays in ejecting the photoelectrons. Between characteristic absorption edges, where fluorescent secondary radiation is excited, the following approximate relation holds:

$$\frac{\tau}{\rho} = k\lambda^m Z^n \qquad (8)$$

where the value of m and n is approximately 3 for $\lambda = 1$ to 10 Å and $k = $ const between one pair of absorption edges. At each absorption edge, there is an abrupt change in the constant k, because at this frequency the radiation is just energetic enough to eject an electron from one of the electron shells of the atom. Tables of the so-called "mass absorption coefficient," found in most x-ray textbooks, give values of the photoelectric coefficient, based on Equation 7. The coefficient τ/ρ is of the order 10^2 to 10^3 for most elements at wavelengths used in microradiography, and even at 1 Å is only < 10 when $Z < 12$. For very hard x-rays in light elements, the loss of energy by scattering becomes comparable, and σ/ρ must then be taken into account. Scattering can also occur as a result of a different cause if particles of a size comparable with λ are present. Values of the true or total mass absorption coefficients for the elements are given by Barrett and Massalski.[16]

It is often necessary to calculate the mass absorption coefficients for an alloy phase or compound from published values for the elements. The mass absorption coefficient depends only on the number of each kind of atom present and is almost independent of their chemical state. Thus, if w_1, $w_2 \cdots$ are the weight fractions of all elements concerned and $(\mu/\rho)_1$, $(\mu/\rho)_2 \cdots$ the appropriate mass absorption coefficients, then:

$$\frac{\mu}{\rho} = w_1 \left(\frac{\mu}{\rho}\right)_1 + w_2 \left(\frac{\mu}{\rho}\right)_2 + \cdots \qquad (9)$$

and the density of the phase is given approximately by

$$\rho = w_1\rho_1 + w_2\rho_2 + \cdots \qquad (10)$$

A convenient index of the penetrating power of x-rays is the half-value thickness, the thickness of an absorber which reduces the intensity' to one-half its initial value. From Equation 1, putting $I_1 = I_0/2$, the half-value thickness is $0.693/\mu$. Some values of half-value thickness as a function of x-ray wavelength and atomic number Z of the absorber are shown in Table 2.

Table 2   Thickness of Absorber to Reduce Intensity to Half Value

Target	K_α-Wavelength (Å)	Absorber (cm)			
		Al	Fe	Cu	Pb
Cr	2.29	0.0017	0.00077	0.00050	0.00010
Cu	1.54	0.0053	0.00027	0.0015	0.00025
Mo	0.71	0.048	0.0023	0.0016	0.00043
Ag	0.56	0.094	0.0044	0.0029	0.00082

2. Choice of Radiation

Although the early microradiographic studies used white radiation, a considerable difference in atomic number is then necessary between two phases to obtain sufficient contrast. Nevertheless, for many qualitative applications, the unfiltered continuous radiation from almost any x-ray tube is adequate if deep absorption steps are avoided. Normally, the radiation used should be fairly monochromatic, in order to permit more quantitative interpretation or estimation of the concentration of an element, and to allow direct comparison of the observed contrast with calculation. For many purposes, once the minimum wavelength necessary to give a desired degree of contrast has been estimated, characteristic radiation can be used with a somewhat higher wavelength. If the wavelength used is between the absorption edges of two elements to be differentiated, good contrast can be obtained between elements of similar atomic number such as the alloying elements in steel. In samples containing more than two elements, a series of microradiographs, taken using radiations of different wavelengths, may give additional information.

The effect of choice of radiation on microradiographs of a 0.001-in. thick slice of nitrided Fe-40% Cr alloy, containing β-Cr_2N constituent particles, is shown in Figure 3 from Davis.[17] The contrast obtained with Co K_α radiation (Figure 3a) is the reverse of that given with Cr K_α (Figure 3b). The reason, as apparent from Figure 4, is that Co K_α radiation is more strongly absorbed by chromium than iron, whereas Cr K_α radiation is more strongly absorbed by iron than by chromium. Thus the Cr-rich particles appear white (weakly absorbing), relative to the matrix in Figure 3a. When the wavelength of the radiation lies between the absorption edges for iron and chromium, there is a much bigger

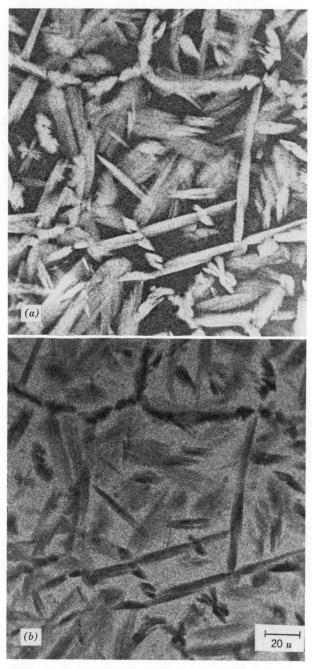

Figure 3 Contact microradiographs of nitrided Fe-40% Cr alloy: (a) Co-K$_\alpha$ radiation; (b) Cr-K$_\alpha$ radiation. From Davis.[17] (Copyright American Elsevier Publishing Company, 1970).

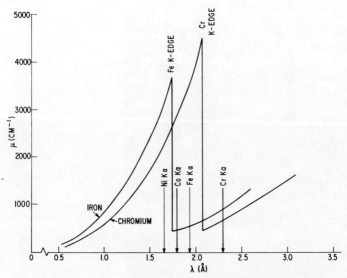

Figure 4 Effect of wavelength on linear absorption coefficient μ for iron and chromium. The arrows indicate the wavelengths of Ni-, Co-, Fe-, and Cr-K_α radiation. From Davis.[17] (Copyright American Elsevier Publishing Company, 1970).

relative difference in absorption by the phases, giving better contrast (Figure 3a).

C. Factors Affecting Resolution

1. Photographic

Since the x-ray magnification is unity in a contact microradiograph, the resolution obtainable is limited by that of the optical system used to enlarge the microradiograph, provided that other sources of loss of resolution and the grain size of the developed emulsion are sufficiently small. Although a photographic enlarger may be used for low resolution work, a good metallograph is desirable, in order to approach the optical resolution limit of about 0.2 μ. According to Lindström,[18] Lippmann-type photographic emulsions seemed to have the smallest grain size ($\sim$0.05 μ), but Kodak High Resolution and Maximum Resolution plates were preferable for microradiographs because of their homogeneity over large areas and freedom from artifacts. Kodak High Resolution plates gave a spatial resolution of about 0.5 μ.[19] Eastman Kodak No. 649 plates also appear suitable and have an average developed grain size of 0.1 μ.[1]

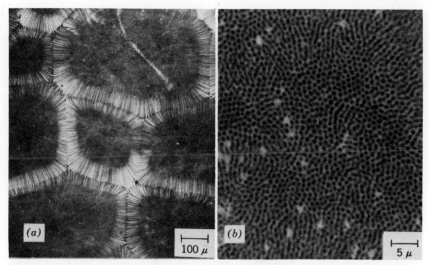

Figure 5   Contact microradiograph (Cr-K$_\alpha$ radiation) of a NiAl-Cr directionally solidified eutectic sample, etched to dissolve the chromium-rich rods; (*b*) is enlarged from the interior of one of the crystals visible in (*a*). From Davis.[17] (Copyright American Elsevier Publishing Company, 1970).

Figure 5*b* from Davis[17] is a contact microradiograph, taken on a Kodak high resolution plate showing a resolution ≤ 0.2 μ, using the smallest dot diameter as a criterion. This is an unusual sample for microradiography, since the black dots represent holes due to etching-out of rodlike particles normal to the 0.001 in. thick slice. The loss of clarity at the edge of the field in Figure 5*b* results from deviation of the hole direction away from the normal slice, as apparent at lower magnification in Figure 5*a*, where the rod direction changes near the grain boundaries. Excellent contrast would be expected, regardless of the type of radiation used. Etched samples are, of course, not normally used in microradiography. Better resolution can be obtained by the contact technique than by the projection technique (Figure 6).

While electron microscopic enlargement of the x-ray image recorded on a photographic emulsion is possible, and has been tried in contact microradiography using a strippable emulsion, the possible gain in resolution is limited. However, progress has been made[20-22] in developing grainless recording media such as plastic films, which may then be combined with optical or electron optical enlargement. Other sources of loss of resolution, namely, geometrical blurring and Fresnel diffraction, will now be considered.

Figure 6 Microradiographs of aluminum-tin alloy with tin phase outlining the aluminum grains. (*a*) Projection microradiograph taken with electrostatic projection microscope; (*b*) contact microradiograph. Private communication from H. J. Beattie, Jr.[15]

2. Geometrical Blurring

The two main sources of geometrical blurring are the formation of a penumbra and relative movement of the object, x-ray beam, and emulsion during exposure. The latter may be minimized in the contact method by firmly fastening the camera to the x-ray tube. Because of the finite size of an x-ray source, the image of an object in a contact or projection microradiograph is never completely sharp but falls off from maximum to minimum over a finite distance, resulting in blurring. The width P of the blurring or penumbra for an object on the front surface of the specimen (Figure 7) is

$$P = f \cdot \frac{b}{a} \qquad (11)$$

where f is the focal spot size, a the distance from the source to the specimen, and b the distance from the specimen to the film, or

$$P = \frac{f(y + z + e)}{a} \qquad (12)$$

where y is the specimen thickness, z the distance from the back of the specimen to the front of the emulsion, and e the emulsion thickness.

The image magnification is

$$M = \frac{a + b}{a} \tag{13}$$

and the geometric resolving distance is

$$r_g = \frac{P}{M} \tag{14}$$

or

$$r_g = \frac{f(y + z + e)}{a + b} \tag{15}$$

For the contact method $M \approx 1$, so that we see from Equation 14 that the geometric resolution is approximately equal to the penumbra width. Adding

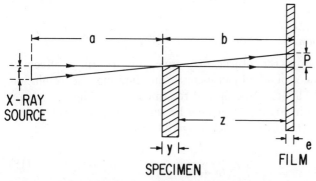

Figure 7 Schematic illustrating formation of penumbra.

the typical values of $f = 1000\ \mu$, $y = 100\ \mu$, $z = 0$, $e = 20\ \mu$, and $a = 12$ cm, we see from Equation 15 that r_g is approximately 1 μ. In order to decrease the penumbral blurring below the resolution limit imposed by the grain size, it would be necessary to decrease the angular size of the source, that is, decrease the source size f, specimen thickness y, emulsion thickness e and/or increase the distance a between the source and the specimen. Decreasing y has the advantage of not increasing the exposure time.

In the projection method, from Equations 11 and 14

$$r_g = \frac{fb}{aM}$$

and from Equation 13

$$r_g = f \cdot \frac{b}{a+b} \tag{16}$$

so that, since $b \gg a$,

$$r_g \approx f \tag{17}$$

Thus the geometric resolution in projection microscopy is approximately equal to the source size. The error in assuming $r_g = f$ is only 10% at a magnification of 10X. It can be shown[3] that, in the projection technique, r_g is also approximately equal to the smallest width of object which can be detected, and for practical purposes to the closest spacing of two objects which can just be resolved. Since, from Equation 16, r_g is largest at the surface of the specimen nearest the source and decreases through the specimen thickness, successive object planes are imaged with increasing sharpness (and reduced magnification). Thus a three-dimensional object can be imaged at high magnification with high definition, unlike the light microscopy in which the depth of focus is of the same order as the resolved distance; yet stereographs can be made that give a good impression of depth.

3. Fresnel Diffraction

Fresnel diffraction in the object may give rise to fringe effects in the microradiograph around object detail, thus limiting the resolution. Applying the usual (light) wave optical treatment and taking an opaque linear object ($y = 0$ in Figure 7), it may be shown[3] that the diffraction-resolving distance r_d (the first Fresnel fringe width) is

$$r_d = \frac{ab\lambda}{(a+b)^{\frac{1}{2}}} \tag{18}$$

where λ is the wavelength of the illumination. In the contact method, where the object is close to its image, $a + b \approx a$ and

$$r_d \approx (b\lambda)^{\frac{1}{2}} \tag{19}$$

In the projection method, if the object is close to the source $a + b \approx b$ and

$$r_d \approx (a\lambda)^{\frac{1}{2}} \tag{20}$$

If the specimen has a finite thickness y, then for both methods

$$r_d \lessgtr (y\lambda)^{\frac{1}{2}} \tag{21}$$

In order to obtain a resolution of 0.1 μ for $\lambda = 1$ Å, then a or $b \not> 100$ μ. In the contact technique this can be satisfied if the emulsion and specimen are in contact and the specimen thickness is less than 100 μ. In the projection technique, it would clearly be difficult to achieve, unless both the target and specimen in Figure 1b are in vacuo, so that they can be placed nearly in contact. In the projection technique, Fresnel diffraction will, in fact, be limiting unless $r_d < r_g$ which as already shown is approximately equal to f, the source diameter. The situation is worsened if soft x-rays ($\lambda \sim 10$ Å) have to be used to get adequate contrast, as in the case of biological material. For practical purposes[3] a disk-shaped feature with a width $< r_d$ will not be detectable because of the Fresnel effect if this is limiting the resolution, and the resolving power for two such objects will be $1.25r_d$. The line resolution will be 2 to 5 times better.

In many instances the nature of the object detail limits the magnification that can be usefully employed; thus a spherical precipitate gives rise to a diffuse contrast effect, since the thickness traversed by the x-rays falls off at its edges. Also, superposition of image detail is likely, unless the specimen is then compared with the detail examined. This confusion can be overcome by stereoscopic viewing of pairs of micrographs, made either with the specimen at different angles to the beam or slightly displaced. In practice, microradiographs of most metallurgical structures cannot be usefully examined at magnifications much above a few hundred diameters. Figure 3 is a good example of the results obtainable by the contact technique.

D. Contact Microradiography

In both the contact and projection techniques, the production of successful microradiographs is dependent on the preparation of thin specimens of uniform thickness. The specimen is reduced in thickness by grinding and polishing to about 100 μ or less for steel or up to 1000 μ for light alloys.

In the contact technique the specimen is placed firmly in contact with a photographic plate, and the whole suitably enclosed to avoid light fogging of the emulsion. Handling is facilitated if the mounted or unmounted specimen is thinned locally to provide a "window" for the x-rays to penetrate. Detailed techniques have been described.[3,17,23] Figure 8 is the type of plate cassette used by Davis.[17] For reasons that will become apparent later, characteristic radiation is normally used and it is necessary to monochromate this if quantitative measurements are required that may increase the exposure time up to many hours from the $\frac{1}{2}$ to 1 hr which is normally sufficient. X-rays from a normal diffraction tube such as

Figure 8 Microradiographic cassette to accommodate 1 in. × 3 in. plate. A 3 in. long, ⅜ in. diameter copper tube is used to attach the cassette to the x-ray port. From Davis.[17] (Copyright American Elsevier Publishing Company, 1970).

the General Electric CA-7 are usually adequate. However, a special ultra-soft x-ray tube was designed by Henke.[24] Scrupulous cleanliness at all stages is necessary to avoid spurious detail on the negative, due to dust; scratches on the sample tend to give rise to unwanted detail. After examination of the developed negative under an optical microscope, selected portions are photographically enlarged which can conveniently be done on a standard metallograph using transmitted light.

Stereo-microradiographs may be made, using a camera that holds the specimen and plate firmly at some variable but known angle to the x-ray beam. A commercial type is sold by Siemens (Erlangen). Alternatively, the sample is simply tilted in the cassette.[17] Williams and Smith[25] successfully used stereoscopy to measure contact angles between three grains in an aluminum-5% tin alloy, in which the tin was segregated to the grain boundaries. Stereoscopic microradiographs of dislocations were published by Jenkinson.[26] For a review of the technique, see Latham.[27]

For reviews of contact microradiography, see Refs. 3, 23, 28 to 30.

E. Point Projection Microradiography

Although, in contact microradiography, the magnification is limited to the secondary magnification that can be obtained by optical enlargement of the initial microradiographic image, in the projection technique, additional magnification is obtained by moving the specimen away from the photographic emulsion toward the source, so that primary enlargement is obtained at the expense of the area of the sample recorded in the image. The greater the primary magnification, the smaller the area imaged. The simplest technique is to place a source close to the object, the magnification being equal to the ratio of source-image to source-object distance.

The definition in the image is principally determined by the source size. Sievert[31] was the first to improve the definition by employing a 5-μ diameter pinhole between the x-ray tube and sample, obtaining a primary magnification of about 5X. More recently, Rovinsky and Lutsau,[32] using a 1-μ diameter pinhole and high intensity x-ray tube obtained images with a primary magnification of 100X using $\frac{1}{2}$ to 1 hr exposures. The specimen and photographic plate were rotated relative to the source to even out the intensity. The pinhole size and pinhole-image distance now replace the source size and source-image distance as critical dimensions, and, as would be expected, Rovinsky and Lutsau obtained a resolution of about 1 μ.

Stereomicrographs may be made, tilting or translating the specimen between exposures.[3] The technique was applied to an aluminum-tin alloy by Nixon.[33]

A second method of reducing the source size involves the use of a *point anode x-ray tube*. A focal spot size of 0.2 μ has been attained,[34] with a primary radiographic magnification of 30 to 500 and reasonable exposure times.

A third method, which has led to the development of several commercially available microscopes, employs electron lenses to focus a beam of electrons onto a thin metal target. The foil target can dissipate several times as much energy as a needle, permitting higher x-ray intensities; and also has the advantage that the specimen can be placed close to the target, while remaining outside the vacuum system and can be manipulated in a stage. The target thickness is a compromise, being thick enough to secure reasonable x-ray intensity, while thin enough to avoid undue scattering of electrons which increases the source size. An image resolution of ≤ 1 μ can be obtained with exposure times of 5 to 10 min. While requiring more costly equipment than the contact technique, the projection method lends itself to stereomicroscopy and gives shorter exposure times. It also provides a convenient x-ray source for microdiffraction, including the Kossel line technique which can be used to study the grain orientation in a thin metal sample.[35] Semiquantitative microanalysis is possible by determining the intensity of fluorescent x-rays emitted by the sample; however, the technique is limited compared with the electron probe microanalyzer.

In the projection microscope once made by the General Electric Company, New York, the electron beam was focused using two electrostatic lenses; a resolution of 1 μ or better was possible. In other commercial instruments such as those built by the Shimadzu Company (Japan), Philips Electronics Company (New York) based on the Philips' EM75 electron microscope, and Canalco (Washington, D.C.) based on the RCA EMU2B electron microscope, magnetic lenses are used. In the Philips instrument, seven different target materials may be rotated into position, the accelerating voltage is continuously variable from 2 to 50 keV; direct magnification of up to 70X can be made on the plate and the useful magnification after optical enlargement is about 500X, with a resolution of about 1 μ. Openshaw[36] obtained a resolution < 1 μ using an AEI microprobe analyzer adapted for x-ray projection microscopy.

Since less optical enlargement of projection microradiographs is necessary for a given resolution than in the case of contact microradiographs, defects in the photographic material are less troublesome, and emulsions of moderate speed and grain such as lantern plates are adequate. The grain size after development should be just fine enough to ensure that, at the selected primary magnification, two resolved points of the image do not fall on the same grain. With present designs of projection microscope and lenses, the exposure time advantage over the contact method disappears as high resolutions better than 1 μ are approached. Whether or not such resolutions can be achieved will again depend, as in the case of contact microradiographs, on the nature of the structural detail in the sample.

Further improvement in the projection technique is likely to result from the utilization of cold cathode x-ray sources that could give higher intensities, permitting reduction in exposure time, and from the incorporation of an image-itensifier system (see page 411) that would permit detailed visual inspection of the image, facilitating adjustment of the instrument and preselection of the area to be recorded.

For reviews of point projection microradiography, see Refs. 3, 23, 28, 29, and 37.

F. Quantitative Microradiography

The absorption of x-rays depends on the atomic number and amount of the elements present, so that the identity and concentrations of elements present may be determined by microradiography (x-ray absorption microanalysis). Analytical information may also be obtained from characteristic line emission from the incident x-ray beam, using a spectrometer to analyze the emitted x-rays, either in the transmission position using a thin sample, or in reflection, in which case we have a microversion of the usual fluorescent x-ray technique. It turns out that a much larger fluorescent x-ray signal can be excited from a microregion by an incident electron beam than by an x-ray beam; this has led to the commercial development of electron probe microanalysers described in Chapter 9. If an average analysis of a region large in extent (up to about 1 in. square) is required, the whole of this can be simultaneously irradiated with an x-ray beam, as in the conventional fluorescent x-ray technique, and a few ppm concentration detected in a volume of approximately 10^{11} μ^3, whereas in the electron microprobe analyzer, in which the beam is usually of 1 to 10-μ diameter, typically 1% concentration can be detected in a volume of 1 to 10 μ^3. It is apparent that there is no real advantage in combining microradiography with fluorescent analysis. Although x-ray absorption microanalysis is capable of a similar sensitivity to emission, amounts of an element of the order of 10^{-10} to 10^{-13} g being detectable in a volume as small as 1 to 10 μ^3, it has not been developed much as a macrotechnique and, as a microtechnique, has been employed principally for biological materials, where the use of longer wavelengths of the order of 10 Å offers an advantage over emission for elements of low atomic number. Recent developments in electron microprobe analysis are tending to remove this advantage.

Since semiquantitative analytical information is readily obtainable from x-ray microradiographs, the basis of absorption analysis will now be considered. The possibility of elemental analysis depends on the fact that the mass absorption coefficient varies rapidly with atomic number Z, being approximately proportional to Z^3, as seen in Equation 8. Monochromatic (line) radiation is normally employed to define the absorption coefficient. If the absorption coefficient and specimen thickness are measured, then, for a binary alloy of two known elements, their concentration may be calculated from Equations 9 and 10. If the densities ρ_1, ρ_2 are equal, these equations simplify to

$$w_1 = \frac{\mu - \mu_2}{\mu_1 - \mu_2} \qquad (22)$$

In principle this method could be extended to ternary or quaternary alloys by determining the absorption coefficient with two or three wavelengths, respectively, to obtain the necessary number of equations. It is desirable that the wavelengths be chosen so that an absorption discontinuity occurs between each pair of wavelengths for one element, in order to realize a substantial change in the absorption coefficient.

The transmitted intensity may be measured photometrically on a negative, or directly with counters which is more readily possible in the projection technique. A specially designed photometer[1] is necessary if areas smaller than about 50-μ diameter are to be examined. Using a step specimen of known absorption coefficient, the emulsion may be calibrated in terms of optical density versus the product of absorption coefficient and thickness, and hence the absorption coefficient of an unknown sample of measured thickness determined from the measured optical density, and the composition calculated as indicated above.

If a series of homogeneous standard alloys is available, an even simpler technique for determining the composition of a binary alloy is to plot a calibration curve of alloy composition versus the optical density for standardized x-ray and phtographic conditions. Synthetic standards may be made by stacking pure metal foils in appropriate proportions.

When the second element is present as an inclusion in a uniform matrix (as in Figure 2) Equation 2 may be applied or, if the intensities differ by no more than 20%, the approximate form may be used:

$$1 - \frac{I_2}{I_1} = \frac{\Delta I}{I} = (\mu_2 - \mu_1)x \tag{23}$$

A direct evaluation is only possible if the specimen contains regions free of the second element, so that I_1 is determined unambiguously. If this is not so, it may be obtained on a reference sample of pure element 1, preferably of the same thickness y (Figure 2). If the thickness is in fact y', giving an intensity I_1', the reference intensity is given by

$$I_1 = I_1' e^{-\mu_1(y-y')} \tag{24}$$

The use of a reference absorber (matrix or pure element) can be avoided by taking measurements on the same region of the specimen at two wavelengths, one above and one below the wavelength of an absorption edge (see Ref. 3 for more detail). This is more conveniently carried out in the projection technique, as it is otherwise necessary to locate and measure identical areas in two contact microradiographs.

The accuracy of the analysis depends on the uniformity of irradiation of the areas measured, on the accuracy of measurement of the transmitted intensities, on the accuracy of measurement and uniformity of specimen thickness (difficult to measure to better than 10%), on the purity of the radiation reaching the recording system (in the projection technique a spectrometer or pulse height analysis can be employed for greater accuracy), and on the accuracy to which the absorption coefficients are known.

The mass per unit area m in Equation 6 is also derivable from measurements made at a single wavelength with a reference absorber, or at two wavelengths about an absorption edge. Thus, beside micromeasurements of the relative and absolute concentration of elements, the thickness may be derived when the density is known, or vice versa, and the mass included within a measured area or the mass of a body of known shape such as a spherical inclusion determined. The microabsorption technique has found important application for determining the dry weight per unit area of biological material[3,38] or the water content if the measurement is made before and after dehydration.[39] A sensitivity of the order of 10^{-16} g is claimed.[39]

Relatively little use of stereograms for displaying internal structure seems to have been made on metallurgical and mineral samples,[26,27] although the technique has sometimes been employed for studies on insect morphology, plant, and other biological structures in conjunction with heavy element staining techniques.[1] One problem is in the presentation of such information which necessitates stereoviewing. One solution would be to apply the medical radiological practice of "planigraphy" (or "tomography"), in which both plate and specimen are moved laterally during exposure, to the projection microradiography method.[3] In this way a sharp image is produced of only one plane of the specimen, that is, where the plate movement equals the product of the specimen movement and the magnification of the desired plane. A series of such planar "sections" may be made through the entire specimen thickness.

III. AUTORADIOGRAPHY

A. Introduction

In autoradiography, a photographic emulsion placed in close contact with a metallographically prepared sample is locally exposed by ionizing radiation emitted from the sample. Subsequent development reveals the qualitative or quantitative distribution of the active element or elements over a thin layer at the sample surface by the distribution of blackening, and a one to one correlation may be possible between features such as grain boundaries in the etched microstructure and local blackening in the autoradiograph. The technique thus has some features in common with sulfur printing, but, unlike the latter which requires only a matter of minutes, may necessitate exposure times of up to several weeks. Autoradiography has proved valuable in biological studies,[40-43] as well as in studies on metal and ceramic materials;[23,44-46] its oldest use was in the detection of natural activity in minerals. An exhaustive review of applications classified by alloy has been given by Condit.[44]

While metallurgical applications include metal transfer during friction or wear, the most important are probably diffusion studies, segregation studies, and determination of the distribution of heavy elements in irradiated nuclear fuel materials. If the radioactive isotope is deliberately added during melting or by diffusion into the solid, rather than produced in situ by (neutron) irradiation, then the resulting autoradiographic indications will be specific and unambiguous. Figure 9 illustrates the use of autoradiography to study grain boundary diffusion in zirconia.[47] The sample was sectioned after diffusing in ^{95}Zr from one surface. Accelerated cation self-diffusion occurred along grain boundaries.

B. Methods of Adding Activity

The radioisotope may be added to a melt, preferably as a master alloy of moderate specific activity. The precautions necessary to avoid radiation hazards depend on the nature of the isotope used. This technique lends itself to segregation studies.

In diffusion studies it is common to employ the radioisotope as one component of a diffusion couple, applied to the sample by electroplating or vaporplating. After diffusion treatment, the interface may be taper-sectioned for autoradiography.

A third technique involves the formation of the active material in situ by neutron irradiation in an atomic pile or by ion (usually deuteron) irradiation in a cyclotron. In order to employ this approach successfully, it is necessary that the activity of one element be considerably greater than that of others present. Since different radioisotopes decay at different rates, it may be possible to make a succession of autoradiographs that reveal the distribution of different elements.[48] Samples may be completely prepared, except for the application of the emulsion, prior to irradiation. The use of short-lived isotopes is possible. High resolutions are unlikely to be possible because of the background activity that is always present.

In biological studies the radioisotope is commonly administered to the living organism, and tissue or other samples later removed for autoradiography.

Ion bombardment activation may also be used[44] and has the advantage that activation can be confined to a thin surface layer, thus improving the resolution.

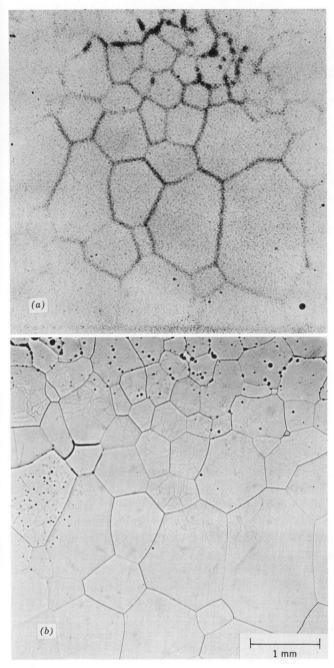

Figure 9 (*a*) Autoradiograph of as-polished sample of calcia-stabilized polycrystalline zirconia containing ⁹⁵Zr radioisotope; (*b*) micrograph of same area after repolishing and etching to show normal appearance of grain boundaries. Private communication from R. E. Carter.

C. Factors Affecting Resolution

With careful technique, resolutions of 2.5 μ (the distance distinguishable between two sources) may be obtained, compared with only 25 μ or worse as commonly practiced. Three main factors limit the resolution obtained: (*a*) the energy and hence the range of the particles, (*b*) geometrical factors, and (*c*) the size of the photographic grains. These will now be considered in order.

1. Particle Energy

Although α, β, and γ radiations are emitted simultaneously by natural radium-bearing minerals, in practice autoradiographs fall into two main classes, namely, those obtained using α emitters and those with β-emitting isotopes. Alpha particles are helium nuclei and are emitted during the disintegration of the heavier unstable nuclides. Alpha particles of only a single energy are emitted during the radioactive decay of any isotope and thus have a fixed range R in any medium given by:[23]

$$R = 0.0003194 \, R_a \frac{\psi_s}{\rho_s} \tag{25}$$

where R_a is the range in dry air (available in standard nuclear reference tables), ρ_s the density of the solid, and ψ_s the permeability of the solid. Then

$$\psi_s = w_1(A_1)^{1/2} + w_2(A_2)^{1/2} + \cdots \tag{26}$$

where w_1, w_2 $\cdots$ are weight fractions, and A_1, A_2 $\cdots$ the atomic weights of the elements in the solid.

Some examples of the range of α particles in selected materials of metallurgical interest, using the naturally occurring isotopes radium F (polonium 210) and thorium X (radium 224), are given in Table 3. Although still used as tracers for elements of similar atomic number, the natural isotopes above have tended to be superseded by the artificial isotopes of atomic number greater than 82.

Beta particles are emitted from nuclei having an excess of protons or neutrons. Beta particles, unlike α particles, are emitted over a range of energies up to a characteristic maximum energy E (MeV) and thus have a range of penetrations in a given material. The half-thickness a, which is the amount of absorber required to reduce the β intensity to one-half, is given by the empirical relation:

$$a = 0.046 \, E^{3/2} \text{ g } cm^{-2} \tag{27}$$

Over 200 artificial radioisotopes are available with half-lives greater than

Table 3   Range 'R' of α Particles from Various Isotopes

		Isotope	
		RaF	ThX
Absorber	α Energy:	5.30 meV	5.68 (and 5.44) meV
Dry air at 15°C		38,400 μ	42,800 μ
Platinum		7.8	8.9
Copper		10.9	12.2
Iron		11.7	13.0
Lead		15.6	17.3
Aluminum		23.6	26.3
Magnesium		34.7	38.7

10 hr and thus of utility in tracer studies. About a dozen are pure β emitters, while the rest also emit electromagnetic radiation. The linear half-thicknesses, given by the ratio a/ρ (ρ = density) for a number of absorbers, are listed in Table 4 for several emitters, after Ward[23] but revised using more recent data.[49]

Table 4   Half-Thicknesses of the Most Energetic β Emission

	Isotope				
	[63]Ni	[14]C	[99]Mo	[32]P	[56]Mn
β energy (meV)	0.067	0.156	1.23	1.71	2.85
Half-life	62 yr	5.370 yr	67 hr	14.3 days	2.58 hr
Half-thickness a ($g/cm^{-2} \times 10^4$)	7.98	28.3	628	1030	2213
Absorber	Linear Half-Thickness $a/\rho(\mu)$				
Tungsten	0.4	1.5	33	53	110
Lead	0.7	2.5	55	91	190
Copper	0.9	3.2	70	120	250
Iron	1.0	3.6	80	130	280
Titanium	1.8	6.3	140	230	490
80% AgBr emulsion	2.2	7.7	170	280	600
Aluminum	3.0	10	230	380	820
Magnesium	4.6	16	360	590	1270
Gelatine	6.2	22	480	790	1700

It will be noted that the linear half-thickness for [99]Mo, ^{32}P, and [56]Mn, is considerably greater in a particular material than the range of α particles (Table 4).

2. Geometrical Factors

Image spread arises from the fact that radioactive materials emit ionizing particles in all directions; thus the thickness of the sample (or in a thick sample of the surface layer contributing to the image), the separation between the sample and the emulsion, and the emulsion thickness are all important factors. The resolution can be improved by reducing the sample thickness to less than the range of α or β particles in the sample. This is particularly important in the case of the more energetic β emitters. Very roughly, the resolution is equal to the linear half-thickness of the specimen for a massive specimen and is about equal to the specimen thickness if this is less than the half-thickness but greater than or equal to the emulsion thickness. Thus if the specimen and the emulsion are each of 3-μ thickness and in intimate contact, the geometrical resolving power will be about 3 μ.

3. Photographic Grain Size

Energy is lost when an ionizing particle crosses an undeveloped photographic grain, and some of this is used to liberate electrons that render the grain developable. The amount of energy lost increases with the grain size, which determines the length of path in the grain, and for a given type of ionizing particle decreases as the particle energy increases. It is desirable to select an emulsion in which the passage of a single particle through a grain will render it developable, in order to optimize the sensitivity. Experience with visible light indicates that absorption of about 35 eV per grain is needed to make it developable.[50] It is estimated[50] that, for ^{35}S or ^{14}C ($E = 167$ and 155 keV, respectively) 0.1 μ is likely to be the limit. For more energetic isotopes, larger grained emulsions would have to be used in order not to sacrifice sensitivity, and the geometrical resolution would be decreased, since it cannot exceed the grain size.

Autoradiographs for visual examination require exposures of at least 10^7 β/cm^2 or 10^7 α/cm^2. However, if the emulsion is to be examined while still in position on the sample surface, much shorter exposures giving about 10^4 β/cm^2 or 10^4 α/cm^2 must be used in order not to obscure the specimen. An image of this density may be invisible to the naked eye; however, the exposed silver grains may be counted under a microscope at high magnification. A lower limit to the exposure is set by the background fog which may amount to 500 grains/cm^2. For greater accuracy in α autoradiographs, thicker emulsions may be employed and tracks counted, instead of random silver grains in thin emulsions (see pages 400–401). The photographic process has been reviewed by Pelc and Welton.[51]

D. Emulsions for Autoradiography

The three types of sensitized material that may be employed are plates, strippable emulsions, and liquid emulsions. Some examples of available emulsions are shown in Table 5. A more exhaustive list and bibliography of applications is given by Condit.[44]

Table 5   Sensitized Materials (Kodak) Available for Autoradiography

Material Designation	Type of Emitter	Nominal Emulsion Thickness μ	Nominal Overcoat Thickness μ	Mean Grain Diam. (Undeveloped) μ
Type no-screen autoradiographic plate	α or β	25	1	0.65
Type A autoradiographic plate	α or β	25	1	1.25
Type NTE liquid emulsion		—	—	0.06
Type NTA plate or liquid	α moderate energy	10, 25, 100	1	0.22
Type NTB plate or liquid	α high energy / β up to 30 KeV	10, 25, 100	1	•0.29
Type NTB 2 plate or liquid	α up to 0.2 Mev	10, 25, 100	1	0.26
Type NTB 3 plate or liquid	All charged particles	25, 100	1	0.34
Fine grain autoradiographic stripping plates AR.10	α or β	5	None	—

NT = nuclear track

Plate materials such as Kodak type A or no-screen have the disadvantage of being coated with a protective layer of gelatine (Table 5) which prevents intimate contact between the specimen and the emulsion and tends to limit the resolution. They are used primarily for low resolution, high contrast work with α or β emitters and as a means for estimating exposure times for the finer grained emulsions. Great care is necessary to avoid relative motion of the plate and specimen during application and exposure. If the specimen is mounted inside a steel cylinder, a magnet may be employed to hold the plate and specimen in intimate contact, permitting later removal for development.

The use of liquid emulsions permits intimate contact with the specimen. Liquid emulsions are gels at room temperature. After heating in a water bath to 40 to 42°C to liquify, a few drops are placed on the specimen and the excess immediately drained off. The film is left on the specimen during subsequent exposure which may last from 2 hr to several months, and during development and microscopic examination. It may be necessary to coat the sample initially with, for example, a 2% solution of vinyl in methyl ethyl ketone to avoid corrosion during photographic processing. In addition, specimens may be coated with a 1% solution of collodion in amyl acetate to improve the adherence of the emulsion.

Special autoradiographic stripping emulsions such as Kodak AR.10 are available without a protective gelatine layer and may be applied to the specimen after stripping from their glass supports. Background fogging, due to γ radiation or cosmic rays, is more severe with Kodak type NTB3 or NTB2 emulsions which have more sensitivity to higher energy particles than the other emulsions listed.

Alpha track autoradiography has been widely used in biological studies as an alternative to the examination of images composed of random silver grains in thin emulsions. In this technique a thick, fine grained, nuclear-track-type emulsion is used, of say $25\text{-}\mu$ thickness, in which the path of an α particle is visible as a straight continuous track of silver grains. Using the liquid emulsion technique the track can be traced under the microscope to the point of entry at the sample emulsion interface. Since α particles form straight tracks, they are readily distinguished from background fog and even a single track may be recognized. If low exposures are used to minimize coincidences, quantitative track counting is possible.

Efforts have been made by biophysicists[50,52] and metallurgists[46] to extend autoradiography for use in the electron microscope, with the aim of improving resolving power and sensitivity. Very thin sections combined with thin, fine grained liquid emulsion,[52] or degelatinized liquid NTE emulsion specially processed to refine the grain size down to 250 to 500 Å have beem employed. The Kodak fine grain autoradiographic stripping plates AR.10 are intended for use in conjunction with electron microscopy.

Recently, it has been discovered[53] that nuclear tracks in a variety of plastic and other materials may be developed by a simple etching technique, so that tracks may be made visible in the electron microscope or after further etching in the light microscope. This technique has now been adapted to the autoradiography of irradiated nuclear fuel materials.[54] Cellulose nitrate was found to be sensitive to α particles but insensitive to β and γ radiation. The tracks were selectively etched with sodium hydroxide, appearing as dark pits when viewed by transmitted light.

The strip of cellulose served as a negative for photographic enlargement. Quantitative track counting was possible. Pit diameters between 2 to 5 μ were found convenient for optical measurements. The technique is at an early stage of development and appears capable of much broader application.

Beta track autoradiography has not been greatly used, since the β particles are deflected repeatedly by collisions in the emulsion and are difficult to trace above the background fog.

IV. X-RAY METHODS FOR DIRECT IMAGING OF DEFECTS

A. Introduction

A number of x-ray techniques for the direct imaging of dislocations and other defects have been developed since 1958 (for reviews see Refs. 56–64) from earlier work on the fine structure of diffraction spots.[65] The detailed theory has been reviewed by Webb,[58] and Amelinckx.[59] Since the contrast width at a dislocation is typically 2 to 10 μ, these imaging techniques are only suitable for crystals containing dislocation densities below about 10^6 cm^{-2}. For higher densities from 10^6 up to about 10^{12} cm^{-2}, transmission electron microscopy is suitable;[59] however, since thin slices must be used, this technique is unsuitable for viewing the spatial distribution of dislocations in relatively perfect crystals with densities below about 10^6 cm^{-2}. Information on densities and types of defects may be obtained by other x-ray methods such as measurements of the Bragg line shape and displacements,[66] distortion of diffraction spots,[67] fine structure of spots,[65] and rocking curves.[68] These are outside the scope of the present discussion, however, which is concerned with intensity-mapping methods rather than methods that measure local variations of the direction of the diffracted beam.

X-ray methods depend for their success on the fact that dislocations and certain other defects produce localized differences in the diffraction of x-rays from an otherwise perfect crystal, which, although small, may be recorded on a photographic film with an x-ray sensitive emulsion with a one to one correlation. Unlike the transmission electron microscope technique which involves the use of relatively high magnification and a correspondingly small region of crystal, the x-ray image may be recorded at a magnification as low as unity over a large area of crystal, and selected regions subsequently enlarged optically by several hundred diameters for detailed examination.

The three principal x-ray methods illustrated in Figure 10 are (a) the back reflection Berg-Barrett method, (b) the Lang transmission method, also referred to as the transmission Berg-Barrett method, and (c) the Borrmann anomalous transmission method. Although there are many individual variations in technqiue, these methods may be described briefly as follows. All use characteristic x-ray irradiation, usually employing the K_α line. Some collimation is desirable, although crystal collimation is only necessary for special purposes. Photographic recording is often convenient, although a diffractometer may be used for some purposes.

B. Back Reflection Berg-Barrett Method

In this method a collimated parallel monochromatic beam of x-rays (Figure 10a) is directed onto a surface of the sample for which μt

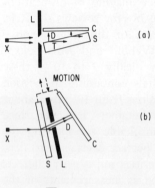

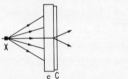

Figure 10 Schematic illustrating techniques for making x-ray topographs. (a) Back reflection Berg-Barrett; (b) Lang transmission; (c) Borrman anomalous transmission. Here X is the x-ray source, S the sample, T the transmitted beam, D the diffracted beam, and C the photographic emulsion.

> 1 (μ = absorption coefficient, t = thickness), which is usually a rather perfect single crystal, and is reflected back from any region that is oriented at the correct-angle Bragg angle θ, given by Bragg's law:

$$n\lambda = 2d \sin \theta \qquad (28)$$

where n is an integer, λ the x-ray wavelength, and d the spacing of the particular set of crystal planes.

Lining-up is facilitated if a slightly divergent, for example, 2°, beam is employed. The diffracted beams are recorded on a film from which light is excluded. In order to avoid the overlapping of detail, and to obtain geometrical correspondence, it is desirable that the film be close to the sample; thus a small angle of incidence of x-rays is generally employed. The exposure of the film to the direct x-ray beam is avoided. A filter may be interposed between the specimen and the film to remove unwanted fluorescent excitation. It is desirable that the film be tilted with reference to the sample surface, in such a way that the signal path through the emulsion is at least as great as the noise path. Multiple dislocation or other images may occur if the crystal is oriented in respect to the beam in such a way that several different Bragg reflections are possible, and should be avoided, using stereographic analysis as a guide. High resolution emulsion film is desirable, since the image has to be studied under a low power microscope, but may necessitate exposures as long as 8 hr. Weissmann[69,70] varied the technique by making a set of Berg-Barrett photographs at successively greater distances from the sample and tracing the diffracted beams back to the sample. In this way he obtained quantitative measurements on the orientation and misalignment of grains, and the size of subgrains and low angle boundary structures. Only a relatively thin surface region of the sample is explored in this technique, since the characteristic x-rays penetrate only to a relatively small depth, typically about 50 μ in lithium fluoride which is a relatively favorable material, and only a micron or less for elements heavier than aluminum.

Dislocations appear as dark lines, due to enhanced diffraction from the region around a dislocation which is stronger for strong, low order reflection. The amplitude of the beam reflected from a perfect crystal results from interference of the once-diffracted and multiply diffracted beams. A phase change of $\pi/2$ occurs on each reflection, so that the twice-diffracted beam, for example, is parallel but out of phase with the incident beam, resulting in destructive interference, the so called "primary extinction." If the strict periodicity is destroyed near an imperfection, the once-diffracted beam is no longer incident at the correct Bragg angle and emerges on second reflection with a larger amplitude, since it is only partly diffracted again into the crystal. This results in enhancement of the diffracted intensity recorded from the region of the imperfection. Hirsch, Howie, and Whelan[71] applied their kinematical theory of diffraction contrast to transmission topographical observations; although it is not strictly applicable, they were able to explain some of the dislocation contrast effects qualitatively.

The back reflection method has been used to study dislocations and subgrain structures,[72-77] and a three-dimensional display of the dislocation "forest" in the crystal interior was reconstructed by Newkirk[78] (Figure 11). Newkirk demonstrated a one to one correlation between dislocation etch pits and dislocation images in lithium fluoride and in silicon[78] (Fig-

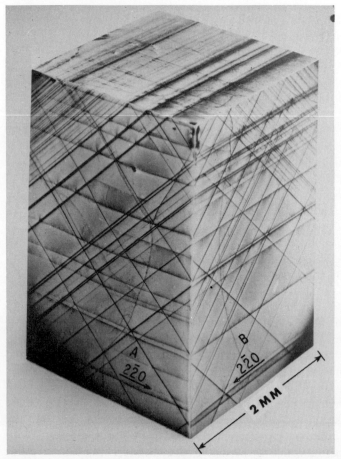

Figure 11 Reconstructed compressed lithium fluoride crystal with {100} faces, as seen by the reflection Berg-Barrett technique, showing subgrain boundaries, slip planes, and grown-in dislocations. From Newkirk.[78] (By permission of The Metallurgical Society of AIME.)

ures 12 and 13). Great care is necessary in the preparation of crystals to avoid confusion between surface damage, for example, latent scratches, and dislocation images.[77,79] Cutting operations such as spark cutting produce a damage layer that must then be removed by chemical and/or electrolytic means. Surface faceting is also highly objectionable,[79] and a smooth electropolished surface is desirable. The criterion for the absence of contrast is, to a first approximation, the same as in transmission electron microscopy, so that by using different reflections and observing charac-

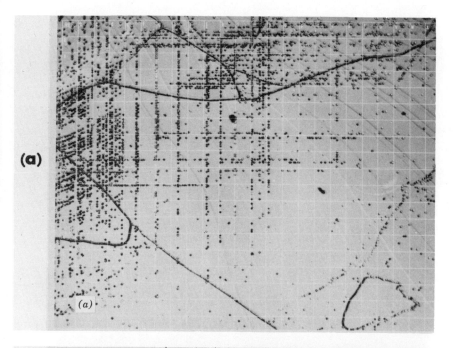

(a)

(a)

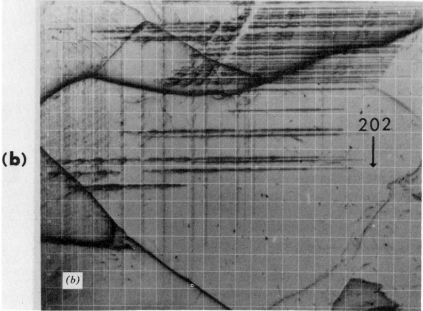

(b)

202

(b)

Figure 12 Etch-pitted, cleaved lithium fluoride crystal. (*a*) Light micrograph with superposed 0.07-mm mesh reference grid, showing orthogonal rows of $< 100 >$ etch pits extending from point of cleavage initiation, and grown-in subboundaries; (*b*) Berg-Barrett reflection topograph of the same area, using 202 reflection and Cr-K$_\alpha$ radiation, showing lines corresponding to the rows of etch pits in (*a*). The horizontal lines were strongest in this reflection, indicating a $[\bar{2}02]$ Burgers vector. Private communication from Newkirk, after Newkirk.[78]

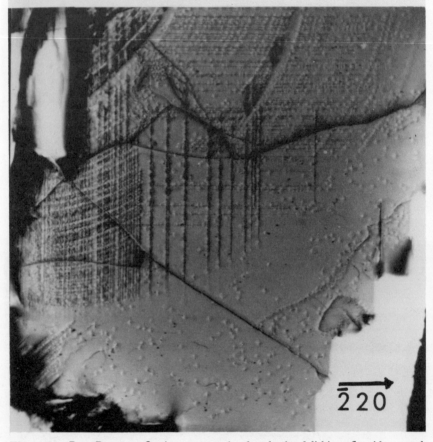

Figure 13 Berg-Barrett reflection topograph of etch-pitted lithium fluoride crystal, demonstrating the resolution possible. Some of the grown-in dislocations associated with etch-pits are seen as black hair lines extending into the crystal. From Newkirk.[78] (By permission of The Metallurgical Society of AIME.)

teristic extinctions it is possible to determine the direction of the Burgers vectors of dislocations.

Yoshimatsu[76] studied the polygonization of bent lithium fluoride crystals, using the Newkirk technique. Dislocations in crystals of BeO were successfully imaged by Austerman et al.[80] Lommel and Kronberg[79] applied the Newkirk technique, modified by using a line microfocus x-ray tube, to sapphire and ruby crystals and obtained a resolution of about 2 μ. They successfully determined the Burgers vectors. The penetration depended upon the crystal orientation and was about 10 μ for an (0001) surface. Dislocations in zinc were successfully resolved by Schultz and Armstrong,[81] using the Newkirk technique.

The technique used by Bonse and Kappler[82] and Bonse[83,84] is also a reflection technique but uses a parallel beam of x-rays, monochromated by Bragg reflection from a rather perfect germanium crystal which falls on the second germanium crystal (the sample) set for the same Bragg reflection. Misorientations of the order of 0.1 sec can be revealed on a plate set as close as possible and nearly parallel to the sample, and corresponding strains ($\Delta d/d$) of only 10^{-8} to 10^{-9} detected. Quantitative observations were possible on the strain field around a dislocation.[82] The arrangement may also be used with the specimen crystal set for transmission. The double-crystal diffractometer technique was combined with x-ray diffraction microscopy (Berg-Barrett) by Weissman[85,86] and Weissman and Kalman,[87] in order to measure the misorientation across selected subboundaries.

C. Lang Transmission Method

If the sample is thin enough, $\mu t \ll 1$ (μ = absorption coefficient, t = thickness), an image may be obtained using either the diffracted beams from the crystal, analogous to dark field in light or electron micrographs (Figure 15), or the transmitted beams analogous to bright field. Variations of this technique have been employed by many workers.[88–123] For additional reviews see Refs. 55 through 64. In the Lang projection technique,[92–94] an aperture is placed behind the sample, permitting passage of a particular Bragg reflection from a set of planes nearly perpendicular to the crystal slab, while shutting off the directly transmitted beam. The distribution of diffracted intensity is recorded on a photographic plate mounted normal to the diffracted beam. The technique provides a sensitive way of imaging defects such as dislocations, stacking faults, strain fields, domain walls,[115] and segregation. Under suitable diffraction conditions, the sense, as well as the direction, of dislocations having an edge component can be determined.[116,117] Stereotechniques are applicable.[94]

Figure 14 Scanning topographic camera based on the Lang principle. (Courtesy of Crystallogenics.)

A commercially available scanning topographic camera, based on the Lang principle, is shown in Figure 14. The goniometer sample holder and film holder can be traversed up to 2 in. by a high precision slide in front of the collimated x-ray beam from a slit. The holder affords 360° rotation of the sample in its own plane. The scintillation counter can be set at twice the Bragg angle on a scale and locked. The sample crystal is set close to the Bragg angle and can be rocked through the Bragg angle by a motor. The crystal and film can be traversed automatically. The camera can be employed with a standard or microfocus x-ray tube and used for either of the three techniques (*a*) through (*c*) of Figure 10; however, much simpler equipment may be used for techniques (*a*) or (*c*). Other cameras based on the Lang design are available from the Jarrell-Ash Company, from Ridaku-Denki of Japan, and from Cie, Generale de Radiologie at Issy-les-Moulineux, France. A high temperature Lang-type camera, operating up to 800°C, is available from Krystallos of Palo Alto, California.

Carlson and Wegener[108] varied the Lang technique to avoid simultaneously moving the specimen and film and used a collimated beam from a line focus source with a relatively large specimen to plate distance, resulting in some loss of definition, and image doubling due to the use of the K_α doublet. A similar technique was employed by Amelinckx, Strumane and Webb[106] and by Webb.[107] The practical resolution of the Lang technique for dislocations is about 5 μ, compared with 2 μ for the reflection methods.

Using the Lang technique, or modifications thereof, dislocations were successfully imaged in silicon,[93,94,99,104,105,111–113] germanium,[94] lithium fluoride,[94] silver chloride,[94] magnesium oxide,[94,96–98] calcite,[94] quartz,[94,99] aluminum,[94,95] diamond,[94,101–103,121] ice,[120] silicon carbide,[106] sodium chloride,[94,107] cadmium sulfide,[119] copper-silver alloy,[122] and iron-silicon.[117] Stacking faults were imaged in silicon,[111,112] and zinc.[109] Segregation was imaged in silicon,[110,114] aluminum,[90] aluminum-zinc alloy,[90] and diamond.[103] Subboundaries were imaged in lithium fluoride,[92] aluminum,[92] magnesium oxide,[96] and iron-silicon.[117] This list is by no means exhaustive.

Figure 15 shows examples of Lang-type transmission scanning topographs of silicon webb dendrite from the work of Merian and Blech.[118] When the topograph of this (111) crystal was made, using the $2\bar{2}0$ reflection (Figure 15a), all dislocations including partials were visible, except those for which the Burgers vector was normal to the [220] diffraction vector **g**. In the second topograph of the same region (Figure 15b) made with the $11\bar{1}$ reflection, stacking fault contrast is observed, since the criterion for the visibility of stacking faults whose fault vector **R** is $\frac{1}{3}$ [111], that is, faults parallel to the web plane, is, as in the case of transmission electron microscopy:

$$\mathbf{g} \cdot \mathbf{R} \neq 0, \pm 1, \pm 2 \tag{29}$$

and so forth. Comparison of the two topographs, indicates that most of the dislocations in Figure 15a separate faulted regions in (b) and so are partial dislocations. Striped contrast is not seen at the faults in Figure 15b because they are parallel to the plane of the sample.

D. Borrman Anomalous Transmission Method

Borrman, Hartwig, and Irmler[124] employed a widely divergent (30°) beam of x-rays from a point focus source and a crystal thick enough to give large normal absorption; for example a 1-mm thick silicon crystal is suitable and gives μt as approximately 15 (μ = absorption coefficient, t = thickness). The arrangement is very simple (Figure 10c). As a result of the beam divergence, Bragg reflection occurs from several different sets of crystal planes, along which there is "anomalous" transmission of

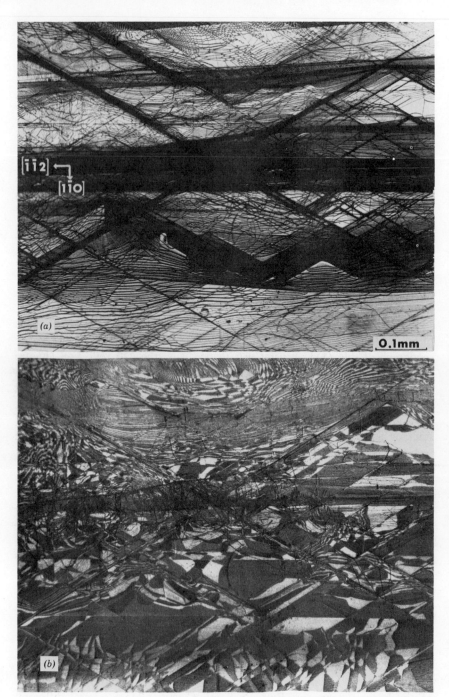

Figure 15 Lang-type transmission topographs of (111) silicon web, taken with Mo-K$_\alpha$ radiation from 100-μ spot focus. (*a*) [2$\bar{2}$0] topograph; (*b*) [11$\bar{1}$] topograph. From Meieran and Blech.[118] (Courtesy of the American Institute of Physics.)

the stationary waves. The latter split into two beams on emerging from the crystal, the diffracted and the transmitted beams. The presence of a dislocation locally destroys the exact periodicity necessary for anomalous transmission, giving a "shadow." Since the plate is in contact with the crystal, the images of the diffracted and transmitted beams coincide.

Barth and Hosemann[99] and Authier[125] used a combination of the Lang and Borrman methods with a stationary crystal, a film separated from the crystal, and a parallel beam of x-rays. Barth and Hosemann recorded the diffracted image only, whereas Authier used a crystal monochrometer and recorded both transmitted and diffracted beams, obtaining double images of the dislocations with complementary contrast. Gerold and Meier[126] demonstrated the Burgers vector determination in germanium by the anomalous transmission technique.

Weissman and Kalman[127] applied the Borrman technique to study microstrained crystals. They showed[64] that the technique can be made more quantitative by combining with back reflection divergent beam x-ray patterns, in order to determine the interplanar spacings.

Newman[128] demonstrated that *moiré fringes* could be produced by a point projection x-ray microscope, using crossed gratings of fine wire and soft x-rays. Bonse and Hart[129,130] exploited the Borrman effect in thick crystals to produce an x-ray interferometer, using as components parts of the same block of a perfect crystal of silicon. Moiré patterns were obtained by deforming,[131] rotating, or translating one part of the interferometer relative to the rest. Displacements of 0.1 Å and rotations of 0.001 sec of arc could be detected, and moiré images of dislocations obtained. The technique was extended to germanium and quartz.[132] Chikawa[133] obtained x-ray topographs, showing moiré patterns, by superposing copper-doped, single-crystal platelets of CdS. The moiré was attributed to variations in lattice spacing, caused by impurity segregation. Authier[134] reported moiré fringes, due to parameter differences, between the proton irradiated and unirradiated portions of a crystal. Lang and Miuscov (see Figure 16 in Ref. 62) observed moiré fringes in x-ray topographs of a quartz crystal, containing a crack that divided a crystal into two, superposed, simultaneously reflecting layers.

E. Instantaneous Video Display of X-Ray Topographs

A very interesting recent development by Rozgonyi, Haszko, and Statile[135] of Bell Laboratories is the instantaneous video display of x-ray diffraction topographs with a resolution < 15 μ. Previous attempts used one of two approaches, either image intensification and pickup, after converting the x-ray to visible light;[136,137] or direct image pickup, using x-ray sensitive Vidicon-type[138],[139] television camera tubes. The best resolution previously obtained was about 30 μ.

Rozgonyi et al.[135] used the silicon diode array target from a Picture Phone camera tube as an image transducer for x-rays in a closed circuit 525 line TV system (Figure 16). The silicon target, which was 15 to

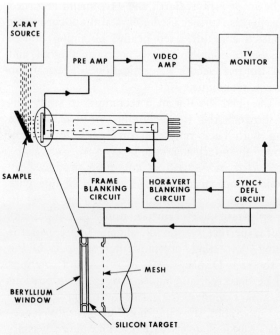

Figure 16   Schematic of video x-ray topographic apparatus. From Rozgonyi et al.[135] (Courtesy of the American Institute of Physics.)

20 μ thick, contained over 750,000 diodes and was incorporated in a magnetically focused and deflected camera tube, provided with a 100 μ thick beryllium x-ray window. The x-ray flux was integrated before it was read out by the electron beam. Using chromium K_α radiation and Berg-Barrett reflection geometry, TV images were successfully obtained from a silicon wafer integrated circuit (Figure 17) and a cleaved surface of a GaAs wafer (Figure 18a). A plot of x-ray intensity versus distance across the cleaved surface of the latter sample was obtained by displaying a single video scan line on a time-delay oscilloscope (Figure 18b).

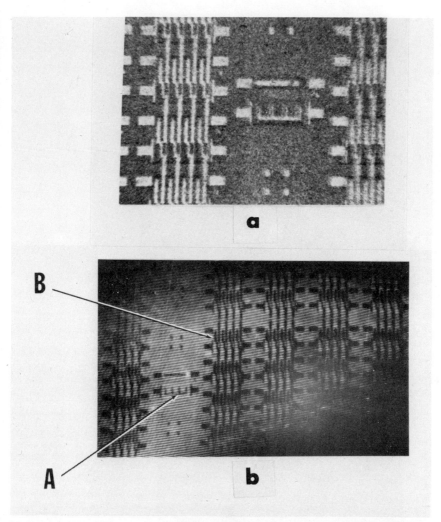

Figure 17 Silicon-integrated circuit. (*a*) Berg-Barrett topograph imaged on Kodak dental film; (*b*) video x-ray topograph, showing 13.5 μ wide metallized strips at *A* and 38-μ strips with 19-μ separation at *B*. From Rozgonyi et al.[135] (Courtesy of the American Institute of Physics.)

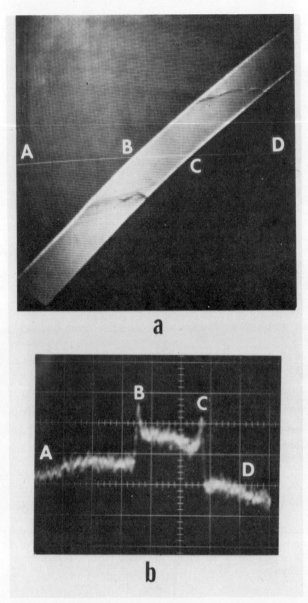

Figure 18 (*a*) Video x-ray topograph of cleaved 0.030 in. thick GaAs wafer *BC;* (*b*) intensity variation across the video scan line *AD* photographed from time-delay oscilloscope. From Rozgonyi et al.[135] (Courtesy of the American Institute of Physics.)

The technique appears to be a major step forward, since instantaneous display of an x-ray topograph can be obtained with electronic control of magnification and results equivalent to dental film. The crystal can be reoriented to optimize the image, while the image is being observed. In addition, a quantitative profile of x-ray intensity can be measured.

V. SCANNING X-RAY DISTRIBUTION MICROSCOPY

Scanning x-ray distribution microscopy, which is the newest of the x-ray microscopic techniques, is now a standard feature on a variety of commercial electron microprobe analyzers and is being added to scanning electron microscopes. A portion of the characteristic x-rays excited by an electron beam which is scanned over the sample, emitted in a particular direction, are picked up by a proportional counter and used to produce a visual image on a cathode ray screen of long persistence time, in which brightness is proportional to the local intensity of the x-ray emission. The cathode ray tube scan is synchronized with the sample scan. The image is most commonly recorded by photographing the screen using a single slow frame scan, to avoid loss of resolution due to registry problems, and it may be advantageous to use exposure times of as much as 30 min. Using a proportional counter, with or without a Bragg spectrometer, the wavelength of the x-rays used to form the image may be selected to correspond to a particular element in the sample, so that a series of such x-ray scanning micrographs may be made on a given area of the sample, giving a semiquantitative measure of the elements present in a surface layer about 25 μ in depth at the usual beam voltages. If the elements present are unknown, they may be identified by a preliminary spectral scan. Recently, semiconducting solid state detectors, in conjunction with a multichannel analyzer, have been successfully used (See page 476).

The segregation of elements at particular features is apparent and can be readily correlated with an optical microscope image or with scanning electron micrographs, made using the back-scattered electron or speciment current signals. The micrograph is, in effect, a *composition map*, visually depicting the composition distribution of a selected element. The magnification of the x-ray micrograph is given by the ratio of the width of the displayed image to the width of the region scanned on the sample and may be varied electronically. The magnification may be as high as 1000X. The electron beam can be stopped on a selected spot or scanned to and fro on a line, and quantitative analysis effected using counter techniques in the usual way. Figure 15 on page 500 is an example of the results obtainable.

Analytical information obtainable from a scanning x-ray image is at the best semiquantitative, since, as in ordinary x-ray emission analysis, the experimental accuracy is limited by statistical considerations, that is, by the number of counts recorded in the time available versus the noise level. This is overcome to some extent by photographing the display over a long period, thus integrating the signal. However, other factors such as the variation of the beam voltage and current, wandering of the probe, and contamination of the area under examination, impose a limitation on the useful exposure time. For more quantitative information, it is necessary to make a microprobe analysis on selected spots.

The quality of x-ray distribution images is limited at present by the relatively high background of random counts and the low signal to noise ratio, due to the unfavorable counting statistics, so that it is impossible to obtain resolutions better than a few microns. An improved alternative approach called "concentration mapping," pioneered by Heinrich,[140] employs computer construction of an image, using as input the x-ray signal produced by an automated point by point x-y scan. The background can be greatly reduced by computer processing, better counting statistics employed, and better resolution obtained. Even so, resolutions comparable with the approximately 200 Å achievable with current commercial scanning electron microscopes cannot be expected, since, even if the beam size is reduced below the 1-μ diameter commonly used in the electron microprobe, the volume of material from which x-rays are excited is still several microns in diameter. This can be improved only by the use of samples a micron or less in thickness.

VI. REFERENCES

1. V. E. Cosslett, A. Engström, and H. H. Pattee, Jr., Ed., X-Ray Microscopy and Microradiography, Proc. 1st Int. Symposium in Cambridge, England, 1956, Academic Press, New York, 1957.
2. A. Engström, V. Cosslett, and H. Pattee, Ed., X-Ray Microscopy and X-Ray Microanalysis, Proc. 2nd Int. Symposium in Stockholm, 1959, Elsevier Publishing Co., Amsterdam, 1960.
3. V. E. Cosslett and W. C. Nixon, X-Ray Microscopy, Cambridge University Press, Cambridge, England, 1960.
4. H. Berger, Non-destructive Testing, 20, 185 (May-June, 1962).
5. H. Berger, J. Appl. Phys. 34, (4), 914 (1963).
6. H. Berger, Int. J. Appl. Radiation and Isotopes, 15, 407 (1964).
7. H. Berger, Neutron Radiography, Elsevier Publishing Co., Amsterdam, 1965.
8. J. L. Hoyt, C. E. Porter, and C. D. Wilkinson, Trans. Amer. Nucl. Soc., 10, (2), 446 (1967).

9. W. C. Roentgen, *Sitzungsber, Phys.-Med. Ges. Würzburg*, Part I, 137–141 (1895); Part II, 11–19 (1896).

10. C. T. Heycock and F. H. Neville, *J. Chem. Soc.*, 73, 714 (1898).

11. C. T. Heycock and F. H. Neville, *Phil. Trans. Roy. Soc.*, *Ser. A*, 194, 201 (1900).

12. V. E. Cosslett and W. C. Nixon, *Nature*, 168, 24 (1951).

13. J. J. Trillat, *Metallurgical Rev.*, 1, 3 (1956).

14. J. J. Trillat, *Exploring the Structure of Matter*, translated by F. W. Kent, Interscience Publishers, New York, copyright George Allen and Unwin Ltd., 1959.

15. H. J. Beattie, General Electric Co., Materials and Processes Laboratory, Schenectady, N.Y., private communication.

16. C. S. Barrett and T. B. Massalski, *Structure of Metals*, 3rd ed., McGraw-Hill Book Co., New York, 1966, p. 621.

17. A. M. Davis, *Metallography*, 3, 165 (1970).

18. B. Lindström, *Acta Radiol., Stockholm, Suppl.*, 125, 27 (1955).

19. B. Lindström, in *X-Ray Microscopy and Microradiography*, V. E. Cosslett, A. Engström, and H. H. Pattee, Eds., Proc. 1st Int. Symposium in Cambridge, England, 1956, Academic Press, New York, 1957, p. 443.

20. H. H. Pattee, in *X-Ray Microscopy and X-Ray Microanalysis*, A. Engström, V. Cosslett, and H. Pattee, Eds., Proc. 2nd Int. Symposium, Stockholm, 1959, Elsevier Publishing Co., Amsterdam, 1960, pp. 56, 61.

21. S. K. Asunmaa, in *X-Ray Microscopy and X-Ray Microanalysis*, A. Engström, V. Cosslett, and H. Pattee, Eds., Proc. 2nd Int. Symposium, Stockholm, 1959, Elsevier Publishing Co., Amsterdam, 1960, p. 66.

22. S. K. Asunmaa, in *X-Ray Optics and X-Ray Microanalysis*, H. H. Pattee, V. E. Cosslett, and A. Engström, Eds., 3rd Int. Symposium, Stanford University, 1962, Academic Press, New York, 1963, p. 33.

23. R. G. Ward, "Microradiography and Autoradiography," in *The Physical Examination of Metals*. 2nd ed., B. Chalmers and A. G. Quarrell. Eds., E. J. Arnold & Son, Leeds, England, 1960, Chapter 17.

24. B. L. Henke, in *Proc. 4th Int. Conf. X-Ray Microscopy*, X-Ray Optics and Microanalysis, Orsay, France, 1965, Hermann & Cie, Paris, 1966.

25. W. M. Williams and C. S. Smith, *J. Metals*, 4, 755 (1952).

26. A. E. Jenkinson, *Philips Tech. Rev.*, 23, (3), 82–88 (1961–1962).

27. R. V. Latham, *J. Roy. Microscop. Soc.*, 85, (3), 255 (1966).

28. P. S. Ong, *Microprojection with X-Rays*, Martinus Nykoff, The Hague, 1959.

29. V. E. Cosslett, *Metallurgical Rev.*, 5, 225 (1960).

30. P. S. Ong. in *Techniques of Metals Research*, Vol. 2, Part 1, R. F. Bunshah, Ed., Interscience Publishers, New York, 1968, pp. 159–176.

31. R. M. Sievert, *Acta Radiol.*, 17, 299 (1936).

32. B. M. Rovinsky and V. G. Lutsau, *X-Ray Microscopy and Microradiography*. V. E. Cosslett, A. Engström, and H. H. Pattee, Jr., Eds., Academic Press, New York, 1957, p. 128.

33. W. C. Nixon, *Proc. 3rd Int. Congress for Electron Microscopy*, London, 1954, Royal Microscopical Society, London, 1956, p. 307.

34. B. M. Rovinsky, V. G. Lutsau, and A. I. Avdeyenko, *X-Ray Microscopy and X-Ray Microanalysis*, Proc. 2nd Int. Symposium in Stockholm, 1959, A. Engström, V. E. Cosslett, and H. H. Pattee, Jr., Eds., Elsevier Publishing Co., Amsterdam, 1960.

35. D. R. Schwarzenberger, *Phil. Mag.*, **4**, 1242 (1959).
36. I. K. Openshaw, in *The Electron Microprobe*, Proc. Symposium, Washington, D.C., 1964, T. D. McKinley, K. F. J. Heinrich, and D. B. Wittry, Eds., John Wiley & Sons, New York, 1966, pp. 439–453.
37. H. Anderton, *Sci. Progr. Oxford*, **55**, 337–356 (1967).
38. B. Lindström, *X-Ray Optics and X-Ray Microanalysis*, H. H. Pattee, V. E. Cosslet, and A. Engström, Eds., Academic Press, New York, 1963, p. 13.
39. A. Engström, *X-Ray Optics and X-Ray Microanalysis*, H. H. Pattee, V. E. Cosslett, and A. Engström, Eds., Academic Press, New York, 1963, p. 23.
40. G. A. Boyd, *Autoradiography in Biology and Medicine*, Academic Press, New York, 1955.
41. S. Dales, *J. Cell Biol.*, **18**, 51 (1963).
42. A. M. Glauert, in *Techniques for Electron Microscopy*, 2nd ed. D. H. Kay, Ed., 1965, F. A. Davis Co., Philadelphia, pp. 254–310.
43. L. G. Caro, *J. Roy. Microscop. Soc.*, **83**, 127 (1964).
44. R. H. Condit, in *Techniques of Metals Research*, Vol. 2, Part 2, R. F. Bunsah, Ed., Interscience Publishers, New York, 1969, pp. 877–952.
45. O. J. Huber, J. E. Gates, A. P. Young, M. Pobereskin, and P. D. Frost, *J. Metals, Trans. AIME*, **209**, 918, (1957).
46. C. B. Gilpin, D. H. Paul, S. K. Asunmaa, and N. A. Tiner, *Advances in Electron Metallography*, Vol. 6, ASTM STP 396, American Society for Testing and Materials, Philadelphia, 1966, p. 7.
47. W. M. Rhodes and R. E. Carter, *J. Amer. Ceram. Soc.*, **49**, 244 (1966).
48. A. Kohn, *Rev. Met.*, **48**, 219 (1951).
49. "Chart of the Nuclides," 9th ed., revised to July 1966, Knolls Atomic Power Laboratory, distributed by Educational Relations, General Electric Co., Schenectady, N.Y., November 1966.
50. L. Bachmann and M. M. Salpeter, *3rd European Reg. Conf. on Electron Microscopy*, Vol. B, Prague, Publishing Company of the Czechoslovak Academy of Sciences, Prague, Czechoslovakia, 1964, p. 15.
51. S. R. Pelc and M. G. E. Welton, in *Advances in Optical and Electron Microscopy*, Vol. 2, R. Barer and V. E. Cosslett, Eds., Academic Press, New York, 1968, pp. 151–166.
52. S. R. Pelc, J. D. Coombes, and G. C. Budd, *Exp. Cell Res.*, **24**, 192, (1961).
53. R. L. Fleischer, P. B. Price, and R. M. Walker, *Phys. Rev.*, **133**, A1443, (1964).
54. J. H. Davies and R. W. Darmitzel, *Nucleonics*, **23**, 86 (1965).
55. P. Kirkpatrick and H. H. Pattee, Jr., "X-Ray Microscopy," in *Handbuch der Physik*, Vol. 30, S. Flügge, Ed., Springer-Verlag, Berlin, 1957, pp. 305–335.
56. G. Hildenbrand, "Grundlagen der Rontgenoptik und Rontgesmikroskopie," *Ergeb. Exakt. Naturw.*, **30**, 1–133 (1958).
57. V. E. Cosslett, *Metallurgical Rev.*, **5**, 225–266 (1960).
58. W. W. Webb, "X-Ray Diffraction Topography," in *Direct Observation of Imperfections in Crystals*, J. B. Newkirk and J. H. Wernick, Eds., Interscience Publishers, New York, 1962, p. 29.
59. S. Amelinckx, *The Direct Observation of Dislocations*, Academic Press, New York, 1964.
60. L. V. Azaroff, *Progress in Solid State Chemistry*, Vol. 1, H. Reiss, Ed., Macmillan Co., New York, 1964, pp. 347–379.
61. C. S. Barrett and T. B. Massalski, *Structure of Metals*, 3rd ed., McGraw-Hill Book Co., New York, 1966, pp. 418–430.
62. S. Weissman, in *Fifty Years of Progress in Metallographic Techniques*, ASTM

Symposium, 1966, Atlantic City, ASTM Spec. Tech. Pub. No. 430, American Society for Testing and Materials, Philadelphia, 1968, pp. 141–191.

63. K. Futagami, *Repts. Res. Inst. for Appl. Mech.*, 17, (57), 49 (1969).
64. S. Weissman and Z. H. Kalman, in *Techniques of Metals Research*, Vol. 2, Part 2, R. F. Bunsah, Ed. Interscience Publishers, New York, 1969, pp. 839–873.
65. P. B. Hirsch, "Mosiac Structure," in *Progress in Metal Physics*, Vol. 6, B. Chalmers and R. King, Eds., Pergamon Press, New York, 1956, p. 236.
66. B. E. Warren, "X-Ray Studies of Deformed Metals," in *Progress in Metal Physics*, Vol. 8, B. Chalmers and R. King, Eds., Pergamon Press, New York, 1959, p. 147.
67. P. Gay, P. B. Hirsch, and A. Kelly, *Acta Met.*, 1, 315 (1953).
68. B. W. Batterman, *J. Appl. Phys.*, 30, 508 (1959).
69. S. Weissman, *J. Appl. Phys.*, 27, 389, 1335 (1956).
70. S. Weissman, in *X-Ray Microscopy and X-Ray Microanalysis*, A. Engström, V. Cosslett, and H. Pattee, Eds., Proc. 2nd Intern. Symp., Stockholm, 1959, Elsevier Publishing Co., Amsterdam, 1960, p. 488.
71. P. B. Hirsch, A. Howie, and M. J. Whelan, *Phil. Trans. Roy. Soc., Ser. A*, 252, 499 (1960).
72. B. Ancker, T. H. Hazlett, and E. R. Parker, *J. Appl. Phys.*, 27, 333 (1956).
73. J. B. Newkirk, *Phys. Rev.*, 110, 1465 (1958).
74. J. B. Newkirk, *J. Appl. Phys.*, 29, 995 (1958).
75. M. Yoshimatsu and K. Kokra, *J. Phys. Soc. Japan*, 15, 1760 (1960).
76. M. Yoshimatsu, *J. Phys. Soc. Japan*, 16, 2246 (1961).
77. A. P. L. Turner, T. Vreeland, Jr., and D. P. Pope, *Acta Cryst.*, A24, (4) 452 (1968).
78. J. B. Newkirk, *Trans. Met. Soc. AIME*, 215, 483 (1959).
79. J. M. Lommel and M. L. Kronberg, in *Direct Observation of Imperfections in Crystals*, J. B. Newkirk and J. H. Wernick, Eds., Interscience Publishers, New York, 1962, p. 543.
80. S. B. Austerman, J. B. Newkirk, and D. K. Smith, *J. Appl. Phys.*, 36, 3815 (1965).
81. J. M. Schultz and R. W. Armstrong, in *Direct Observation of Imperfections in Crystals*, J. B. Newkirk and J. H. Wernick, Eds., Interscience Publishers, New York, 1962, p. 569.
82. U. Bonse and E. Kappler, *Z. Naturforsch*, 13A, 348 (1958).
83. U. Bonse, *Z. Physik*, 153, 287 (1958).
84. U. Bonse, in *Direct Observation of Imperfections in Crystals*, Interscience Publishers, New York, 1962, p. 431.
85. S. Weissman, *J. Appl. Phys.*, 27, 389 (1956).
86. S. Weissman, *Trans. Am. Soc. Metals*, 52, 599 (1960).
87. S. Weissman and Z. H. Kalman, in *Techniques of Metals Research*, Vol. 2, (2), R. F. Bunsah, Ed., Interscience Publishers, New York, 1969, pp. 839–873.
88. R. Smoluckowski, C. M. Lucht, and M. Mann, *Phys. Rev.*, 70, 318 (1946).
89. E. Votava, A. Berghezan, and R. H. Gillette, in *X-Ray Microscopy and Microradiography*, V. E. Cosslett, A. Engström, and H. H. Pattee, Jr., Eds., Proc. 1st Int. Symposium, Cambridge, England, 1956, Academic Press, New York, 1957, p. 603.
90. A. Berghezan, P. Lacombe, and G. Chaudron, *Compt. Rend.*, 231, 576 (1950).
91. C. K. Jackson, in *X-Ray Microscopy and Microradiography*, Proc. 1st Int. Symposium in Cambridge, England, 1956, V. E. Cosslett, A. Engström, and H. H. Pattee, Jr., Eds., Academic Press, New York, 1957, p. 623.

92. A. R. Lang, *Acta Met.*, **5**, 358 (1957).
93. A. R. Lang, *J. Appl. Phys.*, **29**, 597 (1959).
94. A. R. Lang, *J. Appl. Phys.*, **30**, 1748 (1959).
95. A. R. Lang and G. Meyrick, *Phil. Mag.*, **4**, 878 (1959).
96. V. F. Miuscov and A. R. Lang, *Kristallografiya*, **8**, 652 (1963).
97. A. R. Lang and V. F. Miuscov, *Phil. Mag.*, **10**, 263 (1964).
98. A. R. Lang and G. D. Miles, *J. Appl. Phys.*, **36**, 1803 (1965).
99. H. Barth and R. Hosemann, Z. *Naturforsch.*, **A13**, 792 (1958).
100. J. S. Makris and C. H. Ma, *Trans. Met. Soc. AIME*, **230**, 1110 (1964).
101. F. C. Frank and A. R. Lang, *Phil. Mag.*, **4**, 383 (1959).
102. A. R. Lang, *Proc. Roy. Soc. Ser. A*, **278**, 234 (1964).
103. Y. Kamiya and A. R. Lang, *J. Appl. Phys.*, **36**, 579 (1965).
104. A. E. Jenkinson and A. R. Lang, in *Direct Observation of Imperfections in Crystals*, J. B. Newkirk and J. H. Wernick, Eds., Interscience Publishers, New York, 1962, p. 471.
105. A. R. Lang, *Acta Cryst.*, **12**, 249 (1959).
106. S. Amelinckx, G. Strumane, and W. W. Webb, *J. Appl. Phys.*, **31**, 1359 (1960).
107. W. W. Webb, in *Growth and Perfection of Crystals*, R. H. Doremus, B. W. Roberts and D. Turnbull, Eds., John Wiley & Sons, New York, 1958, p. 230.
108. K. A. Carlson and R. Wegener, *J. Appl. Phys.*, **32**, 125 (1961).
109. A. Fourdeux, A. Berghezan, and W. W. Webb, *J. Appl. Phys.*, **31**, 918 (1960).
110. G. H. Schwuttke, in *Direct Observation of Imperfections in Crystals*, J. B. Newkirk and J. H. Wernick, Eds., Interscience Publishers, New York, 1962, p. 497.
111. K. Kohra, M. Yoshimatsu, and I. Shimuzu, in *Direct Observation of Imperfections in Crystals*, J. B. Newkirk and J. H. Wernick, Eds., Interscience Publishers, New York, 1962, p. 461.
112. K. Kohra and M. Yoshimatsu, *J. Phys. Soc., Japan*, **17**, 1041 (1962).
113. E. S. Meieran and K. E. Lemons, *Advances in X-Ray Analysis*, Vol. 8, W. M. Mueller, G. Mallett, and M. Fay, Eds., Plenum Press, New York, 1965, p. 48.
114. M. Yoshimatsu, A. Shibata, and K. Kokra, *Advances in X-Ray Analysis*, Vol. 9, G. R. Mallett, M. Fay, and W. M. Mueller, Eds., Plenum Press, New York, 1966, p. 14.
115. M. Polcarová and A. R. Lang, *Appl. Phys. Lett.*, **1**, 13 (1962).
116. M. Hart and A. R. Lang, *Acta Cryst.*, **16A**, 102 (1963).
117. A. R. Lang and M. Polcarova, *Proc. Roy. Soc., Ser. A*, **285**, 297 (1965).
118. E. S. Meiran and I. A. Blech, *J. Appl. Phys.*, **38**, 3495 (1967).
119. J. Chikawa, *J. Appl. Phys.*, **36**, 3496 (1965).
120. A. Fukuda and A. Higashi, *Japan J. Appl. Phys.*, **8**, 993 (1969).
121. B. R. Lawn and H. Komatsu, *Phil. Mag.*, **14**, 689 (1966).
122. O. Nittono, N. Onodera and S. Nagakura, *Japan J. Appl. Phys.*, **9**, 328 (1970).
123. A. W. Ruff and L. M. Kuehner, in *X-Ray Microscopy and X-Ray Microanalysis*, Proc. 2nd Int. Symposium, Stockholm, Sweden, 1959, A. Engström, V. E. Cosslett and H. Pattee, Jr., Eds., Elsevier Publ. Co., New York, 1960, p. 153.
124. G. Borrmann, W. Hartwig, and H. Irmler, Z. *Naturforsch.*, **13A**, 423 (1958).
125. A. Authier, *J. Phys. Radium*, **21**, 655 (1960).
126. V. Gerold and F. Meier, in *Direct Observations of Imperfections in Crystals* J. B. Newkirk and J. H. Wernick, Eds., Interscience Publishers, New York, 1962, p. 509.

127. S. Weissman and Z. H. Kalman, *Phil. Mag.*, **15**, 539 (1967).
128. S. B. Newman, *J. Res. Natl. Bur. Standards, Phys. Chem.*, **67A**, (2) 149 (1963).
129. U. Bonse and M. Hart, *Appl. Phys. Lett.*, **6**, 155 (1965).
130. U. Bonse and M. Hart, Z. *Physik*, **188**, 154 (1965).
131. U. Bonse and M. Hart, Z. *Physik*, **190**, 455 (1966).
132. M. Hart, *Sci. Progr. Oxford*, **56**, 429–447 (1968).
133. J. Chikawa, *Crystal Growth* (Suppl. to *J. Phys. Chem. Solids*), Proc. Int. Conf. on Crystal Growth, Boston, June 1966, Pergamon Press, New York, 1967, p. 817.
134. A. Authier, discussion to Paper 2.28, Int. Union of Crystallography, 7th Int. Congress, Moscow, July 1966.
135. G. A. Rozgonyi, S. E. Haszko, and J. L. Statile, *Appl. Phys. Lett.*, **16**, 443 (1970).
136. E. S. Meiran, J. K. Landre, and S. O'Hara, *Appl. Phys. Lett.*, **14**, 368 (1969).
137. A. R. Lang and K. Reitsnider, *Appl. Phys. Lett.*, **15**, 258 (1969).
138. A. N. Chester and F. B. Koch, *Advan. X-Ray Anal.*, **12**, 165 (1969).
139. J. Chikawa and I. Fujimoto, *Appl. Phys. Lett.*, **13**, 387 (1968).
140. K. F. J. Heinrich in *"Symposium on X-Ray and Electron Probe Analysis,"* Atlantic City, 1963, ASTM Special Technical Publication No. 349, American Society for Testing and Materials, Philadelphia, 1964, p. 163.

8

SCANNING ELECTRON MICROSCOPY

I. Introduction 423

II. Instrumentation 424

III. Theory 430
    A. Electron Beam Diameter 431
    B. Image Display 433
    C. Operating Modes 435
        1. Emissive Mode 435
        2. Reflective Mode 436
        3. Absorptive Mode 437
        4. Conductive Mode 437
        5. Luminescent Mode 437
    D. Image Contrast 438

IV. Applications 442
    A. Metallurgical Studies 442
    B. Determination of Crystal Orientation 446
    C. Observation of Magnetic Domain Structure 447
    D. Semiconductor Examination 449
        1. Surface Voltage Contrast 449
        2. Surface Field Contrast 451
        3. Induced Current Contrast 451
        4. Combined Surface Voltage and Induced Current Contrast 452
        5. Deflection Modulation 453
        6. Induced Conductivity Contrast 453
        7. Stroboscopic Scanning Microscopy 454
    E. Cathodoluminescence Studies 456

V. References 456

I. INTRODUCTION

Scanning electron microscope systems first became available commercially as adjuncts to electron microprobes that provided a means—in some microprobes, the only means—by which the sample surface could be viewed, while under the electron beam, and an area selected for analysis by means of the characteristic x-ray emission. It was soon realized that improved microscopic performance and versatility could be obtained if the design were not dominated by the analytical needs, in particular, the need for an approximately 1-μ diameter electron beam, sufficient beam current, and the choice of geometry to optimize the sensitivity of x-ray analysis using focusing crystal spectrometers. A new class of instruments, scanning electron microscopes, resulted, which are already finding a multitude of applications in all fields of science and engineering, biological as well as nonbiological, as evidenced in recent symposia.[1-6]

Most scanning microscopes are relatively expensive instruments designed for versatility of application. Already, simpler instruments for specialized applications, notably for control and inspection purposes, are being marketed. Furthermore, the gap between microprobes and scanning microscopes is being closed by improvements in solid state x-ray detectors, which, being nondispersive, can collect a larger fraction of the emitted x-rays, making it possible to work with smaller beams and beam currents and yet retain good x-ray sensitivity. These detectors can be used as adjuncts to scanning electron microscopes with a substantial advantage over focusing crystal spectrometers. It is thus possible that the next generation of instrument will again combine the scanning microscope and microprobe analysis functions, with the addition of facilities for scanning transmission electron microscopy and electron diffraction.

The big advantage of the scanning electron microscope is that sample surfaces can be examined directly with a depth of field very much greater than that of the optical microscope at high magnifications, and in some cases with better resolution. Since surface topography can be examined, the technique is competitive with replication electron microscopy, although the latter is currently capable of superior resolution. Replication artifacts are avoided; however, the interpretation of contrast effects in scanning electron micrographs is sometimes far from straightforward. An important difference between the scanning electron microscope and other microscopes is in the nature and variety of what may be called "information-encoding mechanisms" available. This will become clear when we discuss the signals that may be used and the origin of contrast effects. In some cases, unique information can be obtained.

The instrument normally uses the secondary electron, primary back-scattered electron, or specimen current signal, generated by scanning a finely focused electron beam over the specimen surface to modulate the brightness on a synchronized cathode ray display tube. The magnification obtained is simply the ratio of the distance scanned on the display to that scanned on the sample. Since the signal is averaged over the beam diameter, the resolution is for practical purposes limited by the beam size. An instrument has already been built that gives up to 5 Å resolution (see Chapter 6, page 352) which is much better than that currently realized in commercial instruments. In order to realize high resolution, it is necessary that the sample show suitable detail, bearing in mind the signal used; and that the resolution not be limited by other factors such as vibration, the design of the electronic circuits, the features of the display system, stray fields, charging effects, and so forth. It should also be emphasized that the point-to-point resolution at one place in the image is not an adequate criterion of general image quality. There may be image distortion, background noise, banding effects, and other artifacts present.

Bibliographies to the literature are to be found in Refs. 7 through 12. The historical development of the instrument will be briefly reviewed. Stemming from the work of Knoll,[13] the first true scanning microscope was built by von Ardenne in 1938.[14] This employed thin specimens and the transmitted electron signal. Subsequently, Zworykin, Hillier, and Snyder[15] constructed a similar but improved instrument; using the secondary electron current signal, they achieved a resolution of about 500 Å. Developments in France in the postwar years led to the construction of instruments by Leauté (see Ref. 9) in 1946 and by Davoine[16] in 1957. The latter instrument gave a resolution of about 2 μ and was later used by Davoine, Pinard, and Martineau[17] to investigate cathode luminescence. Work in the Cambridge University group under Oatley, begun in 1948, has been reviewed by Oatley, Nixon, and Pease[9] and more recently by Nixon.[18] The work carried out by successive research students, now numbering over 20, involved the construction, application, and improvement of numerous instruments. In 1965 Pease[19,20] and Nixon[20] succeeded in obtaining a resolution of about 100 Å with a scanning microscope provided with three magnetic lenses. The work at Cambridge led to the first commercial scanning electron microscope—the Stereoscan—marketed in 1965 by the Cambridge Instrument Company.

II. INSTRUMENTATION

The typical components of a scanning electron microscope are illustrated in Figure 1 after Belk.[21] The instrument consists of an electron optical column and sample chamber, furnished with a vacuum system, regulated high voltage, and lens power supplies, plus an appropriate electron collector, signal processing, and handling and display electronics. The function of the electron optical column is to provide a narrow electron beam. This is scanned across the sample, by sets of x and y deflector

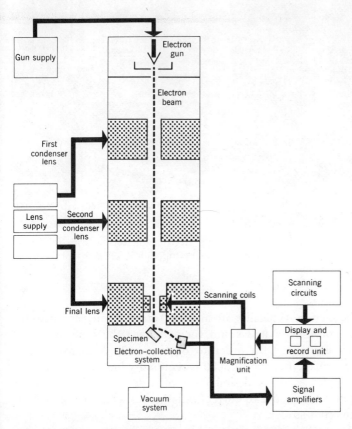

Figure 1 Schematic illustrating the essential components of a typical scanning electron microscope. From Belk.[21]

coils, located inside or just before the final lens, in synchronization with a cathode ray tube display.

The electron source usually consists of a conventional triode electron gun with a tungsten thermionic cathode and a 1 to 30 kV high voltage source. Filament changing is facilitated and column contamination minimized if the gun is provided with a valve to isolate it from the rest of the column. The diameter of the source or crossover of the beam

Figure 2 The Stereoscan Mark IIA scanning electron microscope with solid state x-ray detector (left) and multichannel analyzer (top right). (Photograph courtesy of General Electric Company.)

as it leaves the gun is typically 50 μ. The source is demagnified by a series of electromagnetic lenses and focused onto the surface of the specimen which is inclined. In order to obtain a 100 Å diameter spot at the sample, a demagnification of 5000 is required which can be accomplished by three lenses. It is desirable to provide the final lens with a magnetic stigmator, in order to correct astigmatism which would otherwise increase the spot size.

A variety of specimen stages is desirable, permitting x, y, and z linear motions, tilting, rotation about one or more axes, heating or cooling, and x-ray microanalysis with accommodation for pure standards. Lead-in connections enable working voltages to be applied to microcircuits, while under examination. The specimen is normally insulated and connected to a ground through a suitable high resistance path, so that the specimen current, suitably amplified, can be used to provide a signal if desired. Specimen changing is facilitated if the specimen chamber is provided with a column isolation valve. A stretching stage for the Stereoscan has been designed by Dingley[22] and used to study superplasticity.[23] Lane[24]

Figure 3 Stage of Stereoscan Mark IIA scanning electron microscope, showing specimen platform being inserted. Note the detector for secondary or backscattered electrons and light guide. (Stereoscan-scanning electron microscope is a product of Cambridge Scientific Instruments, London, England, and is represented in the United States by the Engis Equipment Company.)

has built an ingenious wet stage to enable samples to remain moist during examination.

Specimens can be examined mounted or unmounted. If mounted, the use of thermoplastic resins is undesirable, since these tend to decompose in the electron beam and redeposit on the sample. Conducting mounts are desirable and the general practice follows that described in connection with the electron microprobe.

The incident electron beam current on a sample is typically in the range 10^{-9} to 10^{-12} A, corresponding to beam diameters of 1000 to 100 Å. Efficient low noise detectors for electrons are required; these are located inside the specimen chamber. In the Stereoscan Mark IIA Model (Figures 2 and 3), a secondary electron collector is provided which·consists of an electrostatic focusing electrode with continuously variable focusing potential from -30 to $+250$ V. This is suitable for collecting either secondary or backscattered electrons and converting to light by a scintillator. The light travels via a light guide to a photomultiplier and the signal after amplification is used to brightness-modulate the beam of the cathode ray tube. By positively biasing the collector, low energy, that is, secondary electrons, are drawn in curved paths to the collector, thus revealing details inside "shadows" and depressions in the image. By negatively biasing the collector, low energy electrons are deflected away from the collector; whereas primary backscattered electrons, which have high energy, are collected.

When the backscattered electron signal is used, image detail in shadows or depressions is lacking and poorer resolution of topographical detail is obtained than with the secondary electron signal. Since backscattering increases with the atomic number of the target element, the image obtained on a flat surface shows some contrast, related to the compositional variations of an alloy specimen.

A solid state x-ray detector and multichannel analyzer are shown attached to the Stereoscan microscope in Figure 2, making it possible to use the equipment for qualitative or semiquantitative microprobe analysis. The x-ray detector is a lithium-drifted silicon diode. By feeding the appropriate x-ray signal from the analyzer into the brightness input of the cathode ray display, Lifshin, Morris, and Bolon[25] were able to produce x-ray scanning images of good quality. Figure 4, taken from their work, shows the copper and lead distribution in a copper-lead monotectic alloy which may be compared with the backscattered electron image. A twin semifocusing spectrometer attachment for the Stereoscan has been described by Kynaston and Stewart.[26]

Cathodoluminescence, that is light excited from certain materials by the electron beam, can be collected and used for image forming. The light is simply led by a light pipe into a photomultiplier system.

Since the specimen may be tilted by up to about 50°, relative to the incident electron beam, the image is foreshortened. An electronic control can be provided to reduce the probe scan in one direction to correct for this.

The magnification, which is simply the ratio of the distance scanned on the display to that scanned on the sample, which is adjustable, typically ranges from 20 to 100,000 diameters at a working distance of about 10 mm. The use of a greater working distance involves some sacrifice in maximum magnification. The useful magnification is determined by the resolution. Since the eye can readily perceive two points separated by 0.2 mm, if a resolution of 200 Å is desired, the useful magnification which is simply the ratio would be 10,000X. The use of a higher magnification for recording is undesirable, since it will decrease the field size. Somewhat higher magnification is desirable for easy viewing on

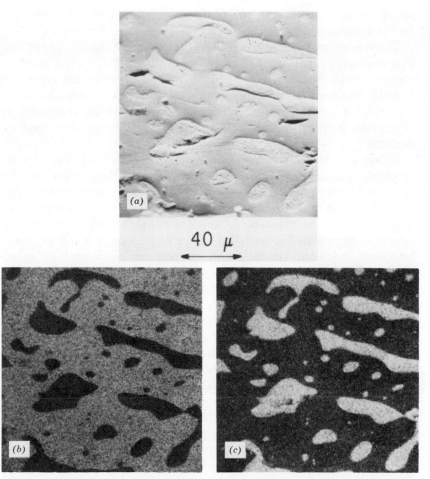

Figure 4 Electron and x-ray distribution micrographs of copper-lead monotectic, taken on a Stereoscan microscope with solid state x-ray detector and multichannel analyzer attachment. (*a*) Backscattered electron image; (*b*) copper-K_α x-ray distribution image (100 sec, 0.5 nA specimen current); (*c*) lead-L_α x-ray distribution image (100 sec, 0.5 nA). From Lifshin et al.[25] (By permission of The Metallurgical Society of AIME.)

the cathode ray display tube. The resolution in the emissive mode is affected by the accelerating voltage used and may be a factor of 5 worse at 1 kV than at 20 kV.

Instead of using the signal to modulate the brightness on the cathode ray display, it may be used to modulate the amplitude of the display, thus providing micrographs from which quantitative measurements of signal intensity can be made. This is also useful in focusing the beam on the sample.

The resolution of the instrument, in common with other types of electron microscope, is liable to be lowered by interference from stray a-c magnetic fields in the region of the column. Such fields are commonly at main frequency or some harmonic thereof and may arise due to the proximity of other electrical equipment—power supplies (including the microscope supply), transformers, motors, switchboards, a-c wiring, and so forth.

Unless used in a vibration-free environment, high resolution may not be obtainable, since the specimen is often of fairly large mass and thus tends to vibrate independently, that is, out of phase with the microscope. To obviate this, it is desirable that the column in a high resolution scanning microscope be mounted on shock absorbers to help isolate it from external sources of vibration. The choice of a location is still of importance and should be remote from reciprocating machinery, air conditioners, and so forth.

The slow scanning speed of the conventional scanning electron microscope makes it difficult to carry out dynamic experiments inside the instrument. Kimoto, Sato, and Adashi[27] developed a standard rapid scan television type display with a $\frac{1}{30}$ sec frame speed to replace the usual 10- to 100-sec speed. While a video recorder can then be used, these devices are expensive. Ciné film gives a permanent record but involves delay in development and is nonerasable. An inexpensive alternative developed by Nixon[28,29] used an audio tape recorder, in conjunction with a pulsed beam, to sample the waveform and record individual scan lines or points on one length of tape, with each line or point properly synchronized.

MacDonald, Marcus, and Palmberg[30] have carried out Auger electron analysis on fracture surfaces in the scanning electron microscope. This is likely to become more useful if a clean high vacuum microscope is employed. Recent developments by Crewe of a high resolution transmission scanning microscope and by Cowley of a high voltage transmission scanning microscope, both incorporating energy analysis, are discussed in Chapter 6, pages 352 and 358, respectively.

III. THEORY

More detailed treatments of the theory of the scanning electron microscope may be found in the literature.[8,9] The electron optical design is similar to that of a conventional electron microscope which is described in standard sources.[31,32]

A. Electron Beam Diameter

Magnetic lenses are normally preferred, since the coils can be mounted outside the vacuum system and the lens surfaces are accessible for cleaning. Cleanliness is necessary to avoid aberrations and charging effects which tend to decrease the resolution. Aberrations are minimized by regulating the constant current lens supplies to better than 50 ppm over, say, 5 min in the first two lenses and 15 ppm in the final lens. The accelerating voltage must be stabilized to avoid chromatic aberration and typical limits are 20 ppm over 10 min at 7.5 to 30 kV, and < 150 mV at 1 to 7.5 kV. If aberrations, stray fields and vibration are all minimized, then the beam diameter d can be approximated after Smith[33] by:

$$d^2 = d_0{}^2 + d_s{}^2 + d_c{}^2 + d_f{}^2 \qquad (1)$$

where d_0 is the diameter of the Gaussian image of the electron source (i.e., the diameter obtained simply from the source diameter and the positions and focal lengths of the lenses), and d_s, d_c and d_f are the diameters of the disks of confusion, due to the spherical aberration, chromatic aberration, and diffraction, respectively.

The origin of spherical and chromatic aberrations in an electron lens is quite analogous to that in a glass lens in a light microscope. The least diameter d_s of the disk of confusion, due to spherical aberration, is

$$d_s = \frac{1}{2} C_s \alpha^2 \qquad (2)$$

where C_s is the coefficient of spherical aberration in the final lens, and α the semiangle of convergence of the beam emerging.

The corresponding least diameter d_c for chromatic aberration is

$$d_c = \frac{\Delta V}{V} C_s \alpha \qquad (3)$$

where $\Delta V/V$ is the energy spread of electrons passing through the final lens, and C_s the coefficient of chromatic aberration.

The diameter d_f (Å) of the Airy disk in angstroms, due to diffraction, is

$$d_f = \frac{15}{\alpha} V^{-\frac{1}{2}} \qquad (4)$$

where V is the electron energy in electron volts.

The diameter d_f can be neglected at high voltages. Now, the brightness, that is, the current density at the sample, j (A/cm^2) of the image of an electron gun source is given by the Langmuir equation which is closely approached in practice by hairpin filaments:

$$j = j_c \left(\frac{eV}{kT} + 1\right) \sin^2 \alpha \tag{5}$$

where j_c (A/cm^2) is the cathode emission current density, e the electronic charge in coulombs, V the accelerating potential in volts, k (J/deg. K) the Boltzmann's constant, and T (°K) the cathode temperature. The semiangular aperture at the cathode is taken to be $\pi/2$.

Pease and Nixon,[20] following Smith,[33] showed that considering only spherical aberration and diffraction, differentiating Equation 1 with respect to the semi-aperture angle α and equating to zero gave an optimum α. By combining this result with j from Equation 5, they obtained an expression relating the theoretical minimum probe diameter d_{min} to the beam current i, as follows:

$$d_{min} = 1.29 C_s^{1/4} \lambda^{3/4} \left(7.92 \frac{iT}{J_c} \times 10^9 + 1\right)^{3/8} \tag{6}$$

where λ (Å) is the electron wavelength,

$$\lambda = \frac{12.4}{V^{1/2}} \tag{7}$$

The theoretical maximum probe current i_{max} was similarly found to be

$$i_{max} = 1.26 \frac{J_c}{T} \left(\frac{0.51 \, d^{8/3}}{C_s^{2/3} \lambda^2} - 1\right) 10^{-10} \tag{8}$$

The semiaperture angle α is simply:

$$\alpha = \left(\frac{d}{C_s}\right)^{1/3} \tag{9}$$

It may be seen from Equation 8 that the x-ray emission, which is directly proportional to the probe current i, is proportional to $d^{8/3}$, so that it falls off very rapidly as the probe diameter is reduced.

Pease and Nixon[20] point out that, in the limit for zero probe current, Equation 6 reduces to

$$d_{min} = 1.29 C_s^{1/4} \lambda^{3/4}$$
$$(i = 0) \tag{10}$$

which is the expression for the resolution of a conventional transmission electron microscope (less than 5 Å). In practice this is not achievable for a variety of reasons.[19] One reason is that V, and hence λ, is generally lower than in a conventional electron microscope, to minimize penetration, resulting in increased chromatic aberration and diffraction as seen from Equations 3 and 4. If a secondary electron signal is employed, the field at the sample, due to the final lens, must be kept small; the design requirements for the lens then result in increased aberration, compared with the conventional electron microscope. The minimum practicable value of the probe current i is determined by such factors as the shot noise, due to the random time intervals between electrons in the beam; the image quality, that is, the signal to noise ratio, there being many sources of noise; and the time necessary to record a micrograph which should not be excessive, either from the convenience or power instability standpoints.

Pease[19] calculated from Equation 6 that $d_{min} = 30$ Å at 30 kV and 50 Å at 15 kV for $C_s = 1$ cm, $C_c = 0.8$, and $i = 10^{-12}$ A, compared with measured values of 50 and 75 Å, respectively. However, the measured point-to-point resolution was inferior, amounting to 100 Å at best and about 200 Å on most specimens which was attributed to the penetration of the primary electron beam into the sample. The secondary electrons are of low energy < 5 eV; therefore only those emitted from within about 100 Å of the surface emerge from the sample. For metal samples, roughly half of these are excited by the primary beam and emerge from an area only slightly larger than this beam. However, the other half are excited by the backscattered electrons and emerge over a much larger area, resulting in diminished resolutions.

B. Image Display

The number of lines per picture has been discussed by Oatley et al.[9] For visual examination of the cathode ray display picture, the line spacing should be just resolvable by eye, that is, should be approximately 0.02 cm. Thus 400 lines would be sufficient on a 7.5-cm square raster. Since smaller detail can be recorded photographically and subsequently enlarged, the choice of the number of lines N would normally lie between 400 and 1000, and the cathode ray tube must be capable of resolving whatever number is chosen. Since the same number of lines N is scanned on the sample, and it is desirable that these be just touching in order to avoid missing sample features, the optimum distance scanned on the sample will be a square of side l which, for a probe diameter d, is

$$l = Nd \qquad (11)$$

A more sophisticated treatment would need to take account of the probe shape and the distribution of intensity within the beam. These depend on the exact shape of the crossover of the electron gun and the distribution of current within it, which in turn vary with the geometrical adjustment of the gun.

If a square of side L is scanned on the final display, the magnification M is then:

$$M = \frac{L}{l} = \frac{L}{Nd} \tag{12}$$

Thus, for a 1000 Å probe diameter, a 7.5-cm square image and 1000X magnification, a 750-line picture would be needed to assure optimum resolution of both the final image and the sample. Since the eye could only then discern about 0.02 cm in the final image equivalent to 2000 Å on the specimen, additional resolution should be obtained by photographically recording and enlarging to 2000X. Further enlargement would result in "empty magnification," that is, magnification with no improvement in resolution. Similarly, increasing the display magnification above 2000X in the usual way by decreasing the distance l scanned on the sample would again result in empty magnification, unless the probe diameter is reduced below 1000 Å.

Shot noise will be present in the final image, since the arrival of electrons at the specimen is a random process that is subject to statistical fluctuations, very small beam currents are used, and the dwell time on a resolvable specimen detail is very small.[34] The brightness of such a detail in the display image is determined by the number of electrons passing from specimen to collector, while the corresponding specimen detail is being scanned. Oatley et al.[9] point out that, if a time t is required to scan an object area of side l, then the number n of electrons reaching a just resolvable, that is, beam-sized specimen detail of side d_0 is

$$n = jd_0^2 \cdot \frac{d_0^2 t}{l^2 e} \tag{13}$$

where j is the current density of the beam (assumed square) at the sample, and e the electronic charge.

The rms fluctuation in n will be $\sqrt{n}$, so that the signal to noise ratio of the element being scanned is $n/\sqrt{n}$ or $\sqrt{n}$. We shall neglect any worsening of this ratio during production of the final image. In order for the eye to distinguish an area of brightness B in the final image from an adjacent one of brightness $B \pm \Delta B$, Rose[35] stated that:

$$\sqrt{n} \geq \frac{5B}{\Delta B} \qquad (14)$$

giving with Equation 13:

$$25 \left(\frac{B}{\Delta B}\right)^2 \leq j \frac{d_0{}^4 t}{1^2 e} \qquad (15)$$

Empirical studies[4] suggest that the value of 25 in Equation 15 is more probably about 100:

$$100 \left(\frac{B}{\Delta B}\right)^2 \leq j \frac{d_0{}^4 t}{1^2 e} \qquad (16)$$

Using conventional electron guns which give a value of current density j close to Equation 5, it may be shown from Equation 16 that, in order to combine maximum resolution (minimum d_0) with distinguishable brightness and obtain photographic recordings, scanning times of up to several minutes may be needed. Problems of stability and scan registry now become important. The scan registry problem can be avoided by using only a single frame for recording. Typical operating (recording) conditions for a resolution of 200 Å would be 10^{-12} A beam current, 5000X magnification, and 100 sec time for one frame; or for 2000 Å, 10^{-8} A, 500X, and 2 sec.

It might be expected that a scan rate of 60 or more frames/sec would be necessary to avoid flicker in visual presentation. Since noise necessitates the use of much slower rates, a long persistence phosphor screen is commonly used for visual purposes and a separate short persistence screen for photographic recording.

C. Operating Modes

The terminology used to describe the operating modes of a scanning electron microscope has not yet become standardized. The physics of the interaction of the electron beam with the sample, which leads to the understanding of image contrast mechanisms, has been considered by Thornton[8] and will not be discussed here in detail.

1. Emissive Mode

In the emissive mode of operation, the current of emitted secondary electrons, which is low in energy compared with the primary incident beam, is used as a signal to modulate the image brightness. Topographical contrast is obtained, and since, as already mentioned, the secondary electrons follow curved trajectories into the collector, details in reentrant holes and behind protrusions may be visible. Since surface potential differences alter the trajectories of emitted secondary electrons, it is possible to observe directly electrical potential differences of 0.5 V or less of individual elements of solid state integrated circuits, while they are in operation,[36] and observe voltage contrast superimposed on topographical contrast. Resolutions of about 200 Å are obtainable in these emissive modes, under ideal conditions.

Since leakage magnetic fields from the specimen surface alter the trajectories of emitted secondary electrons, it is possible to observe directly domain structures in magnetic materials, using special detector arrangements.[37] Domain structures are relatively coarse, so that micrographs so far obtained show resolutions of a few microns.

Compositional contrast may be obtained in the emissive mode, using a characteristic emitted x-ray signal as in the scanning electron microprobe, which is discussed elsewhere. Sufficient beam voltage must be used to excite characteristic emission from the element of interest. The poor resolution ($< 1\ \mu$), theoretically obtainable in such images (from bulk samples), limits their usefulness. Even with a 1-μ probe diameter and using a focusing x-ray spectrometer, the signal from a minor element segregated in the sample may be so weak that photographic exposure times of up to $\frac{1}{2}$ hr are necessary. This situation may be improved, using a solid state detector, to increase the fraction of emitted x-rays picked up, at the sacrifice of some elemental discrimination. For further discussion see page 475 in Chapter 9.

2. Reflective Mode

In the reflective operation mode, backscattered electrons, which have the same energy as the primary beam, are used as a signal. These high energy electrons follow straight line paths into the collector and hence tend to give high contrast images from topographical features, with details lacking in shadows and reentrant features. Resolutions are substantially poorer than when the secondary electrons are used (see pages 440–441). Compositional contrast is also obtained, since the proportion of primary electrons backscattered increases as the average atomic number of the specimen increases. Superposition of compositional contrast on topographical contrast may be avoided by metallographically preparing the specimen surface. Elements two to three atomic numbers apart can usually be distinguished on a flat surface.

3. Absorptive Mode

In the absorptive mode the specimen current is used as a signal. Since this signal principally represents the difference between the incident beam current and the reflected and emitted currents, contrast is a function of both the emitted and reflected signal strengths, and can reveal both topographical detail and compositional variations. The resolution and/or discrimination depend upon the form of the image contrast. The penetration may vary from about 20 Å to 10 μ depending upon the potential of the primary beam and the density of the specimen surface.

Electrons collected by the sample from indirect sources constitute noise; these may be backscattered electrons backscattered again from the lens, secondaries emitted from the sample and then recollected, and secondaries from the lens excited by backscattered or secondary electrons from the sample. The magnitude of these currents has been investigated by Heinrich.[37,38] Secondary electrons from the lens can be suppressed by interposing a suitably biased grid.[39] Backscattered electrons from the lens can be minimized by attention to design.

4. Conductive Mode

If the sample is a semiconducting device suitably biased, subsurface conductive currents may be generated as the primary beam sweeps across certain points in the sample and may be collected as in the absorptive mode, giving unique information in the image. For example, if a reverse bias is applied to a conventional p-n junction it appears mainly across the thin highly resistive "depletion layer" between the layers of P- and N-type semiconductor. A burst of electron/hole pairs is created when a high energy electron beam strikes this region, resulting in a current flow and thus an image of the subsurface depletion layer in the display picture. This may be correlated with surface voltage images. Resolutions of 1000 to 3000 Å are possible. The penetration may vary from about 20 Å to 10 μ, depending upon the incident beam energy and the nature of the sample.

5. Luminescent Mode

In the luminescent mode, light emitted from the specimen as a result of excitation by the primary beam is used as a signal. The phenomenon of cathode luminescence, as it is called, is described in Chapter 9 on electron microprobes. Resolutions of 1000 Å or better are obtainable with a large depth of field.

D. Image Contrast

Scanning electron microscopes are normally operated in the emissive, reflective, or absorptive modes, using secondary electron, backscattered electron, or specimen current signals, most commonly the secondary electron signal. The principal factors determining the contrast in the image are then usually the surface topography and average atomic number.

When a solid target is hit by a beam of relatively high energy electrons, some of the electrons are reflected or backscattered elastically by collisions with target atoms, with their energy unchanged. Backscattered electrons, escaping from the target, may therefore come from a considerable depth of up to several microns, depending on the energy of the original beam and the penetrability of the target which decreases with an increase in the average atomic number. Other primary electrons lose their high energy by ionizing target atoms (inelastic collisions), but the electrons thus stripped from the target atoms may retain sufficient energy to cause further ionization. Electrons thus diffuse outward from the track of the primary electrons, producing a cascade of ionizations, until their energy is dissipated, or until they reach the target surface. If their energy exceeds the surface barrier energy, they can escape from the target and be picked up as a secondary electron signal. In silicon, for example, 3.3 eV is needed to create a free electron by this ionization process, so that a single 10-keV primary electron can create up to 3000 secondary electrons before its energy is dissipated. (This calculation ignores the fact that there are other competitive energy dissipation processes operating.) However, only secondary electrons produced within about 100 Å of the surface have any chance of escaping. This has important consequences on the specimen contrast, as illustrated in Figure 5 after Thornton.[40] Primary or secondary electrons that do not escape constitute the specimen current which may be led to ground via a suitable resistance path to provide a signal.

It will be seen from Figure 5 that, as the angle between the incident beam and the specimen surface is decreased, the proportion of secondary electrons escaping from the surface, that is, the yield, is increased. Thus local surface tilts on the sample will give contrast differences. Small, approximately beam-sized asperities, or edges, give enhanced emission (Figure 5b); conversely, depressions will tend to give decreased emission. No secondary emission may be obtained from an insulating inclusion, such as an oxide particle in a metal sample, if this charges up and repels the primary beam (Figure 5c), although the backscattered electron signal will be greatly enhanced if these electrons reach the detector. Finally, the primary penetration will be decreased as the average atomic number of the target is increased and, in general, the secondary yield will be increased, particularly at low primary beam energies (Figure 5*d*).

The secondary electron yield goes through a maximum at about 1

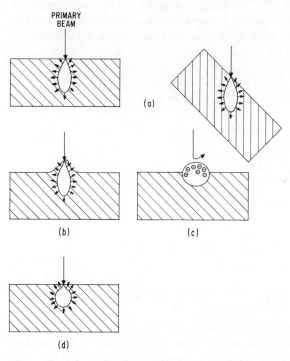

Figure 5 Schematic illustrating the origin of a number of contrast effects, using the secondary electron signal, after Thornton.[40] The area penetrated by the primary electron beam is shown unshaded. Small arrows indicate excited secondary electrons. Note how tilting the sample away from the normal incidence in (*a*), or the presence of a small protrusion on the sample surface in (*b*), increase the secondary yield. The charging of an insulating region in (*c*) has repelled the primary beam, so that no secondary electrons are obtained. In (*d*) the target has a higher atomic number than in (*a*) and (*b*), so that the primary beam energy is dissipated more rapidly, resulting in the generation of more secondary electrons near the surface and a bigger secondary signal.

kV, and may be as high as 1.0 for the lighter elements and 1.7 for the heavier elements, compared with about 0.15 at 30 kV (see Figure 6 after Morris, Lifshin, and Bolon[41]). At 30 kV, which is at the upper end of the operating range of the scanning microscope, the secondary yield is relatively insensitive to change in atomic number above about 30, whereas the backscattered yield increases with atomic number and is substantially greater than the secondary yield, particularly at higher atomic number (Figure 6). The backscattered yield is fairly independent of the operating voltage over the range normally used. These backscattered electron images show better compositional discrimination than secondary electron images, with specimen current images intermediate.

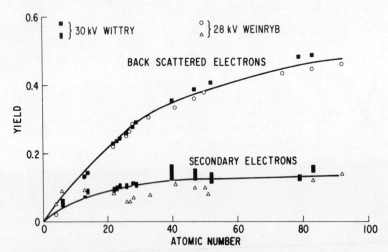

Figure 6 Variation of primary backscattered and secondary electron yield with atomic number of target at about 30 kV. After Morris et al.,[41] using data from Wittry[42] and Weinryb.[43]

In view of the high yields obtainable, the use of secondary electrons at low accelerating voltage offers a considerable advantage in signal to noise ratio, when the specimen is of low average atomic number. Unfortunately, although a resolution approaching 200 Å may be realized at 20 kV, it is difficult to achieve better than 1000 Å at 1 kV, since the spot size is increased by such factors as increased lens aberrations, increased diffraction effects at the longer wavelength, decreased theoretical current density, and increased sensitivity to stray fields.

The backscattered electron signal emanates from a surface layer 1 to 3 μ thick and represents the average properties of this layer. Furthermore, these electrons emerge from an area on the surface of several microns diameter, so that the resolution is limited by this effective source size, rather than the actual beam diameter. The range of these electrons within the sample can be reduced by reducing the electron energy. A reduction from 30 to 20 kV should nearly double the resolution. The secondary electron signal emanates from a surface layer about 100 Å thick and mainly emerges from an area equal to that of the incident electron probe. The specimen current signal combines a good measure of the depth-averaging property of the backscattered signal with an effective source size, approaching that of the secondary signal.

The nature of surface topographical contrast depends on the signal used and the position of the electron detector, as well as on the surface detail. Images formed using specimen current are free from those shadows in the backscattered image which result from the fact that electrons travel in a straight line to the detector which is located obliquely. Since secondary electrons can follow curved paths to the detector, detail in shadows tends to be visible in secondary electron images, although absent in backscattered images. Since, as already discussed, the secondary yield is much less sensitive to composition, in general, than the backscattered yield, most of the detail in the secondary electron images relates to the surface topography. On the other hand, if backscattered detectors are located 180° apart and their signals added or subtracted, then Kimoto and Hashimoto[44] showed that images based primarily on composition or topography, respectively, can be obtained.

The surface detail depends a great deal on the nature of the sample. In the case of a fracture surface, the detail is predetermined. It is usually desirable, however, to make stereoscopic pairs of micrographs, recording the same field twice but at different inclinations to the beam, obtained by tilting the sample a few degrees or more about a single axis normal to the beam. Since the surface is usually very rough, superior detail will probably be obtained if secondary electron or specimen current imaging is employed, rather than backscattered imaging. In the case of a standard metallographic sample, this may either be examined unetched, or after etching to produce topographical detail such as boundaries, etch pits, and so forth, related to the structure of interest. In the case of a polished unetched sample, topographical detail is absent and it is usually advantageous to employ specimen current or backscattered imaging to bring out compositional contrast detail.

Materials with a resistivity exceeding about 10^8 ohm/cm tend to charge up under the incident beam and may develop a potential great enough to cause dielectric breakdown. The resulting varying surface potential may cause loss of resolution. If breakdown cannot be avoided by reducing the beam potential, the problem can usually be overcome by evaporating a thin ($\sim$100 Å) layer of carbon or aluminum onto the sample. Nonconducting samples are normally coated as a matter of course.

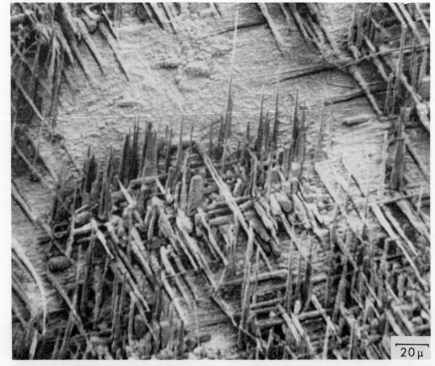

Figure 7 Scanning electron micrograph of aluminum oxide capacitor dielectric. The aluminum foil was etched with a "tunneling etch," oxidized, and the matrix dissolved, leaving the oxide formed inside the tunnels. From Lifshin et al.[25] (By permission of The Metallurgical Society of AIME.)

In the case of semiconductors and semiconductor devices, a number of other types of contrast are available, namely, surface voltage, surface field, induced current, and induced conductivity. (These are discussed on pages 449 to 454.)

IV. APPLICATIONS

A. Metallurgical Studies

No attempt will be made to review the myriad applications already found to all types of materials in many fields of science (see Refs. 1 to 6 on symposia and the bibliographies in Refs. 7 to 12). Figures 7 through 13 show a few examples, mainly from the field of metallurgy, taken using the secondary electron signal by Lifshin et al.[25] Scanning

Figure 8 Scanning electron micrograph of directionally solidified Cu-Pb mono-tectic alloy, etched to remove copper-rich matrix phase at the surface. From Lifshin et al.[25] (By permission of The Metallurgical Society of AIME.)

microscopy, combined with selective etching, has a unique capability for revealing the shape, size, and distribution of second phase networks (Figures 7 and 8). Figure 9 is one of a collection of remarkable micrographs, taken on lunar material brought back from the Apollo 11 mission by Commander Neil A. Armstrong. The samples were gathered 12 miles south southwest of crater Sabine D in the southwestern part of the Mare Tranquillitatis. The spherical shape of the glassy spherule is indicative of prior melting which was one of the key findings of this mission.

The potentiality of scanning electron microscopy as an alternative to replication electron microscopy for studying deformation markings on deformed single crystals is indicated in Figure 10 which shows basal slip in a crystal of cobalt. Kotval[3] has studied the surface shears on a polished surface, resulting from martensitic transformation. The surface topography of fibers used in making composite materials is of considerable

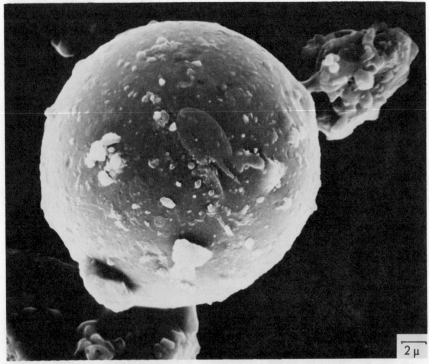

Figure 9 Scanning electron micrograph of 23-μ diameter lunar rock from Apollo 11 mission. The spherule's shape indicates that it underwent melting. Micrograph by Roger B. Bolon. (Courtesy of General Electric Company.)

importance. Figure 11 shows the grooved surface of carbon fibers. The topographical features of scale layers on oxidized or corroded metals are of importance in understanding both the growth mechanism and kinetics. Figure 12 shows the sulfide scale on a nickel-chromium alloy.

Scanning electron microscopy is a natural tool for fractography (see Refs. 3 and 5). Whereas samples such as fractures will normally be examined directly, standard metallographic samples in the form of sections may be examined unetched or after etching chemically, cathodically or otherwise. In the unetched condition, detail is revealed by compositional contrast only; thus, the comparison of scanning micrographs of the same area before and after etching may be useful in distinguishing compositional from topographical contrast. Scanning microscopy shows considerable promise as a tool for phase identification in complex alloys, offering

20 μ

Figure 10   Scanning micrograph of slip bands on deformed single-crystal whisker of cobalt. From Lifshin et al.[25]   (By permission of The Metallurgical Society of AIME.)

an alternative to selective etching procedures, and is applicable where the second phase regions are too small for electron microprobe analysis in situ. Figure 13 shows the structure of an unetched nickel-base superalloy.

Undoubtedly, there is a great future for dynamical experimentation inside the scanning microscope, parallelling that which has already taken place in conventional transmission electron microscopes. Dynamical observations can be made in present commercial instruments, provided that they occur slowly with respect to the frame scanning rate. See Wells[7] for a bibliography of such applications. Special specimen stages will be used permitting deformation, cyclic straining, oxidation, heating, cooling, and so forth, inside the scanning microscope, combined with teletape, ciné, or magnetic tape recording.

The application of the scanning microscope to computerized quantitative metallographic measurements such as grain size determination, volume fraction analysis, and size and shape distribution is another important area under development. Progress has been reported by White et al. (see Ref. 1), White et al.,[45] Dorfler and Russ,[46] and Boyde.[47]

Figure 11   Scanning micrograph of carbon fibers. From Lifshin et al.[25] (By permission of The Metallurgical Society of AIME.)

B. Determination of Crystal Orientation

At very low magnifications of 20 to 100 diameters, crystallographically oriented band patterns may be observed in images of the surface of thick, single-crystal specimens.[48-54] The Coares' patterns were superimposed on the normal topographical image and were obtained with both the backscattered electron and specimen current signals. The bands were qualitatively interpreted[55] in terms of anomalous absorption effects, similar to those observed for x-ray emission, and may be considered as "inverse" channeling patterns. Recently, a quantitative many-beam dynamical theory has been formulated.[56] The band patterns have been observed on GaAs, Cu-10 at.%Al, Ge, Si, Cu, and Co,[48-55] and are probably quite general in occurrence. They may be used to determine the crystal orientation.[48-50,52] The technique is probably limited to single or coarse grained samples (> 1-mm diam. grain size), since the bands are hard to recognize

Figure 12   Scanning micrograph of surface of chromium sulfide scale on Ni-Cr alloy, sulfided in a mixture of hydrogen and hydrogen sulfide gases. From Lifshin et al.[25] (By permission of The Metallurgical Society of AIME.)

at high magnifications. The bands normally terminate at a grain boundary. Subboundaries are revealed as a consequence of a change in band direction.[52] If a crystal is exactly oriented for Bragg reflection, it is predicted[55] that contrast should be observable at dislocations of appropriate Burgers vector running to the surface, although this may necessitate instrumental resolutions < 100 Å. The technique has been used to study 60-keV cadmium ion damage in gallium arsenide.[54]

Tilt or rotation of the specimen moves the bands exactly as occurs for Kikuchi bands in electron diffraction patterns; however, their origin is different.

C. Observation of Magnetic Domain Structure

The presence of magnetic domains inside a ferromagnetic sample gives rise to small variations in the surface magnetic field near the sample.

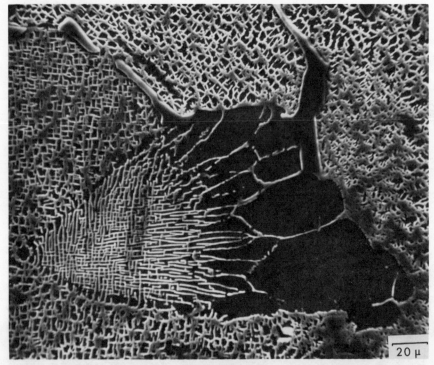

20 μ

Figure 13 Scanning micrograph of as-polished microsection of nickel-base super-alloy, showing low-melting eutectic phase at the center, massive carbides at the grain boundaries, and fine γ' particles (dark contrast) in the grains. From Lifshin et al.[25] (By permission of The Metallurgical Society of AIME.)

This results in a deflection of the secondary electrons, due to the Lorentz force of the magnetic field, enabling the domains to be made visible in a scanning image under suitable conditions.[57-64] In addition to the internal domain structure, there is normally present near the surface a different pattern of "closure-domains" which minimize the leakage field from the sample. An interesting feature of the scanning microscopy technique is that not only the surface domain pattern but also the underlying internal domain pattern can simultaneously be imaged.[58] The first successful attempts at observing magnetic contrast employed samples consisting of magnetic tape, on which a saturated sine wave had been recorded.[65,66] Domains in single and polycrystalline cobalt have now been imaged.[57,58]

Typically, a rather low beam voltage ≤ 5 kV is employed. The magnetic contrast is sensitive to both the secondary electron collector geometry and to the voltage applied to the collector grid, disappearing when the voltage is too low. In order to obtain magnetic contrast, some electrons must have undergone a Lorentz deflection large enough for them to avoid the collector. The efficiency of the collector must be a function of the direction in which the electrons are traveling, as well as of their energy. Thus, instead of employing an open collector, it is necessary to employ an apertured collector. The resolution for magnetic domains is inherently high (better than 1000 Å), but the sensitivity is at present less than for the conventional colloid technique.[62] However, a wide temperature range can be studied and time-varying fields examined, using a stroboscopic technique as for scaler potentials.[62]

D. Semiconductor Examination

Applications of the scanning electron microscope to semiconductors have been discussed by Thornton.[8] The surfaces of integrated circuits and semiconductor devices can readily be examined with or without application of external bias.[67,68] Faults such as poor registration, surface scratches, poorly evaporated interconnections, improperly marked diffusions, and defective oxide layers can be detected as a result of topographical and compositional contrast. Additional information may be obtained by making observations while a voltage is applied to the device. These techniques will now be discussed.

1. Surface Voltage Contrast

If only those secondary electrons within a certain energy range are collected and used to modulate the image, a variation in signal can be obtained dependent on the surface potential at a reverse-biased P-N junction. In this way the junctions in germanium and gallium phosphide were detected.[69] Surface voltages as small as 0.25 volts are apparent.[70] Figure 14 from Matta[71] illustrates voltage contrast obtained in this way on an integrated circuit, when a $+2$-V bias was applied to the common collector region. Two of the three transistor base regions now appear white as a result of the -2-V bias on these regions. Voltage contrast enables a fault, indicated by an arrow in Figure 14a, to be seen clearly.

In certain instances, potential contrast may be induced by the scanning beam without bias applied to a device. Thus Chang and Nixon[72] observed potential contrast in planar transistors, attributed to the movement of electron hole pairs generated near a P-N junction across the junction, thus changing the potential distribution. Shaw and Booker[73] described an alternate method applicable to semiconductors without the need for a P-N junction.

Saparin, Spivak, and Stepanov[74] detected voltage contrast, due to a

450 Scanning Electron Microscopy

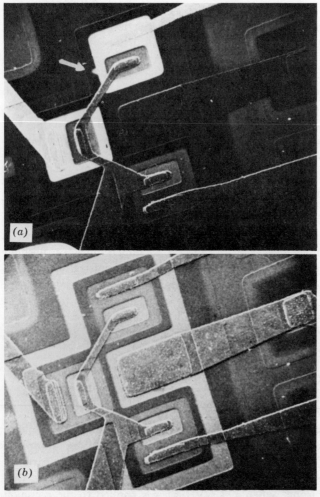

Figure 14 Scanning electron micrographs of integrated circuit. (*a*) 2-V bias applied; (*b*) unbiased, obtained using secondary electron signal. Illustrating voltage contrast. From Matta.[71] (Courtesy of Illinois Institute of Technology Research Institute.)

bias of only 10 mV across a *P-N* junction, using a novel technique in which an a-c voltage of known frequency was applied and a video amplifier employed tuned to this frequency. Thus the image was formed only by the component of the secondary electron signal which was modulated at this frequency.

Driver[75] used a spherically symmetrical detector in a scanning electron microscope, in order to facilitate observation of surface voltage and other forms of contrast by removing topographical and atomic number contrast from the image.

Recent developments have been aimed at quantitative potential mapping or measurement, rather than just the qualitative observation of voltage contrast. MacDonald[76] has developed a computerized system, using the shift in energy of an Auger electron peak to measure the surface potential. Wells[77] incorporated an electrostatic energy analyzer in a commercial scanning microscope between the specimen and electron collector to determine the energy of the secondary electrons, and hence the surface voltage. Banbury and Nixon[78,79] developed a multipurpose cylindrical electrostatic detector, capable of distinguishing between the horizontal and vertical components of the vector field, whether electric or magnetic. Flemming and Ward[80] employed two supplementary electrodes, placed close to the specimen, and electronic circuitry to display a quantitative contour map of the voltage distribution. Gaffney[81] extended the technique of Saparin et al.[74] by using a phase lock amplifier and a time-varying applied potential and were able to map potential gradients quantitatively.

2. Surface Field Contrast

Most observations of voltage contrast have been made on P-N junctions having narrow depletion layers. Spivak, Saparin, and Pereverzev[82] showed that surface fields, as distinct from differences in surface voltage, can contribute to the secondary electron signal if the depletion layer is wide, compared with the scanning spot size. Thornton, Culpin, and Drummond[83] observed similar effects on bulk samples. It is not certain whether the surface fields change the secondary electron emission or whether they simply change the emitted electron trajectories.

3. Induced Current Contrast

The electron hole pair flow, mentioned above in connection with surface voltage contrast, also implies an induced current flow across a P-N junction if the minority carriers cross the junction before recombining. The maximum induced current results when the primary beam impinges in the vicinity of the depletion region. The P-N junction may thus be imaged if the induced current amplitude is used to modulate the displayed image.[84] Figure 15a, after Matta,[71] shows a planar transistor imaged in this way; the leads and bonding pads are visible because they reduce the beam penetration. Considerably more detail and a three-dimensional effect are obtained if the induced current gradient is used as a signal (Figure 15b), that is, if the induced current is differentiated, thus elimi-

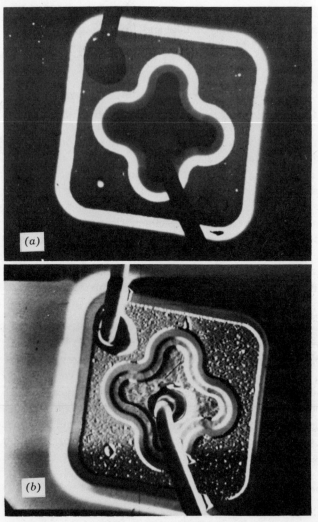

Figure 15 Scanning micrograph of planar transistor, illustrating (*a*) induced current image; (*b*) differentiated induced current image. From Matta.[71] (Courtesy of IITRI.)

nating the d-c component and permitting the use of the amplifier at a higher gain.

4. Combined Surface Voltage and Induced Current Contrast

The electronic processing of signals permits a great deal of flexibility in the image presented. Thus Everhart, Wells, and Matta[85] demonstrated the effect of mixing an induced current signal with a secondary electron

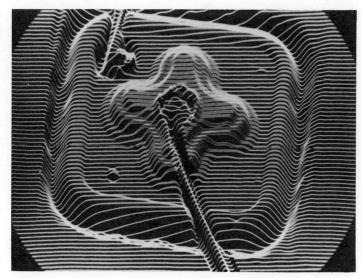

Figure 16 Scanning micrograph, illustrating deflection modulation. From Matta.[71] (Courtesy of IITRI.)

signal from an integrated circuit, with and without inversion of one of the signals which reverses the contrast. An optimum choice may thus be made of the contrast and of the features to be displayed.

More quantitative information is obtained[71] if a line scan is made across an integrated circuit, for example, and the surface voltage and induced current separately displayed or chart-recorded. Information about minority carrier diffusion lengths and lifetimes is obtainable by analysis of the slopes and amplitudes of the signals. The depletion layer widening in silicon diodes has been measured as a function of applied voltage in a similar manner.[86,87]

5. Deflection Modulation

Instead of using a signal to modulate the brightness of the beam in the cathode ray display, a constant amplitude beam may be employed, and the signal used to deflect the beam perpendicular to the scan direction by an amount proportional to the signal. Figure 16 shows an example of this deflection modulation after Matta.[71] Striking pseudo three-dimensional images are obtained.

6. Induced Conductivity Contrast

Silicon devices such as integrated circuits are commonly passivated, that is, are coated with an insulating oxide layer a few hundred to a few thousand angstroms thick. When exposed to the scanning microscope beam, this insulating surface assumes very nearly the same potential as the underlying semiconductor surface, so that surface voltage contrast, as discussed above, is observed at the underlying detail as if the passivating layer were absent. The device shown in Figure 14 was, in fact, passivated. The relationship between the semiconductor potential and that of the oxide surface has been attributed to electron beam induced conductivity in the oxide.[87-90] Energetic incident electrons generate electron hole pairs in the oxide, as well as in the semiconductor layer below. These carriers move freely through the oxide, until they are trapped or recombine. Green[91] coated the oxide layer with aluminum and used the current flowing through an external circuit to the semiconductor substrate to modulate the scanning image, thus obtaining a conductivity map of the oxide. The image revealed defects in the oxide not detectable by other means.

7. Stroboscopic Scanning Microscopy

Since the voltages used by most microelectronic devices vary rapidly and periodically with time, two sorts of information are needed to describe these high frequency operations, as distinct from the static voltage contrast distribution already described. The first can be obtained by stopping the probe at a given point on the surface and monitoring the voltage variation with time. The second, the voltage distribution at a chosen point in time, that is, at a chosen phase, may be obtained by a stroboscopic imaging technique developed by Plows and Nixon.[92,93] If the voltage distribution on the specimen surface varies periodically with time and the electron collector is sensitive to the specimen voltage, then the signal is the sum of the voltage, due to the surface properties plus a periodic variation. By suitably gating the video chain, for a time short compared with the period of a fluctuation, at the same phase during each cycle, a display image is obtained, made up of an array of evenly spaced dots whose intensity is modulated by the voltage. Using a 1-sec frame scan, a continuous image is obtained at specimen voltage frequencies greater than about 250 kHz.

The most suitable arrangement is to pulse the incident electron beam at a pulse repetition frequency equal to the specimen voltage frequency. The time duration of the pulse is made short compared with the period. The progress of a voltage wave through a MOST ladder device can be made visible in this way, as in Figure 17 after Plows and Nixon.[92] At a given frequency, the spatial resolution decreases as the time resolution is improved. However, it appears that, with further refinements, spatial resolution of approximately 0.1 μ may be achievable with sampling

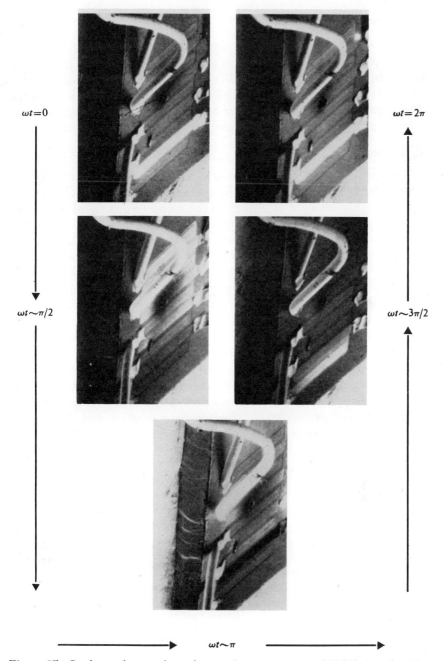

$\omega t = 0$

$\omega t \sim \pi/2$

$\omega t = 2\pi$

$\omega t \sim 3\pi/2$

$\omega t \sim \pi$

Figure 17 Stroboscopic scanning micrograph sequence on a MOST (metal-oxide-semiconductor-transistor) ladder device, after Plows and Nixon.[92] A 5-V peak to peak, distorted, 7-MHz sine wave was applied to gate 1 and the displayed phase varied through the sequence in the direction indicated by the arrows. (Courtesy of The Institute of Physics and The Physical Society.)

455

times of $\frac{1}{100}$ of a period. The absolute time resolution may be < 0.1 nsec. A plot of the time variation of the voltage may be made by stopping the probe at a selected point and sweeping the sampling phase through the specimen voltage period in a similar manner to a sampling oscilloscope. The technique should be useful for analysis of very high speed switching circuits, since it causes no distortion of the waveform under observation.

E. Cathodoluminescence Studies

The field of cathodoluminescence studies has been reviewed by Thornton.[8] Studies fall into two groups, (*a*) those using a scanning beam and employing the light excited from the specimen as a signal to produce a scanning image, and (*b*) those employing a stationary beam and making an analysis of the spectral distribution in the light emitted from a given location on the sample.[94] The former technique has been applied to ionic crystals,[95,96] phosphor powders,[97,98] GaAs,[99–102] and minerals.[103] The presence of dislocations and growth striations in GaAs gave rise to observable effects and the position of the *P-N* junction was revealed.[99–102,104]

The cathodoluminescent signal after conversion to a voltage may be electronically gated, so that only regions of constant cathodoluminescent efficiency are displayed.[94] The electrons emitted from the sample and converted to light by a scintillator may be combined with the cathodoluminescent signal and fed into a photomultiplier to obtain a multimode scanning image displaying surface topography, as well as variations in cathodoluminescent efficiency.[8] Surface field contrast may be usefully combined with luminescent contrast to study electroluminescent devices.[105]

V. REFERENCES

1. *Scanning Electron Microscopy, 1968*, Proc. 1st Annual Scanning Electron Microscope Symposium, Chicago, IIT Research Institute, Chicago, 1968.
2. Proc. 1st Annual Stereoscan Scanning Electron Microscope Colloquium, Chicago, 1968, Engis Equipment Co., Morton Grove, Ill., 1968.
3. *Scanning Electron Microscopy, 1969*, Proc. 2nd Annual Scanning Electron Microscope Symposium, Chicago, IIT Research Institute, Chicago, 1969.
4. Proc. 2nd Annual Stereoscan Scanning Electron Microscope Colloquium, Chicago, 1969, Engis Equipment Co., Morton Grove, Ill., 1969.
5. *Scanning Electron Microscopy, 1970*, Proc. 3rd Annual Scanning Electron Microscope Symposium, Chicago, IIT Research Institute, Chicago, 1970.
6. Proc. 3rd Annual Stereoscan Scanning Electron Microscope Colloquium, Chicago, 1970, Kent Cambridge Scientific, Inc., Morton Grove, Ill., 1970.
7. O. C. Wells, in Record IEEE 10th Annual Symposium on Electron, Ion and Laser Beam Technology, held at Gaithersburg, Md., May 1969, L. Marton, Ed., San Francisco Press, pp. 509–541.

8. P. R. Thornton, *Scanning Electron Microscopy*, 1st ed., Chapman and Hall, London, 1968.
9. C. W. Oatley, W. C. Nixon, and R. F. W. Pease, "Scanning Electron Microscopy," *Advances in Electronics and Electron Physics*, Vol. 21, L. Marton, Ed., Academic Press, New York, 1965, pp. 181–247.
10. W. G. Morris, *Scanning Electron Microscopy of the Surface Structure on Iron Whiskers*, D.Sc. thesis, Massachusetts Institute of Technology, Cambridge, Mass., 1965.
11. V. Johnson, in *Scanning Electron Microscopy, 1969*, Proc. 2nd Annual Scanning Electron Microscope Symposium, Chicago, IIT Research Institute, Chicago, 1969, pp. 483–525.
12. O. C. Wells, in *Scanning Electron Microscopy, 1970*, Proc. 3rd Annual Scanning Electron Microscope Symposium, Chicago, IIT Research Institute, Chicago, 1970, pp. 509–524.
13. M. Knoll, *Z. Tech. Phys.*, **16**, 467 (1935).
14. M. Von Ardenne, *Z. Phys.*, **109**, 553 (1938).
15. V. K. Zworykin, J. Hillier, and R. L. Snyder, *ASTM Bull.*, **117**, 15 (1942).
16. F. Davoine, Dissertation, University of Lyons, France, 1957.
17. F. Davoine, P. Pinard, and M. Martineau, *J. Phys. Radium*, **21**, 121 (1960).
18. W. C. Nixon, in *Scanning Electron Microscopy, 1968*, Proc. 1st Annual Scanning Electron Microscope Symposium, Chicago, IIT Research Institute, Chicago, 1968, pp. 55–62.
19. R. F. W. Pease, Ph.D. Dissertation, Cambridge, University, England, 1964.
20. R. F. W. Pease and W. C. Nixon, *J. Sci. Instrum.*, **42**, 81 (1965).
21. J. A. Belk, *Techniques of Metals Research*, Vol. 2, Part 1, R. F. Bunshah, Ed., Interscience Publishers, New York, 1968.
22. D. J. Dingley, *Micron*, **1**, 206 (1969).
23. D. J. Dingley, in *Scanning Electron Microscopy, 1970*, Proc. 3rd Annual Scanning Electron Microscope Symposium, Chicago, IIT Research Institute, Chicago, 1970, p. 329.
24. W. C. Lane, in *Scanning Electron Microscopy, 1970*, Proc. 3rd Annual Scanning Electron Microscope Symposium, Chicago, IIT Research Institute, Chicago, 1970, p. 41.
25. E. Lifshin, W. G. Morris, and R. B. Bolon, *J. Metals*, **21**, 1 (1969).
26. D. Kynaston and A. D. G. Stewart, *Scanning Electron Microscopy, 1969*, Proc. 2nd Annual Scanning Electron Microscope Symposium, Chicago, IIT Research Institute, Chicago, 1969, p. 465.
27. S. Kimoto, M. Sato, and T. Adashi, in *Scanning Electron Microscopy, 1969*, Proc. 2nd Annual Scanning Electron Microscope Symposium, Chicago, IIT Research Institute, Chicago, 1969, p. 67.
28. W. C. Nixon, in Proc. 2nd Annual Stereoscan Scanning Electron Microscope Colloquium, Chicago, 1969, Engis Equipment Co., Morton Grove, Ill., 1969, p. 165.
29. W. B. Drayton and W. C. Nixon, Proc. 27th Annual Meeting Electron Microscopy Society of America, St. Paul, Minn., 1969, C. J. Arceneaux, Ed., Claitor's Publishing Division, Baton Rouge, La., 1969, p. 18.
30. N. C. MacDonald, H. L. Marcus, and P. W. Palmberg, *Scanning Electron Microscopy, 1970*, Proc. 3rd Annual Scanning Electron Microscope Symposium, Chicago, IIT Research Institute, Chicago, 1970, p. 25.
31. C. E. Hall, *Introduction to Electron Microscopy*, McGraw-Hill Book Co., New York, 1953.

32. M. E. Haine and V. E. Cosslett, *The Electron Microscope*, E. and F. Spon, London, 1961.
33. K. C. A. Smith, Ph.D. Dissertation, Cambridge University, England, 1956.
34. T. E. Everhart, Ph.D. Dissertation, Cambridge University, England, 1958.
35. A. Rose, *Advances in Electronics and Electron Physics*, Vol. 1, Academic Press, New York, 1948, p. 131.
36. T. E. Everhart, *The Electron Microprobe*, T. P. McKinley, K. F. J. Heinrich, and D. B. Wittry, Eds., John Wiley & Sons, New York, 1966, p. 480.
37. K. F. J. Heinrich, *Advances in X-Ray Analysis*, Vol. 7, Plenum Press, New York, 1964, p. 325.
38. K. F. J. Heinrich, *4th Int. Congress of X-Ray Optics and Microanalysis*, Orsay, 1965, R. Castaing, P. Deschamps, and J. Philibert, Eds., Hermann & Cie, Paris, 1966, p. 159.
39. J. R. Banbury and W. C. Nixon, *J. Sci. Instrum.*, **44**, 889 (1967).
40. P. R. Thornton, *Science J.*, **1**, 66–71 (November 1965).
41. W. G. Morris, E. Lifshin, and R. B. Bolon, private communication.
42. D. B. Wittry, *4th Int. Congress of X-Ray Optics and Microanalysis*, Orsay, 1965, R. Castaing, P. Deschamps, and J. Philibert, Eds., Hermann & Cie, Paris, 1966, p. 168.
43. E. Weinryb, Thesis, University of Paris, France, 1965.
44. S. Kimoto and H. Hashimoto, *The Electron Microprobe*, T. P. McKinley, K. F. J. Heinrich, and D. B. Wittry, Eds., John Wiley & Sons, New York, 1966, p. 480.
45. E. W. White, H. Görz, G. G. Johnson, Jr., and R. E. McMillan, *Scanning Electron Microscopy*, 1970, Proc. 3rd Annual Scanning Electron Microscope Symposium, Chicago, IIT Research Institute, Chicago, 1970, p. 57.
46. G. Dorfler and J. C. Russ, *Scanning Electron Microscopy, 1970*, Proc. 3rd Annual Scanning Electron Microscope Symposium, Chicago, IIT Research Institute, Chicago, 1970, p. 65.
47. A. Boyde, *Scanning Electron Microscopy, 1970*, Proc. 3rd Annual Scanning Electron Microscope Symposium, Chicago, IIT Research Institute, Chicago, 1970, p. 105.
48. D. G. Coates, *Phil. Mag.*, **16**, 1179 (1967).
49. D. G. Coates, *4th European Reg. Conf. on Electron Microscopy*, Rome, 1968, Vol. 1, D. S. Bocciarelli, Ed. Tipografia Poliglotta Vaticana, Rome, 1968, p. 81.
50. D. G. Coates, *Scanning Electron Microscopy, 1969*, Proc. 2nd Annual Scanning Electron Microscope Symposium, Chicago, IIT Research Institute, Chicago, 1969, p. 27.
51. E. D. Wolf and T. E. Everhart, *Scanning Electron Microscopy, 1969*, Proc. 2nd Annual Scanning Electron Microscope Symposium, Chicago, IIT Research Institute, Chicago, 1969, p. 41.
52. E. M. Schulson, C. G. Van Essen, and D. C. Joy, *Scanning Electron Microscopy, 1969*, Proc. 2nd Annual Scanning Electron Microscope Symposium, Chicago, IIT Research Institute, Chicago, 1969, p. 47.
53. G. R. Booker, *Scanning Electron Microscopy, 1970*, Proc. 3rd Annual Scanning Electron Microscope Symposium, Chicago, IIT Research Institute, Chicago, 1970, p. 489.
54. E. D. Wolf and R. G. Hunsperger, *Scanning Electron Microscopy, 1970*, Proc. 3rd Annual Scanning Electron Microscope Symposium, Chicago, IIT Research Institute, Chicago, 1970, p. 457.

55. G. R. Booker, A. M. B. Shaw, M. J. Whelan, and P. B. Hirsch, *Phil. Mag.*, 16, 1185 (1967).

56. P. B. Hirsch and C. J. Humphreys, *Scanning Electron Microscopy, 1970*, 3rd Annual Scanning Electron Microscope Symposium, Chicago, IIT Research Institute Chicago, 1970, p. 449.

57. J. R. Banbury and W. C. Nixon, *J. Sci. Instrum.*, 44, 889 (1967).

58. D. C. Joy and J. P. Jakubovics, *Phil. Mag.*, 17, 61 (1968).

59. G. V. Saparin, G. V. Spivak, G. T. Sbezhnev, and N. F. Pesotskii, *Bull. Acad. Sci. USSR, Phys. Ser.*, 32, 896 (1968).

60. G. V. Spivak, G. V. Saparin, N. N. Sedov, and L. F. Komolova, *Bull. Acad. Sci. USSR, Phys. Ser.*, 32, 891 (1968).

61. D. C. Joy and J. P. Jakubovics, *4th European Reg. Conf. on Electron Microscopy*, Rome, 1968, Vol. 1, D. S. Bocciarelli, Ed., Tipografia Poliglotta Vaticana, Rome, 1968, p. 85.

62. J. R. Banbury and W. C. Nixon, *4th European Reg. Conf. on Electron Microscopy*, Rome, 1968, Vol. 1, D. S. Bocciarelli, Ed., Tipografia Poliglotta Vaticana, Rome, 1968, p. 93.

63. D. C. Joy and J. P. Jakubovics, *Brit. J. Appl. Phys.*, (*J. Phys. D*) Ser. 2, 2, 1367 (1969).

64. A. J. Speth, *Rev. Sci. Instrum.*, 40, 1636 (1969).

65. J. R. Dorsey, presented at 1st National Conf. on Electron Probe Microanalysis, University of Maryland, College Park, Maryland, 1964, paper No. 10.

66. V. N. Vertsner, R. I. Lomunov, and Y. V. Chentsov, *Bull. Acad. Sci. USSR, Phys. Ser.*, 30, 778 (1966).

67. J. R. Devaney, Proc. 2nd Annual Stereoscan Scanning Electron Microscope Colloquium, Chicago, 1969, Engis Equipment Co., Morton Grove, Ill., 1969, p. 61.

68. R. Anstead and G. Jacobs, Proc. 2nd Annual Stereoscan Scanning Electron Microscope Colloquium, Chicago, 1969, Engis Equipment Co., Morton Grove, Ill., 1969, p. 67.

69. C. W. Oatley and T. E. Everhart, *J. Elec. Cont.*, 2, 568 (1957).

70. T. E. Everhart, O. C. Wells, and C. W. Oatley, *J. Elec. Cont.*, 7, 97 (1959).

71. R. K. Matta, *Scanning Electron Microscopy 1968*, Proc. 1st Annual Scanning Electron Microscope Symposium, Chicago, IIT Research Institute, Chicago, 1968, p. 131.

72. T. H. P. Chang and W. C. Nixon, *Solid State Electronics*, 10, 701 (1967).

73. A. M. B. Shaw and G. R. Booker, *Scanning Electron Microscopy, 1969*, Proc. 2nd Annual Scanning Electron Microscope Symposium, Chicago, ITT Research Institute, Chicago, 1969, p. 459.

74. G. A. Saparin, G. V. Spivak, and S. S. Stepanov, *Proc. 6th Int. Congr. Electron Microscopy*, Kyoto, 1966, Vol. 1, R. Uyeda, Ed., Maruzen Company, Tokyo, 1966, 609.

75. M. C. Driver, *Scanning Electron Microscopy, 1969*, Proc. 2nd Annual Scanning Electron Microscope Symposium, Chicago, ITT Research Institute, Chicago, 1969, p. 403.

76. N. C. MacDonald, *Scanning Electron Microscopy, 1970*, 3rd Annual Scanning Electron Microscope Symposium, Chicago, IIT Research Institute, Chicago, 1970, p. 481.

77. O. C. Wells, *Scanning Electron Microscopy, 1969*, 2nd Annual Scanning Electron Microscope Symposium, Chicago, IIT Research Institute, Chicago, 1969, p. 397.

78. J. R. Banbury and W. C. Nixon, *J. Sci. Instrum.*, *Ser. 2*, **2**, 1055, (1969).
79. J. R. Banbury and W. C. Nixon, *Scanning Electron Microscopy, 1970*, Proc. 3rd Annual Scanning Electron Microscope Symposium, Chicago, IIT Research Institute, Chicago, 1970, p. 473.
80. J. P. Flemming and E. W. Ward, *Scanning Electron Microscopy, 1970*, Proc. 3rd Annual Scanning Electron Microscope Symposium, Chicago, IIT Research Institute, Chicago, 1970, p. 465.
81. D. P. Gaffney, *Scanning Electron Microscopy, 1970*, Proc. 3rd Annual Scanning Electron Microscope Symposium, Chicago, IIT Research Institute, Chicago, 1970, p. 433.
82. C. V. Spivak, G. V. Saparin, and N. A. Pereverzev, *Izv. Acad. Nauk USSR*, *Ser. Fiz.*, **26**, 1339 (1962).
83. P. R. Thornton, M. J. Culpin, and I. W. Drummond, *Solid State Electronics*, **6**, 523 (1963).
84. T. E. Everhart, O. C. Wells, and R. K. Matta, *J. Electrochem. Soc.*, **111**, 929 (1964).
85. T. E. Everhart, O. C. Wells, and R. K. Matta, *Proc. IEEE*, **52**, 1642 (1964).
86. P. R. Thornton, K. A. Hughes, D. V. Sulway, and R. C. Wayte, *Microelectronics and Reliability*, **5**, 291 (1966).
87. N. C. MacDonald and T. E. Everhart, *Appl. Phys. Lett.*, **7**, 267 (1965).
88. K. G. McKay, *Phys. Rev.*, **74**, 1606 (1948); **77**, 816 (1950).
89. L. Pensak, *Phys. Rev.*, **75**, 472 (1949).
90. F. Ansbacher and W. Ehrenberg, *Proc. Phys. Soc.*, **A64**, 362 (1951).
91. D. Green, *Proc. 8th Annual Electron and Laser Beam Symposium*, 1966, G. I. Haddad, Ed., sponsored by The University of Michigan, Ann Arbor, and The Institute of Electrical and Electronics Engineers, pp. 375–384.
92. G. S. Plows and W. C. Nixon, *J. Sci. Instrum.*, *Ser. 2*, **1**, 595 (1968).
93. W. C. Nixon, *4th European Reg. Conf. on Electron Microscopy*, Rome, 1968, Vol. 1, D. S. Bocciarelli, Ed., Tipografia Poliglotta Vaticana, Rome, 1968, p. 67.
94. D. A. Cusano, *Solid State Comm.*, **2**, 353 (1964).
95. F. Davoine, P. Pinard, and M. Martineau, *J. Phys. Radium*, **21**, 121 (1960).
96. F. Davoine, R. Bernard, and P. Pinard, *Proc. European Reg. Conf. on Electron Microscopy*, Delft, 1960, Vol. 1, A. L. Houwink and B. J. Spit, Eds., De Nederlandse Vereniging Voor Electronenmicroscopie, Delft, 1960, p. 165.
97. R. Bernard, F. Davoine, and P. Pinard, *Compto Rindui Acad. Sci.*, **248**, 2564 (1956).
98. D. A. Shaw, R. C. Wayte, and P. R. Thornton, *Appl. Phys. Lett.*, **8**, 289 (1966).
99. D. B. Wittry and D. F. Kyser, *J. Appl. Phys.*, **35**, 2439 (1964).
100. D. F. Kyser and D. B. Wittry, *The Electron Microprobe*, T. D. McKinley, K. F. J. Heinrich, and D. B. Wittry, Eds. John Wiley & Sons, New York, 1966, p. 691.
101. H. C. Casey, Jr., and R. H. Kaiser, *J. Electrochem. Soc.*, **114**, 149 (1967).
102. H. C. Casey, Jr., *J. Electrochem. Soc.*, **114**, 153 (1967).
103. G. Remond, S. Kimoto, and H. Okuzumi, *Scanning Electron Microscopy, 1970*, Proc. 3rd Annual Scanning Electron Microscope Symposium, Chicago, IIT Research Institute, Chicago, 1970, p. 35.
104. D. A. Shaw, D. V. Sulway, R. C. Wayte, and P. R. Thornton, *J. Appl. Phys.*, **38**, 887 (1967).
105. D. V. Sulway, Htin Kyaw, and P. R. Thornton, *Solid State Electronics*, **10**, 545 (1967).

9

ELECTRON MICROPROBE ANALYSIS

I. Introduction 462

II. Basic Instrumental Features 463

III. The Cameca MS46 Microprobe 464

IV. Specimens for Microprobe Analysis 469

V. X-ray Optics 470
   A. Wavelength Dispersion 470
   B. Counters 473
   C. Energy Dispersion 475
   D. Combined Wavelength and Energy Dispersion 477

VI. Qualitative Analysis 478

VII. Quantitative Analysis 483
   A. Empirical Approach 485
   B. Theoretical Approach 486
     1. Absorption Correction 487
     2. Secondary Fluorescence Correction 488
     3. Atomic Number Correction 490
       a. Evaluation of Stopping Power S 491
       b. Evaluation of the Backscatter Coefficient R 492
     4. Errors due to Contamination Buildup 493

VIII. Light Element Analysis 493

IX. Use of Signals Other Than X-ray for Analysis 494

X. Applications 496
   A. Phase Identification 496
   B. Determination of Concentration Gradients 498
   C. Phase Equilibria 499

    D. Film Thickness Determination. 500
    E. Kossel Line Technique 502
    F. Microradiography 504
    G. Miscellaneous 504

 XI. Combined Electron Microscope and Microprobe. 506

XII. Secondary Ion Emission Microanalysis 511

XIII. References 513

I. INTRODUCTION

Electron microprobe analysis, or electron probe microanalysis as it is often called, in which the x-ray spectra excited from the sample by electrons are used for chemical analysis, has won rapid acceptance as an important metallurgical tool, when used in conjunction with other metallographic techniques. Instrumentally, it has led to the development of a variety of scanning imaging systems, including scanning electron microscope systems, now being marketed as separate instruments which permit higher resolutions to be obtained. Microprobe attachments have been developed for several commercial electron microscopes, permitting transmission electron microscopy and electron diffraction to be combined with microprobe analysis on the same sample. Prototypes of improved dual purpose instruments, combining these techniques, have been built. A recent commercial development is the addition of nondispersive solid state detectors to scanning electron microscopes, which, in conjunction with suitable electronics and a multichannel analyzer, enable the complete spectrum of the emitted x-rays to be displayed. X-ray scanning imaging and x-ray measurement channels may be added. Thus we have a microprobe attachment on a scanning electron microscope. Clearly, a large variety of instruments may be envisaged optimized for specific functions.

Since the signals obtained from the electron microprobe may be collected in digital form, they lend themselves to computer analysis. Important further applications to quantitative metallography are already under development which permit the computerized analysis of grain size, particle size, particle shape and distribution, and volume fraction, using scanning electron systems. Since contrast expansion systems can be incorporated, and a variety of signals employed, this approach offers the promise of substantial advantages over the analysis of light microscopic images, received on a television pickup tube, which has already been exploited commercially. In the latter, one is dependent on etching to control the contrast, and the measurement accuracy is limited by subjective errors, as well as by variations in the pickup tube response in different parts of the field. Attempts to circumvent the etching problem by the use of micrographs, in which the contrast can be enhanced by photographic techniques, suffer from the disadvantages of being indirect and also of the conflicting requirements of resolution and field size. The use of a large field is desirable, in order to minimize sampling and statistical errors in making quantitative metallographic measurements. If a reasonably high magnification is used to permit desired resolution, the print size becomes prohibitively large.

The success of the electron microprobe has encouraged the development of a variety of other instruments for the analysis of microregions of solids such as the laser microprobe, the spark source mass spectrometer, and the plasma jet mass spectrometer. None of these has the microcapability of the electron microprobe which can analyze a volume of about 10^{-11} cc, located with a precision of $< 1\ \mu$.

II. BASIC INSTRUMENTAL FEATURES

Electron excitation of characteristic x-radiation from a sample has proved to be a most versatile method of analysis of microregions. Although the electron microprobe analyzer was only invented about 1951 by Castaing,[1] a large variety of first and second generation instruments has now been marketed. The principal features of these instruments may be listed as follows.

1. An electron optical system consisting of an electron gun, followed by lenses, whose function is to produce at the surface of the sample an electron beam typically about 1 μ in diameter.

2. A stage for holding and mechanically translating the sample in a precision manner.

3. Viewing arrangements, commonly a light microscope and also a scanning electron microscope system, for accurately selecting and recording the region to be analyzed.

4. X-ray detection and measuring equipment for determining the wavelength and intensity of the emitted x-rays. Normally, one or more x-ray crystal spectrometers are provided, each with its own detection system.

5. Facilities for computer interfacing and data processing in a computer.

In advanced equipment, the whole analysis, for example, the determination of concentration gradients in a diffusion couple and calculation of the diffusion coefficients may be carried out in an almost completely

Figure 1 CAMECA Type MS46 electron microprobe, equipped with four channels of Hamner readout electronics. (Courtesy of General Electric Company.)

automated manner once the equipment is set up, using automatic step scanning of the sample and digital readout of the x-ray intensities on-line to a computer or via an intermediary data storage system.

Since a variety of microprobes are commercially available, the choice of instrument here for purposes of illustration is necessarily arbitrary.

III. THE CAMECA MS46 MICROPROBE

The principal features of the instrument are shown in Figures 1 and 2. Two electromagnetic lenses are employed to focus a beam of electrons from the self-biasing, triode-type electron gun toward the specimen surface. The beam strikes the specimen at right angles. Acceleration potentials from 2 to 40 kV may be used, and the filament may be changed without reducing the whole column to atmospheric pressure. A six-pole electrically operated stigmator enables the astigmatism in the objective

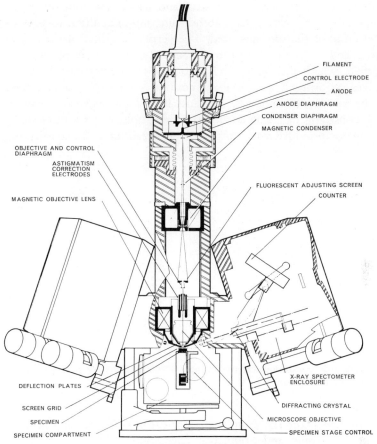

FILAMENT

CONTROL ELECTRODE

ANODE

ANODE DIAPHRAGM

CONDENSER DIAPHRAGM

MAGNETIC CONDENSER

OBJECTIVE AND CONTROL DIAPHRAGM

ASTIGMATISM CORRECTION ELECTRODES

FLUORESCENT ADJUSTING SCREEN

MAGNETIC OBJECTIVE LENS

COUNTER

DEFLECTION PLATES

X-RAY SPECTOMETER ENCLOSURE

SCREEN GRID

DIFFRACTING CRYSTAL

SPECIMEN

MICROSCOPE OBJECTIVE

SPECIMEN COMPARTMENT

SPECIMEN STAGE CONTROL

Figure 2   Principle components of the CAMECA Type MS46 electron microprobe. (Courtesy of Consolidated Electrodynamics Corporation.)

lens to be corrected, thus concentrating all of the electrons into a small circular spot at the specimen surface. The beam is regulated by means of a beam-defining, current-sampling aperture (labeled "objective and control diaphragm" in Figure 2). This provides a feedback signal to the condenser lens supply which automatically maintains a constant beam intensity. Typical beam currents are 2.5×10^{-8} A for heavy and 1×10^{-7} A for light elements. The vacuum system is fully automated.

The specimen may be observed while it is under electron bombardment by means of a coaxial optical microscope system. This consists of a mirror-type, objective lens (focal length 12 mm, numerical aperture 0.48), vertical illuminator, and eyepiece system and has a resolving power of 0.7 μ (Figure 3). This system has another function in addition to providing for selection of the spot analyzed. The vertical position of the objective is fixed, and it has a small depth of focus $< 1 \mu$. The act of focusing the specimen, which can be moved vertically by a micrometer adjustment, serves to bring it into a fixed position on the Rowland circles of the four x-ray spectrometers. A transmitted light accessory may be employed, consisting of a thin section specimen holder and a transmitted light illuminator provided with normal or polarized light. By substituting

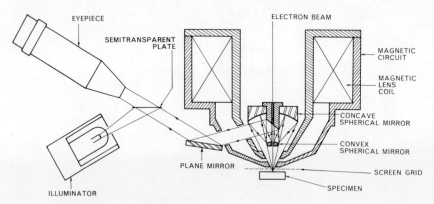

Figure 3   Details of coaxial optical microscopic system. (Courtesy of Consolidated Electrodynamics Corporation.)

a fluorescent screen for the sample, the optical microscope may be used for alignment, focusing, and astigmatism correction.

The four spectrometers are of the fully focusing type, each equipped with a curved crystal which is positioned under Johansson focusing conditions, with respect to the x-ray source (see Fig. 5). The four spectrometers are located in pairs inside vacuum enclosures 180° apart. One spectrometer is equipped with a dual crystal holder, giving a total of five crystals. Each spectrometer is provided with sealed or gas flow proportional counters. Five motor-driven spectrometer scan speeds are provided. The motion of a counter is such that it always receives x-rays reflected by the crystal, excluding all direct or scattered radiation. X-rays are taken off at an angle of 18° to the specimen surface. Different crystals are employed in the four spectrometers which enable a range of wavelengths to be covered, and permit up to four elements to be analyzed simultaneously if the necessary electronics are in place. All elements down to, and including, boron may be analyzed.

X-ray scanning images may be obtained with the aid of an x-y scanning system, incorporating electrostatic deflection in the x direction parallel to the line focus of the analyzer crystal and electromechanical scanning of the sample in the y direction (Figure 4). A sample area up to 0.4

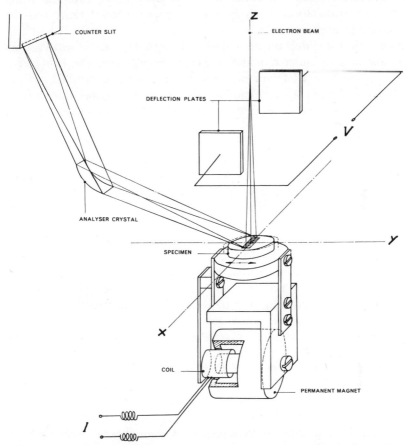

Figure 4 Details of scanning system. (Courtesy of Consolidated Electrodynamics Corporation.)

mm square may be scanned. The former motion is synchronized with the line scan in the image displayed on a cathode ray tube, while the latter, which may be slower, is synchronized with the sweep scan. The x-ray intensity is used to modulate the oscilloscope signal. Since the beam is never deflected in the y direction, the x-ray source is maintained on the Rowland circle during scanning. A scanning electron microscope image may be obtained, using the specimen current to ground as a signal. The specimen is placed in an insulated mount which is provided with a hole acting as a Faraday cup to measure the incident electron current. An attachment permits automatically step scanning the sample in programmed discrete steps in either the x or y direction, counting being initiated automatically at each spot for a predetermined time, and the total count printed out.

Since carbon buildup on the sample during light element analysis can be a problem, an anticontamination device is provided, consisting of a liquid nitrogen cooled finger near the sample and a controlled oxygen leak. A Kossel camera stage can be used to record Kossel transmission or reflection patterns, in order to orient a crystal or determine precision lattice parameters.

A typical data-handling system would consist of preamplifier, counting channel, linear amplifier, scalar, rate meter, and pulse height analyzer. Typical readout modes are (a) scaler and printer (b) rate meter and strip chart recorder, and (c) rate meter, or amplifier and oscilloscope (with Polaroid camera attachment).

Quartz crystals are used in spectrometers 1 and 2 which are designed for the analysis of hard and medium x-rays, respectively. Quartz, gypsum, or mica crystals may be used in spectrometer 3 for progressively softer x-rays, while KAP or a multilayer lead-stearate pseudocrystal is used in spectrometer 4 for the very soft, that is, long wavelength x-rays. The wavelength range from 0.90 to about 80 Å is covered by the range of monochromator crystals provided. Wide overlapping of the wavelength ranges of the various crystals permits the simultaneous analysis of several elements.

It should be noted that more than one of the characteristic K, L, or M lines may be used for the analysis of some elements; thus M lines are excited at a lower voltage than L lines, and L lines at a lower voltage than K lines. In a particular alloy it may be desirable to keep the excitation voltage low, in order to avoid secondary fluorescence of one element, due to x-radiation from another element, which would result in an increase in the effective volume analyzed and an additional correction. In general, the choice of spectra is governed by the desirability of optimizing the line to background ratio of x-ray intensity for maximum sensitivity of analysis.

The literature on electron probe analysis is already too extensive to be reviewed here and the reader is referred to the following sources.[1-11] Reference 6 contains a bibliography up to January 1965, compiled by Heinrich.

IV. SPECIMENS FOR MICROPROBE ANALYSIS

The stages in modern microprobes are usually capable of accepting standard metallographic samples of 1 or $1\frac{1}{4}$ inches diameter by $\frac{1}{4}$- to $\frac{1}{2}$-in. thickness. Larger samples may be machined down to this size. It is often convenient to mount samples in a conducting mount of bakelite plus 5 wt. % graphite or copper/dialyl phthalate. The conducting mount provides a means of avoiding local charge accumulation on the specimen surface, which would result in deflection of the electron beam and image distortion. If a nonconducting mount is employed, it is desirable to provide a path to ground after preparation, using aluminum or silver paint. Electrically insulated samples are coated with a thin evaporated film of carbon or aluminum, avoiding the use of an element that is to be analyzed.

Standard metallographic preparation is employed, taking care to avoid smearing of the structure or rounding-off of the edges if these are to be analyzed. It is important that the sample be flat, in order to maintain the x-ray geometry. Light etching may be necessary to remove flowed material or to reveal structural features of interest such as grain boundaries which would not otherwise be apparent. The use of filming etchants must be avoided. Since etching often preferentially removes the material of interest such as inclusions, it may be necessary to put on fiduciary markers (microhardness impressions or scratches) and repolish before microprobe analysis. Surface roughness, whether caused by residual scratches, porosity, or etching, while giving contrast in the image, changes the geometry of x-ray emission and may, particularly if the x-ray takeoff angle is small, place obstacles in the way of the characteristic x-ray emission, resulting in absorption and analytical errors (see Yakowitz in Ref. 11).

Because of the geometrical considerations above, it is necessary to mount and section irregular samples such as fractures or powders if quantitative analysis is desired. As the sample size approaches the beam size, quantitative analysis will no longer be possible because of the impossibility of calculating corrections. For qualitative analysis of small particles it is often sufficient to attach a particle to a metal plate of a different material. Certain microprobes such as the Geoscan (Cambridge Instrument Co.) and Cameca MS46, designed with the needs of geologists in mind, have provision for the examination of petrographic thin sections by transmitted normal or polarized light. The samples would then be prepared by standard petrographic methods. Since such samples are likely to transmit some of the incident electrons, serious problems of correction again arise.

In the case of electron microprobe attachments for the electron microscope, extraction replicas or electrothinned, partially electron transparent thin films are common types of samples. Analysis will again be qualitative rather than quantitative, due to the nature of the sample, problems due to etching, and so forth.

V. X-RAY OPTICS

In order to carry out a chemical analysis in the electron probe, it is necessary to measure the wavelength and intensity of the characteristic x-rays generated in the specimen. The superimposed background of x-rays from the continuous spectrum must be subtracted out. The characteristic emission may be separated, either by wavelength (crystal) dispersion or by energy dispersion, the former being more commonly used, since the detectors used for energy dispersion are at present less discriminating. These techniques will now be described.

A. Wavelength Dispersion

Wavelength dispersion is accomplished in an x-ray crystal spectrometer, where the various component wavelengths are diffracted according to Bragg's law:

$$n\lambda = 2d \sin \theta \tag{1}$$

where n is an integer representing the order of diffraction, λ the x-ray wavelength in angstroms, d the interplanar spacing of the diffracting crystal in angstroms, and θ the angle of diffraction.

Since the volume emitting x-rays is so small, it is usual to employ focusing optics in the x-ray spectrometer, in order to collect sufficient x-rays to give good sensitivity. Either Johansson (Figure 5) or Johann (Figure 6) focusing conditions are often employed [Ogilvie[4]]. For perfect focusing, the x-ray source, diffracting element, and receiving slit must all be on the arc of a circle (Rowland focusing circle). In Johansson focusing, the crystal is bent with a radius equal to the diameter NM (Figure 5) of the focusing circle and then the crystal is ground to the radius of the focusing circle. Since any normal to the diffracting planes passes through M, any ray of the appropriate constant wavelength from the source S, diffracted from the crystal to the x-ray counter slit at P, forms an angle that subtends the same angle $(2\pi - 4\theta)$ on the focusing circle. The perfection of the crystal is important. It is difficult to grind soft crystals without introducing a substructure too deep to remove by etching; hence some spectrometers employ a crystal such as mica which is bent only—the Johann arrangement. Here only the rays diffracting from the point of tangency pass through the true focus (Figure 6).

Both the collection efficiency, that is, the angular spread of the emitted x-rays, and the resolution, which is dependent on the width of the reflected beam at the counter slit, are important. In order to analyze the elements present, it is necessary to have a scanning spectrometer in which

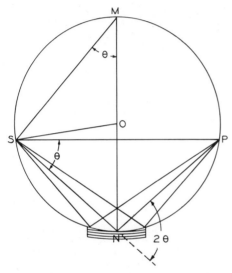

Figure 5 Johansson focusing conditions. From Ogilvie.[4]

the geometry of Figures 5 or 6 is maintained during scanning. A mechanical arrangement is usually used to couple the rotation of the crystal to that of the detector which is kept pointing in the direction of the crystal. The angular velocity of the detector must be twice that of the crystal to keep the angles of incidence and reflection equal. A number of arrangements are possible. In one, the crystal and detector are rotated about the center of the focusing circle, while maintaining $SN = NM$ $\sin \theta$ (Figure 5) to ensure fully focusing conditions. For other arrangements, including semifocusing optics, see Refs. 2 through 5. For a given crystal, the lattice spacing, d, is fixed, so that only wavelengths shorter than $2d$ can be analyzed. A series of crystals is therefore necessary to cover the full wavelength range. Although a crystal-changing device may be incorporated to avoid dismantling a spectrometer to change the crystal, it is convenient to provide several spectrometers equipped with different crystals, particularly since other design factors enter into optimization. This also permits the simultaneous analysis of several elements.

While spectrometers for short wavelengths may be operated in air, using a window to maintain the vacuum in the column, absorption losses become considerable at longer wavelengths, so that it is desirable to evacuate the spectrometer to about 10^{-3} torr.

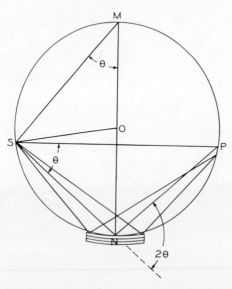

Figure 6   Johann focusing conditions. From Ogilvie.[4]

The line breadth from a fully focusing spectrometer depends only on the size of the x-ray source, and the perfection and accuracy of manufacture of the curved crystal. With a source size of the order of 1 μ, it is the crystal that is limiting. Line breadths $\geq 0.5°$ are common, giving a resolution of < 0.01 λ. The intensity from a curved crystal is nearly independent of the wavelength λ, so that the counter receives a nearly correct spectral distribution of intensity.

Analysis of elements lower than atomic number 12, that is, below sodium, presents special difficulty, due to the long wavelengths involved, necessitating a crystal of large d spacing. Duncumb and Melford (see Ref. 7) have shown that accurate intensity measurements can be made of the emission from elements down to boron (K emission 67 Å) in concentrations well below 1%, using stearate crystals and thin window proportional counters. The crystal consisted of about 100 layers of lead stearate, deposited on mica by the method described by Henke.[12] The window was a lightly aluminized 2000-Å-thick collodion film, supported on a copper grid. Braybrook et al. (see Ref. 7) have successfully used concave gratings of 2 meters radius to analyze carbon. These were made of gold-coated epoxy resin, mounted on glass. Light element microanalysis has also been discussed by Scott.[13]

B. Counters

The function of the x-ray detector or counter is to convert each x-ray photon into an electrical pulse which can then be counted electronically with a scaler. One type (proportional, Geiger-Muller, or ionization counters) records the pulses produced by the ionization of a gas in an electric field; the other (scintillation) counts the flashes of light in a phosphor crystal using a photomultiplier. Scintillation counters generally have poorer energy resolution and are limited to wavelengths below 3 Å because of the excess dark currents present at longer wavelengths. The proportional counter has several advantages over the Geiger-Muller counter and is the type most commonly employed for microprobe analysis.

Gas ionization counters consist of a cylindrical metal shell cathode filled with gas which contains a fine metal wire anode along its axis. One end or the side of the tube contains a thin window made of an x-ray transparent material such as beryllium. A constant potential is applied between the anode and cathode. Photons passing through the window collide with gas atoms and are absorbed, accompanied by the emission of electrons and ionization of the gas. The electrons move toward the anode and the ions toward the cathode. The collection of the electrons and ions results in a current pulse through the counter which leaks away through a resistor, but not before the charge added to a storage capacitor has been detected and converted to a voltage pulse in an attached circuit. The average arrival rate of photons, and hence the x-ray intensity, are measured by a rate meter.

The counter behavior for a given geometry, gas, and gas pressure in the counter depends only on the operating voltage (Figure 7). If the voltage is high enough, all the gas in the tube is ionized by an incoming photon as a result of a corona discharge, the basis of the *Geiger-Muller counter*. A high amplification factor of approximately 10^{10} is achieved, that is, high sensitivity, but the sensitivity is independent of the wavelength of the photon. Counting rates are limited to < 100 counts per sec by the long dead-time required for recovery, for both the ions and electrons produced to be collected. Since the counts from the fluorescent background radiation are often of similar order, this type of counter is, in general, unsuitable for microprobe use.

In the intermediate voltage range we have the *proportional counter* which is commonly used in microprobe work. The amplification factor is still fairly large ($10^3 - 10^6$) but depends sensitively on the voltage. Amplification occurs because the electrons ejected from the gas atoms

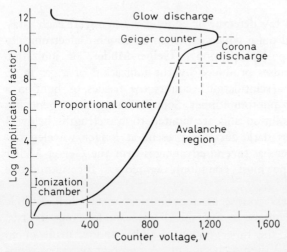

Figure 7   Counter sensitivity as a function of voltage. From Brandon.[14]

by the photons are accelerated so much by the applied voltage that they can knock electrons out of other atoms. The applied voltage is set, so that for a single ionization a fixed number of positive ion-electron pairs are produced. The dead-time is shorter than the Geiger-Muller counter, permitting better resolution between the energy of the photon and the output pulse size. When used with a crystal spectrometer set for a particular characteristic wavelength, the background level can be reduced by amplifying the output pulse and feeding through a pulse height analyzer which limits the pulse energies counted to those between set values of voltage adjacent to the characteristic wavelength. The output is fed into a rate meter and scaler and/or chart recorder which records the peak intensity and location.

At low voltages of about 200 V, no amplification occurs and the height of the pulse produced depends linearly on the photon energy. *Ionization chambers* that work in this range are of little interest here because of their low sensitivity.

Scintillation counters utilize the fluorescent emission of visible light from a phosphor, as a result of x-ray excitation. The photon/electron interaction results in the raising of valence electrons into the conduction band, from whence they return to the ground state. The amount of light emitted is proportional to the incident x-ray intensity, and is converted to electrons and amplified by a photomultiplier tube. Thermionic emission from the photocathode, used to convert the light to an electron signal, results in a high background, or dark current, which is a disadvantage. The relation between the pulse size and phonon energy is much less sharply defined than in the proportional counter. The efficiency of absorption of x-rays is better and approaches 100%. The dead-time is only about 10^{-6} sec. While the scintillation counter is useful at high counting rates, its principal application in the microprobe is for the detection of the electrons backscattered from the sample.

C. Energy Dispersion

Instead of using a crystal or grating to disperse the emitted x-rays, according to wavelength, and counting each wavelength interval separately, the x-rays may be picked up directly by a counter such as a proportional counter which converts them to pulses with an energy proportional to the wavelength of the x-rays. If desired, the energy distribution may be displayed on an oscilloscope screen and recorded photographically. The pulses are then sorted electronically by pulse height analyzers, according to voltage. Since the counter may be placed near the sample and thus command a relatively large solid angle, a much larger fraction of the emitted x-rays may be collected than in the case of an x-ray crystal spectrometer, giving greater sensitivity, perhaps by a factor of 100 or more. Unfortunately, the resolution amounts to only about 0.2 Å for the best energy dispersion counter, compared with 0.002 Å or better for a crystal analyzer. The resolution is limited by the detector characteristics. The number of ionizations is not fixed by the characteristic photon energy but only the most probable number, so that a distribution of pulse energies is obtained which may overlap the distribution from the characteristic emission of another element. The separation of elements closer than three atomic numbers may only be possible by mathematical analysis, assuming a distribution function, and leads to complexities.

It is probable that detectors giving better discrimination will be developed and will tend to replace crystal spectrometers because of the large possible sensitivity gain, discussed above. Until such detectors are available, the use of energy dispersion is likely to be more limited than wavelength dispersion. Energy dispersion has been utilized for rough quantitative analysis in a number of microprobe attachments for electron microscopes, where it may be desired to identify particles 0.1 μ or less in diameter, so that the emitted x-ray intensity is very small. Since the photon energy decreases with increasing wavelength, it is desirable to use energy dispersion for the determination of elements of low atomic number such as carbon, oxygen, nitrogen, and beryllium, using the K lines. However, complications may arise as a result of interference from the L and M characteristic lines of heavier elements if these are also present.

Considerable progress has been made in developing solid state detectors of the cooled, lithium-drifted, silicon diode type, with significantly improved resolution. Descriptions of the design and operation of such detector systems are given in Refs. 15, 16. The x-ray signal passes through a thin window into a vacuum chamber which contains a cooled, reversed bias, lithium-drifted silicon crystal of the P-I-N type (P-type/intrinsic/N-type).[17] X-rays, traveling through the intrinsic region, lose energy both to the crystal and as a result of the generation of electron hole pairs. The latter are swept away by the applied bias and the resultant charge pulse is converted by a charge-sensitive preamplifier to a voltage pulse. Following Lifshin,[17] a beam of monoenergetic x-ray photons produces a statistical distribution of pulses of average energy, proportional to the photon energy, with a standard deviation σ (in eV) of

$$\sigma = (\epsilon E F)^{\frac{1}{2}} \qquad (2)$$

where ϵ is the average energy (in eV) needed to produce a single electron hole pair, E the incident x-ray photon energy, and F the Fano factor.

We see from Equation 2 that F is the variance σ^2, divided by the ion pair yield E/ϵ. If the pulse energies follow a Poisson distribution, E_D the full width at half-maximum ($FWHM$) at the detector is

$$E_D = 2.35\sigma \qquad (3)$$

A further energy spread E_N results from electronic noise in the detector and preamplifier:[17]

$$E_N = E_0 + MC_D \qquad (4)$$

where E_0 is the FWHM without a detector. Hence, at zero capacitance, M is the energy versus the external capacitance curve slope, and C_D the capacitance of the detector. The measured FWHM is then obtained[17] from Equations 3 and 4:

$$E_{\text{FWHM}} = (E_N^2 + E_D^2)^{\frac{1}{2}} \qquad (5)$$

After further amplification and shaping, the pulses are sorted by amplitude, using a multichannel analyzer. The energy spectrum is conveniently displayed on a cathode ray tube for visual examination and photography, or may be recorded using an x-y recorder, or computer processed via appropriate interfacing.

Fitzgerald, Keil, and Heinrich[18] used a detector of this type, in conjunction with a beryllium window, multichannel pulse-height analyzer, cathode ray display, x-y recorder, and high speed printout. Although the peak to background ratios obtained on pure element samples were considerably lower than those usually obtained with crystal spectrometers, they were appreciably better than expected by energy dispersion with gas proportional detectors. Russ and Kabaya[19] and Lifshin[17,20] have used detectors of this type for microprobe analysis, in conjunction with the scanning electron microscope.

Energy dispersion spectrometers of the solid state type clearly have considerable potential for use in conjunction with instruments such as electron microprobes, scanning electron microscopes, and electron microscopes, or in neutron activation analysis. For qualitative analysis a complete spectral display of an unknown alloy is obtained in a matter of seconds for visual inspection. The spectrum may be recorded photographically or in digital form for computer analysis. Many of the standard readout techniques may be applied, for example, concentration mapping in conjunction with electron beam scanning. In view of the higher x-ray collection efficiency, sufficient counts may be obtained at lower incident electron beam currents, or higher speeds employed. This type of spectrometer is thus useful as an additional channel on microprobes equipped with crystal spectrometers. It should be advantageous to use when the x-ray signal is weak because of the small sample volume excited, as in an electron microscope, or in a high resolution scanning electron microscope, where the beam current is typically 4 to 5 orders of magnitude smaller than in the microprobe or conventional electron microscope.

D. Combined Wavelength and Energy Dispersion

The line to background ratio can usually be improved by using a proportional counter and pulse height analyzer, in conjunction with a crystal spectrometer. By setting the discriminating circuits for the wavelength satisfying Bragg's law for the crystal from the element of interest, continuous background radiation is reduced; any fluorescent characteristic radiation from the components of the analyzing crystal is eliminated; and other wavelengths, such as the 2λ and 3λ components diffracted in the second and third orders from other elements at the same crystal setting, are eliminated. Since the line to background ratio is a measure of the composition, and the background cannot be measured directly at the line position, it is necessary to move just off the line and reset the pulse height discriminator for the new position, in order to measure the appropriate background level.

VI. QUALITATIVE ANALYSIS

When the electron beam strikes the sample, it interacts as illustrated schematically in Figure 8, after Lifshin and Hanneman.[21] The electrons

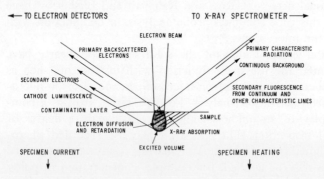

Figure 8   Schematic illustrating interactions that may occur at the sample. From Lifshin and Hanneman.[21] (Courtesy of General Electric Company.)

interact with the atoms by both elastic and inelastic collisions, resulting in some backscattering of the primary electrons, in some cases the production of visible light (cathodoluminescence), the emission of secondary electrons, local heating of the sample, a flow of electrons through the specimen to ground (specimen current), and the emission of x-rays in the form of both characteristic (line) and continuous spectra.

In qualitative analysis the main objective is to optimize the peak to background ratio for the characteristic x-ray signal of interest. The ratio varies from element to element and depends on a variety of operating conditions such as the beam voltage and current, characteristic line used, design of the spectrometer, and nature and performance of the x-ray detector. The nature of the x-ray spectrum will first be briefly discussed. The reader interested in a fuller discussion may refer to Refs. 22 to 24.

The spectral distribution of x-rays emitted by a rhodium target is illustrated in Figure 9, after Webster [see Taylor[25]]. Characteristic radiation is only generated above the critical excitation voltage. Below this,

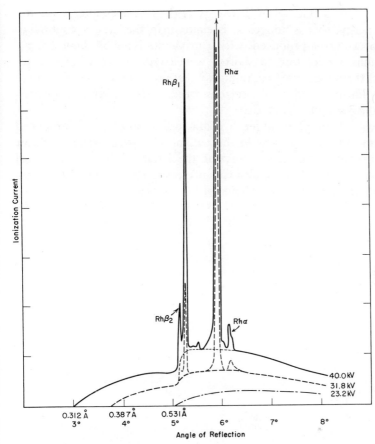

Figure 9   X-ray spectrum of rhodium, bombarded by electrons, as a function of electron voltage. Webster, from Taylor.[25]

the spectrum consists solely of continuous radiation which shows a sharp cutoff, corresponding to the complete deceleration of an electron by a single collision with an atomic nucleus at a minimum wavelength λ_{min} given by:

$$\lambda_{min} = \frac{hc}{eV} \approx \frac{12,400}{V} \tag{6}$$

where h is Planck's constant, c the velocity of light, e the charge on the electron, and V the accelerating voltage. The energy E of the incident electron is simply the product eV.

If the electron undergoes a glancing blow with an atomic nucleus and only gives up a fraction of its energy at the first encounter, a photon of lower energy, that is, longer λ, is emitted. In this way a continuous spectrum of x-rays is produced, with a maximum lying at about $1.5\lambda_{min}$. The maximum moves toward shorter wavelength as the voltage is increased and the intensity of continuous radiation also increases. Less than 1% of the kinetic energy of electrons striking the target is converted to x-rays; the rest appears as heat.

In the case of an alloy target, a characteristic spectrum is produced for each kind of atom present in the volume irradiated by the incident electrons, provided that the energy of the latter is above the critical excitation voltage of all of the elements present. Clearly, some discrimination may be possible by proper choice of the incident electron energy.

The characteristic spectrum can be explained on the basis of the energy level diagram for an atom (Figure 10). The K excitation state corresponds to

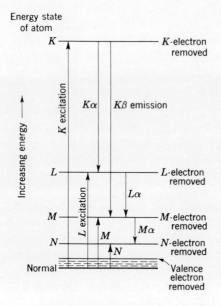

Figure 10 Energy level diagram for an atom (schematic). Arrows indicate excitation and emission processes. From Barrett and Massalski.[26] Copyright 1966 by McGraw-Hill Book Company; used with permission.)

the work required to eject a K electron from the atom. Likewise, the L, M, and N states correspond to the ejection of electrons from the outer shells. An excited atom returns to the ground state by discrete energy transitions (Figure 10), each corresponding to the emission of a particular x-ray photon, as an electron from an outer shell falls into the inner shell which has lost an electron. The energy states are very sharp, so that a photon has a well-defined characteristic wavelength. The K_α radiation is emitted when a K shell vacancy is filled by an adjacent L shell electron, K_β when it is filled by an M electron. Similarly, L lines are emitted when L vacancies are filled. Since many atoms are excited simultaneously, all lines of the K, L, or M, and so forth, series are emitted whenever the voltage is sufficient to eject an electron from a K, L, or M, and so forth, shell respectively. The minimum excitation potential necessary to excite K, L, or M spectra increases with atomic number (Figure 11).[21]

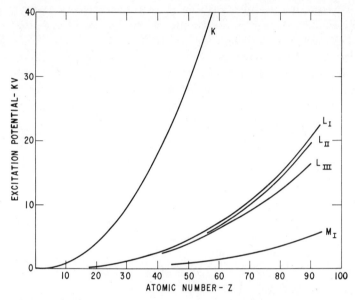

Figure 11   Minimum excitation potentials for K, L, and M spectral series as a function of atomic number. From Lifshin and Hanneman.[21] (Courtesy of General Electric Company.)

For a given element, the difficulty of excitation increases in going from M to L to K lines.

Moseley[27] showed that the wavelength λ of a photon is related to the atomic number Z of an atom as follows:

$$\lambda = P(Z - \sigma)^{-2} \tag{7}$$

where P is a constant for a particular transition, and σ a screening constant that is unity for K lines.

It should be noted that the same characteristic emission occurs when an atom is excited by an incident beam of x-rays, the basis of fluorescent x-ray analysis. If the wavelength of the incident x-rays is steadily decreased, an abrupt increase in absorption occurs when the beam energy becomes sufficient to eject an electron from one of the shells. The process of K absorption produces K excitation and the discontinuity is called the K absorption edge. Radiation due to the excitation of an atom by x-rays is known as *fluorescent radiation* by analogy with light and consists almost entirely of line radiation. In the case of a heavy atom such as uranium, while there is only one K absorption edge, there are three L absorption edges, five M and many N and O edges. A common complication in electron microprobe analysis is that characteristic x-rays, emitted by one element present in the sample, excite fluorescent radiation from a second element, necessitating corrections.

The intensity I of a characteristic x-ray emission line excited ·by an electron beam increases with both the beam current i and the overvoltage. That is, the amount by which the voltage V_0 of the incident electrons exceeds the critical value V_E for excitation of the line. The intensity I after Jönsson[28] is approximately

$$I = ik \, (V_0 - V_E)^n \tag{8}$$

where k is a constant that depends upon the atomic number and spectral series and n is approximately 1.67 for $V_0/V_E \geq 3$ or 4, as commonly employed in microprobe analysis. At higher voltage, n approaches unity.

The intensity I_λ of the continuous background emission at wavelengths between 1 and 2.8 Å was shown by Kulenkampff[29] for eight representative metals (aluminum to platinum) to be approximately

$$I_\lambda = \left(\frac{\bar{Z}}{\lambda^2}\right)\left(aV - \frac{b}{\lambda}\right) + \frac{N\bar{Z}^{-2}}{\lambda^2} \tag{9}$$

where a, b, and N are constants, and $\bar{Z}$ the average atomic number of the target material.

The background intensity at a fixed wavelength and Z value is directly proportional to V when the last term is small. Thus, considering Equation 8, it is desirable to have large overvoltages, in order to increase the peak to background ratio. Since a large overvoltage increases the penetration and lateral diffusion of the electron beam, that is, decreases the size resolution, the best compromise is to use an intermediate voltage giving V_0/V_E between 2 to 5.

Fisher[4] and Ogilvie[4] showed that the theoretical maximum value of the probe current i_{max} is

$$i_{max} \approx \left(\frac{3\pi^2}{16}\right) B \, \frac{eV}{kT} \frac{D^{8/3}}{C_s^{2/3}} \tag{10}$$

where B is the filament brightness, V the applied voltage, kT the thermal energy of the electron, D the diameter of the electron beam at the sample, and

C_s the spherical aberration coefficient. It will be noted that $i_{max} \propto D^{8/3}$, so that the emitted x-ray intensity falls off rapidly as the probe size is reduced, leading to low counting rates and low accuracy in analysis. For this reason, and also because the penetration is usually of the order of 1 μ at operating voltages, it is seldom desirable to reduce the beam size below 1-μ diameter, although this may be advantageous in increasing the resolution in scanning images, obtained using either the backscattered electron, specimen current, or secondary electron signals.

The efficiency of production of characteristic x-rays within the sample is another important factor governing the sensitivity obtainable. The minimum detectability limit will also depend on the degree of statistical confidence placed upon a given intensity measurement [see Liebhafsky et al.[24] for a detailed discussion]. It turns out that the statistics for emission of x-ray quanta are the same as for radioactive decay, provided that ideal operating conditions are maintained, so that other variables do not come in. Under these conditions, if a sufficient number of counting samples are gathered, a plot of frequency of occurrence versus number of counts follows a Poisson distribution. The standard deviation s is thus the square root of the total number of counts N', that is,

$$s = (N')^{1/2} \qquad (11)$$

If a 95% confidence limit is required for stating that an element is detected, the theoretical limit of detectability is then three times the standard deviation.

VII. QUANTITATIVE ANALYSIS

Although in qualitative analysis the problem is to identify the wavelengths present in the emission spectrum from the sample, in quantitative analysis, one is concerned with converting the precise measured x-ray intensities to weight percent. If homogeneous standards of known composition are available for the alloy system measured, the most reliable procedure is usually to construct experimental calibration curves of relative intensity versus composition for fixed operating conditions—the empirical approach in which corrections are avoided. In the absence of adequate standards, it is necessary to use a theoretical approach based on the use of pure metal standards and calculated corrections.

Provided that reasonable count rates are obtained, accuracies of about 1% of the amount of an element present may be obtained by the empirical approach. The problem of the preparation of suitable standards is by no means trivial, as pointed out by Birks.[3] It is not sufficient to use an inhomogeneous standard of which the average chemical composition is known precisely, and to rely on increasing the beam size to average out the local fluctuations in composition, because the probe does not then necessarily measure an arithmetic average composition. This is because effects such as absorption and secondary fluorescence are not linear functions of concentration. It is necessary to employ standards that are homogeneous on a micron scale. Similar problems may arise if the volume of the sample in which x-rays are generated is not uniform in concentration which will be the case if steep concentration gradients exist or inclusions are present.

Whichever approach is used, the best accuracy is normally obtained if the analytical conditions are chosen to minimize factors such as secondary fluorescence and absorption. The accuracy with which an absorption correction can be calculated may be limited by the accuracy with which the fundamental absorption coefficients are known [see tables by Heinrich et al.[6]].

In both approaches, the raw x-ray data for the selected characteristic radiation for the element i being analyzed must first be corrected for (a) counter dead-time, (b) instrumental drift, and (c) background. If the counter dead-time is $< 10\%$, the following simple expression may be used:

$$N = \frac{N'}{1 - N' \tau} \tag{12}$$

where N and N' are the true and observed intensities, respectively, in counts per sec, and τ is the largest value of the resolving time of the components used in the x-ray detection system. Usually, τ is determined by the counter dead-time. See Wittry[4] and Elion[5] for a more detailed discussion.

Instrumental drift may be determined as a function of time by measuring the intensity of the standards before, after, and if necessary, during measurements from the sample. If the drift is other than linear with time, the correction becomes increasingly inaccurate. Electron feedback systems and good probe design enable drift to be reduced to negligible proportions.

If the continuum x-ray intensity I_b is small, compared with the measured characteristic intensity I_T (similarly corrected for counter dead-time and for drift), a constant background may be assumed and subtracted out. The corrected intensities $I_i{}^j$ and $I_i{}^0$ for the ith element in an alloy j or a pure standard are then, respectively,

$$I_i^j = I_T^j - I_b^j \tag{13}$$

and

$$I_i^0 = I_T^0 - I_b^0 \tag{14}$$

However, in the case of dilute alloys, where the background is an appreciable fraction of the measured characteristic intensity, the background intensity must be measured for each characteristic radiation and for each element in the specimen. The background intensity I_{bi}^j for the ith element is then, after So and Potts:[30]

$$I_{bi}^j = \sum_j c_j I_{ji}^0 \tag{15}$$

where c_j is the weight fraction of the jth element, and I_{ji}^0 the background intensity of the ith element, measured with the spectrometer at the i setting, due to the jth element measured on the pure jth element. The summation is made over all the elements present in the specimen.

The resultant intensity value I_i^j is normally expressed as a ratio to I_i^0 for the pure metal standard, and may now be converted to composition by the empirical or theoretical approach.

A. Empirical Approach

In the empirical approach, the composition is simply read off from the calibration curve, determined experimentally using identical experimental conditions and a set of alloy standards spanning the compositions of interest. The calibration curve (see the example in Figure 12) is a plot of the ratio of the characteristic x-ray intensity generated in the alloy standards to that generated by a pure standard of the same element against composition for the same alloy system.

Most binary calibration curves may be closely approximated, as shown by Ziebold and Ogilvie,[31] by the relation

$$c_i^j = \frac{k_i^j a_{il}}{[1 - k_i^j(1 - a_{il})]} \tag{16}$$

where c_i^j is the weight fraction of element i, k_i^j the intensity ratio for element i $(= I_i^j/I_i^0)$ and a_{il} is a constant which gives the best least squares fit to the experimental values of intensity from the series of standard alloys of elements i and l. If it is desired to convert directly to atomic percent, the constant a_{il} is replaced by b_{il}:

$$b_{il} = a_{il}\left(\frac{A_l}{A_i}\right) \tag{17}$$

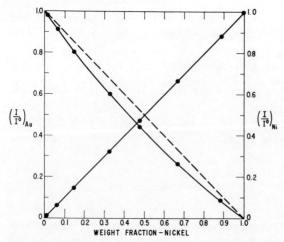

Figure 12 Experimental calibration curve (I/I^0 versus weight fraction) for the gold-nickel system. Data from at $\alpha = 20°$ and $V_0 = 30$ keV. From Lifshin and Hanneman.[21] (Courtesy of General Electric Company.)

where A_l/A_i is the ratio of the atomic weight of element l to that of element i. The values of a_{il} and b_{il} depend on both the voltage and takeoff angle employed.

An expression similar to Equation 16 has been developed for ternary alloys by Ziebold and Ogilvie.[31] The compositions may thus be computed, rather than read off from the calibration curve.

B. Theoretical Approach

In the absence of suitable alloy standards, it is necessary to calculate the concentration of an element A in a sample from the ratio of its characteristic x-ray intensity to that generated by pure element A under the same experimental conditions. Worthington and Tomlin,[32] Archard,[33] Green and Cosslett,[34] and Brown and Ogilvie[35] have treated the intensity of characteristic x-ray production. After Ogilvie,[9] the number N_K of K quanta (frequency ν_K) per incident electron radiated into the solid angle $\delta\omega$ at a takeoff angle α into the x-ray spectrometer is given by:

$$N_K = RW \left[\exp(-\psi_K \bar{x} \operatorname{cosec} \alpha)\right] \int_0^{x_K} \frac{\partial n}{\partial x} \, dx + N_{K_f} \qquad (18)$$

where R is a correction factor for the loss of ionization, due to backscattered electrons; W the fluorescent yield factor; ψ_K the linear absorption coefficient for the K photons; and $\bar{x}$ the average depth of K photon production. Also, x_K represents the path distance at which the electron energy decreases below the energy for K shell ionization; $\partial n/\partial x$ is the number of K shell ionizations per unit electron path; and N_{K_f} is the contribution resulting from secondary ionization such as that produced partly by a strong characteristic line from a second element and partly by the continuous spectrum.

The term in square brackets in Equation 18 is the absorption correction after Castaing[2] which we shall now call $f(x)$. Then $\partial n/\partial x$ is given by the product of the number (N) of atoms per unit volume and the average cross section $(\bar{Q})$ for K shell ionizations. Hence Equation 18 becomes

$$N_k = RWf(x)N\bar{Q}x_K + N_{K_f} \tag{19}$$

It follows from Equation 19 that K_A, defined as the ratio of the intensity from the alloy $I_{A+B+\cdots i}$ to the intensity of pure A is

$$K_A = \frac{RWf(x)N\bar{Q}x_K + N_{K_f}}{R^0Wf^0(x)N^0\bar{Q}x_K{}^0 + N_{K_f}{}^0} \tag{20}$$

where the superscripts indicate the pure element A.

Ogilvie[9] points out that Equation 20 includes all the factors required for an exact solution of the intensity ratio, but is more usefully reduced to Equation 21 below because of the difficulty in evaluating the average ionization $(\bar{Q})$ and the ionizations lost due to the electrons backscattered:

$$K_A = \left[\frac{f(x)}{f^0(x)}\right]\left[1 + \frac{I_{sf}}{I_{pf}}\right]\left[\frac{\alpha_A}{\sum_i \alpha_i c_i}\right] c_A \tag{21}$$

where c_A is the weight fraction of the component A, and the three square-bracketed terms represent respectively Castaing's[2] absorption correction, Castaing's[2] secondary fluorescence correction, and the atomic number correction of Poole and Thomas.[36] In dealing with these three corrections below, we follow Ogilvie[9] in employing the Castaing[2] absorption correction, as modified by Philibert[37] and Duncumb and Shields,[6] replace Ogilvie's[9] modification of Castaing's[2] secondary fluorescence correction by an improved modification due to Reed,[38] and replace Poole and Thomas's atomic correction[36] by an improved correction after Duncumb and Reed [see Ref.[10]]. The merits of alternative combinations of computerized correction procedures have been critically reviewed by Beaman and Isasi.[39]

1. Absorption Correction

Castaing's[2] absorption correction is better expressed in analytical form, after Philibert:[37]

$$f(x) = \frac{1 + H}{(1 + x/\sigma)[1 + H(1 + x/\sigma)]} \tag{22}$$

where $H = 1.2 \Sigma \alpha_i A_i / (\Sigma \alpha_i Z_i)^2$ (in which α_i is the atomic concentration, A the atomic weight, and Z the atomic number), $x = (\psi/\rho)_{A+B+\cdots i} \operatorname{cosec} \alpha$ (in which ψ is the linear absorption coefficient, ρ the density), and σ is the cross section for absorption of electrons, as given by Duncumb and Shields:[6]

$$\sigma = \frac{2.39 \cdot 10^5}{(V^{1.5} - V_{EK}^{1.5})} \tag{23}$$

where V and V_{EK}, both in kilovolts, are the incident electron beam potential and critical excitation potential for the characteristic (K) line involved, respectively.

Equation 23 enables the progressive loss of energy of electrons, as they penetrate into the alloy sample, to be taken into account. Since the excitation potentials of the elements present differ, this means that the emission from one component may come from nearer the surface than another and undergo less absorption.

A tabulated form of Equation 22, covering a wide range of voltages, was published by Adler and Goldstein[40] and makes calculations of the absorption correction a simple matter.

Brown and Ogilvie[35] showed that the $f(x)$ curve could be calculated for any element having any particular tracer. Experimentally, it has been found that a tracer layer of zinc within a copper sample, for instance, since it has a higher photon excitation energy, shifts the x-ray distribution towards the surface. Ogilvie[9] gives a set of such curves for the copper-aluminum system, covering Cu K emission in pure copper and pure aluminum, and Al K emission in pure aluminum and pure copper. In the absence of adequate information, linear interpolation between each pair of limiting curves is assumed for a particular Cu-Al alloy.

The success of the Philibert and other absorption corrections has been reviewed by Poole and Thomas.[6] As Heinrich pointed out,[6] unless accurate values of the mass absorption coefficient are available for systems studied and the wavelength used, the proper evaluation of the various proposed absorption-correction functions is impossible. Disregard for this fact has led to much confusion. Comparative tests are also difficult if other corrections to the data are necessary that are not well known such as the atomic number effect. Current values of the absorption coefficient have been tabulated by Heinrich.[6]

2. Secondary Fluorescence Correction

The secondary fluorescence correction, which is the second square-bracketed term in Equation 21, will now be considered. A correction for secondary fluorescence is necessary when either the characteristic radiation or part of the continuous spectrum, excited by the electron beam from one species of atom in the sample, has sufficient energy to excite characteristic emission from a second series. Since it is usual to employ two or three times the beam voltage necessary to excite the characteristic radiation required in order to increase the line to background ratio, some fluorescence normally occurs in an alloy sample, and a correction should be calculated. When secondary fluorescence occurs, the spatial resolution is reduced, since the volume in which the x-rays are being generated is increased. The basic emission concentration proportionality is, in general, no longer true for the secondary emission. This is because an element that strongly absorbs the exciting radiation undergoes a selective excitation which increases its apparent concentration, unlike primary emission in which all elements possess the same mass absorption coefficient for the beam electrons.

An expression to correct for secondary fluorescence, due to the continuous spectrum, was given by Castaing and Deschamps.[41] It is usual to neglect this correction, since it is difficult to evaluate and usually small. In order to decide whether fluorescence due to the characteristic radiation will occur, it is necessary to know the relationship between the characteristic wavelengths and the absorption edges. Expressions for this purpose have been given by Castaing,[1] Wittry,[42,43] Birks,[3] Lifshin and Hanneman,[21] and others.

Castaing[2] derived a fluorescence correction for K lines exciting K lines (K-K fluorescence), giving the intensity ratio I_f/I_A of the fluorescent intensity to that of primary radiation. These intensities are corrected for absorption in emerging from the sample. Castaing's relation may be written:

$$\frac{I_f}{I_A} = c_B \left[\frac{1}{2} W_K{}^B \left(\frac{r_A - 1}{r_A} \right) \left(\frac{A_A}{A_B} \right) \left(\frac{\lambda_B}{\lambda_A} \right) \left(\frac{\mu_B{}^A}{\mu_B{}^B} \right) \right] X \qquad (24)$$

where

$$X = \left[\frac{\ln (1 + u)}{u} + \frac{\ln (1 + v)}{v} \right]$$

Here c_B is the weight fraction of the element B, $W_K{}^B$ the K shell fluorescent yield of element B, and r_A the K absorption edge jump ratio of element A. Also, A_A and A_B are the atomic weights of elements A and B, respectively; λ_A and λ_B are the wavelengths of the absorption edges of elements A and B; $\mu_B{}^A$ is the mass absorption coefficient of element A for K radiation from element B which is equivalent to the linear absorption coefficient ψ divided by the density ρ; and $\mu_B{}^B$ is the mass absorption coefficient of element B for K radiation from element B. The factor 0.5 comes from the assumption that, if primary x-ray production occurs on the surface of the specimen, one-half of the total intensity will be adsorbed in the specimen. u and v are given by Equations 25 and 26:

$$u = \left(\frac{\mu_A}{\mu_B}\right) \csc \alpha \qquad (25)$$

$$v = \frac{\sigma}{\mu_B} \qquad (26)$$

where σ is the cross section for absorption of electrons given by Equation 23 and α is the x-ray takeoff angle.

While Castaing's Equation 24 has been found satisfactory for K-K fluorescence, modification is necessary, for example, for K-L, L-K or L-L fluorescence. This is necessary because, when the characteristic emission of the jth element is of shorter wavelength than the absorption edge of the ith element, it is able to enhance the characteristic emission of the ith element. Reed[38] modified Castaing's Equation 24 as subsequently expressed in more general form, applicable to a multicomponent system by So and Potts,[30] to include all cases involving K and L characteristic radiations, giving Equation 27. Let element h fluoresce element i in an alloy j, then the contribution to i intensity $\epsilon_{i,h}^j$ is

$$\epsilon_{i,h}^j = \frac{I_{i,h}^j}{I_i} = c_h{}^j \left[\frac{1}{2} W_h \left(\frac{r_i - 1}{r_i}\right)\left(\frac{A_i}{A_h}\right)\left(\frac{\lambda_h}{\lambda_i}\right)\right]\left[\frac{\mu_h{}^i}{\mu_h{}^j}\right] X \qquad (27)$$

where the symbols have the same general meaning as in Equation 24. The coefficient μ_A in Equation 25 becomes $\mu_i{}^j$; μ_B in Equations 25 and 26 becomes $\mu_h{}^j$. The second square-bracketed term in Equation 21 is then written as $[1 + \epsilon_{i,h}^j]$, where $\epsilon_{i,h}^j$ is given by Equation 27.

3. Atomic Number Correction

Finally, we come to the third term in Equation 21, namely, the atomic number correction after Duncumb and Reed (see Ref. 10). This is a two-part correction, consisting of backscatter and penetration factors. The former accounts for the difference in the number of electrons backscattered from the sample, as compared with the standard. This fact means that different fractions of electrons are available for x-ray production in the sample and standard, respectively. The penetration factor accounts for differences in the depth distribution of x-ray production between the sample and the standard. Duncumb and Reed's relation (see Ref. 10) may be written:

$$\frac{L_i{}^j}{L_i{}^0} = \left[c_i{}^j \frac{R^j}{R_i{}^0}\right]\left[\frac{\int_{E_{BK}}^{E_0} (Q/S^j)\ dE}{\int_{E_{BK}}^{E_0} (Q/S_i{}^0)\ dE}\right] \qquad (28)$$

where the first square-bracketed term is the backscatter factor, and the second is the penetration factor. Here $L_i{}^j$ is the intensity of element i determined on alloy j, $L_i{}^0$ the intensity of i determined on pure i, $c_i{}^j$ the weight fraction of i in alloy j, R^j the backscatter factor for alloy j, $R_i{}^0$ the backscatter factor for pure i, Q the ionization cross section, S^j the stopping power of the alloy j, and $S_i{}^0$ the stopping power of pure i. Also, E_0 is the original energy of the back-scattered electrons, that is, is equal to the incident electron energy; and E_{EK} the critical excitation energy for the characteristic K line involved. The backscattered electrons possess a range of energies from E_0 down. In order to evaluate Equation 28, we need to evaluate S and R.

a. EVALUATION OF STOPPING POWER S

The stopping power S, which is defined as the ability of a given material to slow down electrons, is

$$S \equiv -\frac{1}{\rho} \cdot \frac{dE}{dx} \tag{29}$$

where ρ is the density, and dE/dx the loss of energy per unit distance.

In practice for a pure element, S may be determined by the Bethe relation:[44]

$$S = \text{const} \frac{Z}{A} \cdot \frac{1}{E} \ln\left(\frac{1.166E}{J}\right) \tag{30}$$

where Z is the atomic number, A the atomic weight, E the electron energy, and J the mean ionization potential which may be calculated from a table of J/Z as a function of Z given by Duncumb and Reed.[10]

Duncumb and Reed showed that the penetration factor in Equation 28 was approximately

$$\frac{S_i{}^0}{\bar{S}^j} \approx \frac{\int_{E_{EK}}^{E_0} (Q/S^j)\, dE}{\int_{E_{EK}}^{E_0} (Q/S_i{}^0)\, dE} \tag{31}$$

where $\bar{S}_i{}^0$ is the mean stopping power of pure i, and $\bar{S}^j$ the mean stopping power of the alloy j. This simplification is possible because $S_i{}^0/S^j$ varies only slowly with E and can therefore be evaluated at an energy midway between E_0 and E_K and taken with Q outside the integrals which then cancel. The terms $\bar{S}_i{}^0$ and $\bar{S}^j$ can be evaluated from Equation 30 and the relation,

$$\bar{S}^j = \sum_K c_K{}^j S_K{}^0 \tag{32}$$

The term $1/E$ then appears in both the numerator and denominator and cancels, as does the constant. The mean energy $\bar{E}$ is approximated by:

$$\bar{E} = \frac{1}{2} (E_0 + E_K) \tag{33}$$

where E_0 is the incident electron energy, and E_K the critical K line excitation potential.

b. Evaluation of the Backscatter Coefficient R

The backscatter coefficient R is the ratio of the actual x-ray intensity to that which would have been obtained if no backscattering of electrons occurred. The coefficient R is given by

$$R = 1 - \frac{\int_{W_K}^{1} \eta(W) \, Q/S \cdot dW}{\int_{E_K}^{E_0} Q/S \cdot dE} \tag{34}$$

where $W = E/E_0$, $\eta(W)$ is the number of backscattered electrons with energy greater than WE, Q the ionization cross section, and S the stopping power evaluated from Equations 29 and 30.

For an alloy, experiments show that the number of electrons of all energies backscattered per incident electron, that is, the backscattered fraction η, is given quite well by the mass concentration average:

$$\eta = \sum_i \cdot c_i \cdot \eta_i \tag{35}$$

In practice R_i^0 in Equation 28 may be evaluated from a table of R as a function of Z and $1/U$ ($U = E_0/E_K$), computed by Duncumb and Reed,[10] using Equation 27. The R^j in Equation 28 may be determined from:

$$R^j = \sum_k c_k^j R_k^0 \tag{36}$$

where c_k^j is the weight fraction of k in alloy j, and R_k^0 the backscatter factor for pure k.

Thus Equation 28 may be rewritten in the form:

$$\frac{L_i^j}{L_i^0} = c_i^j \left(\frac{\sum\limits_k c_k^j R_k^0}{R_i^0} \right) \left(\frac{\bar{S}_i^0}{\sum\limits_K c_K^j S_K^0} \right) \tag{37}$$

Equation 21, incorporating the three corrections, may be rewritten as follows:

$$\frac{I_i^j}{I_i^0} = c_i^j \left[\frac{f(x)}{f^0(x)} \right] [1 + \epsilon_{i,h}^j] \left[\frac{L_i^j}{L_i^0} \right] \tag{38}$$

where the absorption correction (the first square-bracketed term) is given by Equation 22, the secondary fluorescence correction (the second square-bracketed term) by Equation 27, and the atomic number correction (the third term) by Equation 37.

So and Potts[30] and Beaman and Isasi[39] have discussed the computerization of correction procedures. The use of a computer is almost mandatory, since a number of the factors entering into Equation 38 themselves depend on the composition that one is trying to calculate. An iterative procedure is necessary to solve, starting with the normalized relative x-ray intensities as a first approximation for the weight fractions in calculating the correction terms. The corrected composition is then used to recalculate the correction terms for successive approximations.

4. Errors Due to Contamination Buildup

Finally, the effect of contamination build up when the electron beam strikes the specimen or standard surface, due to the breakdown of hydrocarbon vapors derived from the vacuum system, should be mentioned. The deposit tends to decrease the energy of the incident electron beam, leading to a progressive decrease in x-ray intensity. The contamination thickness depends on the time of exposure, the probe current, the nature and composition of the surface, and the quality of the vacuum. The rate of contamination may be greatly reduced by inserting a cold finger near the specimen surface involved. When alloy standards similar in composition to the sample are used, the effect will tend to cancel if the time of exposure, probe current and voltage are the same. If pure metal standards and calculated calibration curves are employed, particularly if the probe voltage is near the critical excitation voltage for the element measured, the decrease in intensity with time should be measured on the standard and the specimen. If this is appreciable and differs from one to the other, the true intensities can be determined by extrapolation to zero time.

VIII. LIGHT ELEMENT ANALYSIS

Considerable progress has been made in extending the range of the electron microprobe to elements below sodium in the periodic table, although the sensitivity for elements such as boron, oxygen, nitrogen, and to a lesser extent carbon still leaves much room for improvement. The principal problems are associated with the long wavelength of the characteristic x-ray emission which leads to absorption losses in the windows and to problems in energy discrimination. Attenuation has been reduced by the provision of external means for removing the window between the column and the spectrometer and by the development of very thin (1000 to 2000 Å thick) counter windows of relatively transparent material such as nitrocellulose, although these are extremely fragile. Discrimination has been explored by a number of methods, notably the use of crystals of large d-spacing such as multilayer stearate films, ruled diffraction gratings, and pulse height analysis. The first approach is employed in a number of commercially available electron microprobes.

A further problem arises as a result of overlapping of characteristic lines of heavy elements, which may be present, with lines from the light elements; for example, L lines from chromium and vanadium lie close to the K line from oxygen. A preliminary survey for the heavy elements is therefore necessary.

The carbon contamination layer, which gradually builds up on the sample where the electron beam impinges, tends to attenuate the emitted x-rays and, of course, gives misleading results in the case of carbon analysis. The provision of a liquid nitrogen-cooled finger near the sample in current commercial instruments reduces this problem to an acceptable level, although the ultimate solution lies in the employment of a clean high vacuum system. The usual sources of contamination are the backstreaming of oil vapor from the vacuum pumps, the presence of traces of grease and oil in the system, and the use of rubber for seals. The presence of oxide or nitride layers on the surface of reactive phases can similarly lead to problems.

A further difficulty is that both line shapes and peak positions of the soft x-ray spectral lines are far more dependent than those of the heavier elements on the state of chemical combination or even on the crystallographic structure. Thus the peaks of the diamond and graphite lines are displaced by nearly 0.5 Å.

Finally, the present correction procedures, applied to the measured x-rays intensities, tend to be less satisfactory at the lower kilovoltages employed in light element analysis [see Springer[7]]. Strong atomic number effects may occur in the case of metallic oxides and carbides. Information on appropriate values of x-ray mass absorption coefficients is still lacking.

The latest information in this rapidly moving area is to be found in proceedings of recent electron microprobe conferences (see Refs. 6, 7, 45, and 46).

The ability to analyze for low atomic number elements, even qualitatively, is particularly valuable. If a given phase can be identified, for example, as an oxide or nitride, then the amount of oxygen or nitrogen may be more accurately estimated by difference. Even an approximate analysis may be sufficient to decide between different oxide phases in a stoichiometric series.

IX. USE OF SIGNALS OTHER THAN X-RAY FOR ANALYSIS

When an electron microprobe beam is incident on a sample, a variety of signals is produced (Figure 8) which may in principle carry analytical information. The *backscattered electrons* are made up of two types, referred to as primary and secondary backscattered electrons. The latter comprising those with energies less than 50 eV are generally used in scanning electron microscopy. The fraction of electrons backscattered has been shown by Sternglass[47] and others to increase with the atomic number, suggesting the possibility of using this signal as a means of chemical analysis, with the aid of suitable standards as discussed by Ogilvie.[9] Alternatively, the *specimen current*, which is the difference between the incident beam current and the backscattered current, may be used to measure the composition of binary alloys, as first proposed by Poole and Thomas[36] and discussed by Heinrich.[7] This can be passed to ground via a resistor to provide a signal voltage. Very little use has so far been made of either of these signals for quantitative analysis; however, they have been used to establish whether a second phase has an average atomic number lower or greater than the matrix.

Analysis by *Auger electron emission*[48] is currently attaining increasing use (see discussion in Chapter 4).

Cathodoluminescence, that is, the emission of visible light, is characteristic of a few materials within the classes of minerals, oxides, phosphors, semiconductors, and insulators when exposed to an electron probe. Since, in principle, both the characteristic wavelengths (spectral lines) and intensities may be measured, useful analytical information may be obtained. The technique has as yet been little exploited in commercial instruments. Qualitative use may be made of cathodoluminescence in the electron microprobe if the instrument is equipped with an optical microscope for viewing the specimen, while it is exposed to the electron beam in the scanning mode. We have found that thoria particles in tungsten may be recognized by a characteristic blue luminescence, although too small to give detectable thorium x-ray emission. Identification was then possible by a comparison with larger particles which gave a detectable x-ray signal. The size limit of detectability was limited to about 1-μ diameter by the resolution of the optical system provided. Information on the particle size and distribution is thus obtainable.

Variations in composition within, for example, a single crystal of calcite were shown by Long and Agrell[49] to give rise to variations in color and intensity, due to fluctuations in the Mn/Fe ratio. Cathodoluminescence in the visible part of the spectrum is sometimes an intrinsic property of a mineral, as in the case of scheelite,[50] but is more often imparted by traces of impurities such as rare earths or manganese, known as "activators." The light emitted from minerals may be polarized, as reported as early as 1879 by Crookes.[51] The complexities of cathodoluminescence in gallium arsenide have been studied by Kyser and Wittry.[6]

McMullan[52] and Smith[53] employed a photomultiplier without wavelength selection, in conjunction with a scanning electron microscope, to obtain scanning images showing cathodoluminescence. Wavelength selection was incorporated by Bernard, Davoine, and Pinard[54,55] in a scanning microscope and by Davey[7] in a scanning electron microprobe. Pruden et al.[45] modified a commercial Stereoscan scanning electron microscope by the addition of light guides and a photomultiplier, and claimed a resolution of about 750 Å in a scanning micrograph made of a phosphor particle using the light signal. Comparable resolution may not be obtainable on materials exhibiting a lower efficiency of conversion of electron energy into light, as pointed out by Davey.[7] Typically, 10^{-4} to 10^{-12} mass concentration of an activator is required to attain maximum characteristic light emission. At low concentration the relation between intensity and concentration is approximately linear.[7]

The use of cathodoluminescence appears likely to be limited to qualitative analysis because of its variation with the specimen temperature, time of excitation, sensitivity to the presence of other impurities or lattice defects, and variation with the nature of the matrix in which the activator is present. The subject of cathodoluminescence has been reviewed at some length by Thornton.[56]

X. APPLICATIONS

Only a brief review of some of the multifold applications is given here. For further information see Refs. 2–5, 7, 11, 21, 37, and 46.

A. Phase Identification

The electron microprobe has proved invaluable for the analysis and identification, particularly in situ of microscopic inclusions in metallurgical samples. These often produce detrimental effects; identification is a necessary precursor to their control or elimination. The inclusions present may be classified into various types, using x-ray distribution images or one of the several types of display images, and then a point analysis or line scan analysis made across the inclusions. Figure 13 is a good example, taken from a study[57] of the sulfidation of a nickel-20% chromium alloy, the line distribution of nickel, chromium, and sulfur being represented by superposition on the scanning microscopic image. The scale formed near the surface is clearly seen to consist of chromium sulfide. Concentration gradients around inclusions may be studied.

Unfortunately, it is difficult to make reliable in situ quantitative point analyses on a particle if the particle is smaller than approximately 10-μ

diameter. There is usually no advantages in reducing the beam size $< 1\ \mu$, since this results in loss of x-ray intensity; furthermore, the penetration, which is a function of the excitation voltage, is still usually $\geq 1\ \mu$. The minimum volume that can be analyzed is also limited as a consequence of electron scattering, secondary fluorescence, and the takeoff

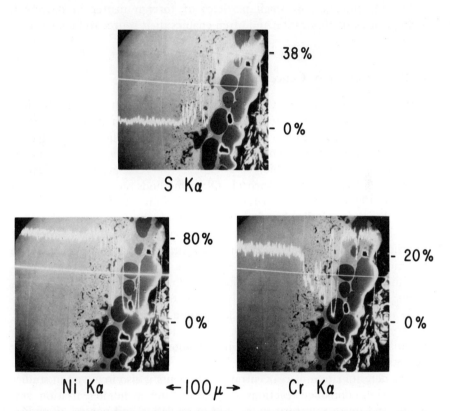

Figure 13 X-ray line scans across sulfide scale, formed on a Ni-20% Cr alloy in an H_2S-H_2 atmosphere at 700°C ($H_2S/H_2 = 6 \times 10^{-3}$). Horizontal white line is superimposed on specimen current image to show scan position. (From Lifshin.[57] (Courtesy of General Electric Company.)

angle. While the resolution can be improved by decreasing the sample thickness to $\leq 1\ \mu$, new complexities are introduced into quantitative analyses by the transmission of electrons through the sample, necessitating an accurate knowledge of the thickness and additional corrections. It is unwise to rely on a quantitative analysis of a single 10-μ diameter inclusion in a metallographic section, since the depth of the inclusion is usually unknown. In analyzing the matrix, it should be remembered that there may be an invisible inclusion just below the surface which will be averaged into the analysis; thus reliance should not be placed on a single-point analysis on the matrix. These problems are much less serious if only qualitative or semiquantitative analyses are required.

The identification of small particles of foreign matter in devices is another obvious application and often enables their source to be identified.

B. Determination of Concentration Gradients

Since microvolumes may be analyzed, the electron microprobe is a powerful tool for the determination of concentration differences from point to point, and thus is extremely useful for studies of segregation and in determining concentration gradients. An obvious application is in diffusion studies, in which it is capable of very general application using the diffusion couple approach. Older methods require the machining of 50 to 100-μ-thick slices parallel to a diffusion interface, followed by the analysis of each slice by chemical, tracer, or x-ray emission techniques. The size resolution of the microprobe beam is one to two orders of magnitude better, permitting the use of shorter diffusion treatments. The technique is nondestructive and applicable to both metallic and non-metallic systems, since machinability is no longer a factor. The measurements and the correction and analysis of data may be automated. Figure 14 is an example of the concentration profile in a Au versus Au-6.1 wt.% Ni couple diffused for 1 week at 875° C determined in this way using automatic step scanning, automatic readout of data and computer processing.[21,58]

The determination of concentration gradients is also involved in studies of diffusion controlled reactions of all kinds such as homogenization, sintering, and phase transformations, and in oxidation and corrosion studies. Figure 15 is an example of microprobe application to a study of the sulfidation of a Ni-20 at. % Cr alloy.[57] The sample was exposed to an H_2S-H_2 atmosphere at 700° C, then sectioned normal to the surface and metallographically polished. The surface scale visible in the specimen current scanning image is shown by the x-ray distribution images to be enriched in chromium and sulfur and denuded of nickel, while a region free of chromium is apparent below the surface scale. As is readily apparent, the layers have a complex morphology. The line scans made on the same area (Figure 13) show greater detail, since the counting statistics are more favorable. More accuracy still is obtainable by point counting for appropriate time periods.

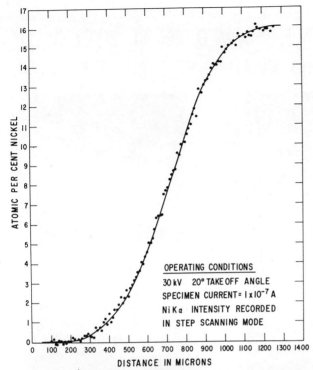

OPERATING CONDITIONS

30 kV   20° TAKE OFF ANGLE

SPECIMEN CURRENT = 1 x 10⁻⁷ A

Ni Kα INTENSITY RECORDED

IN STEP SCANNING MODE

Figure 14 Concentration profile in a Au versus 93.9wt.%Au-6.1%Ni couple diffused for 1 week at 875°C. From Lifshin and Hanneman.[21] (Courtesy of General Electric Company.)

C. Phase Equilibria

Much time-consuming work in the determination of phase diagrams or phase equilibria is avoided in the approach, applicable to binary or pseudobinary systems which employs series of couples of known compositions diffused to equilibrium at various controlled temperatures, subsequently sectioned and analyzed with a microprobe. A smooth concentration gradient appears in single-phase regions, with discontinuities at the boundaries of the two-phase fields, which may thus be determined as a function of composition and temperature. Because of its good resolution, even relatively thin phase layers and steep concentration gradients may be studied with the microprobe technique. In a study[21] of this type on the gold-nickel system (Figure 16), good agreement was found with the established phase diagram. When the resolution is decreased as a result of secondary fluorescence, it may be advantageous to reduce the incident beam voltage and measure the backscattered electron fraction,

SULFIDE SCALE FORMED ON Ni-20Cr ALLOY IN A H₂S-H₂ ATMOSPHERE AT 700°C

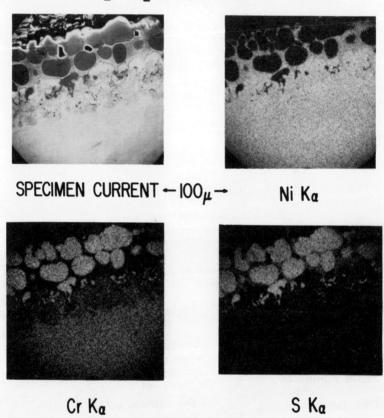

SPECIMEN CURRENT ←100μ→   Ni Kα

Cr Kα   S Kα

Figure 15 Specimen current and corresponding scanning x-ray distribution images of sulfide scale, formed on Ni-20%Cr alloy in an H₂S-H₂ atmosphere at 700°C. From Lifshin.[57] (Courtesy of General Electric Company.)

instead of the emitted x-rays, in order to obtain finer resolution. See Ogilvie[9] for a more detailed discussion.

D. Film Thickness Determination

If a metal film is thinner than the depth of penetration of the electrons, then, for normal incidence, the film thickness may be determined from the intensity of x-ray emission. Until satisfactory calibration curves can be calculated theoretically, it is necessary to prepare a calibration curve of corrected relative x-ray intensity versus film thickness, using a number of standard films whose thickness has been determined by other techniques such as interferometry. The microprobe technique is usually ap-

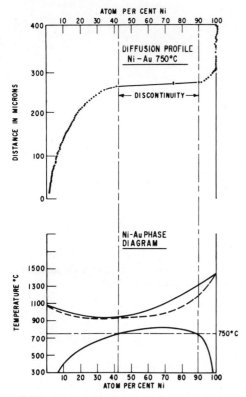

Figure 16 Pure gold-pure nickel couple, diffused at 750°C. Concentration profile showing concentration discontinuities, corresponding to the two-phase region in the gold-nickel phase diagram at 750°C. From Lifshin and Hanneman.[21] (Courtesy of General Electric Company.)

plicable in the thickness range from about 20 Å to 2000 Å. Figure 17 is an example of such calibration curves for Au L_α emission versus film thickness of gold on Si and SiO_2 substrates.[21] The values of corrected relative intensity were obtained by taking the ratio of the x-ray intensity measured on the thin film to that measured on a bulk pure standard of gold, both corrected for background, counter dead-time, and drift.

The ratio was plotted against film thickness and extrapolated to zero thickness. The points marked "current data" and "unknown sample" are from Lifshin and Hanneman[21] for gold on a silica substrate, the remainder from Hutchins[6,59] for gold on silicon. Hutchins[6,59] has discussed the thickness determination of thin films in some detail.

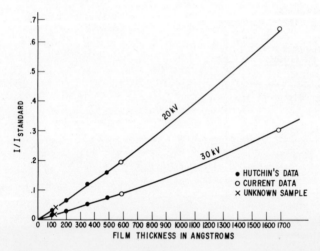

Figure 17 Showing a calibration curve for thin film thickness determination. Corrected relative intensity of Au L_α versus Au thickness on Si and SiO_2 substrates. From Lifshin and Hanneman.[21] (Courtesy of General Electric Company.)

E. Kossel Line Technique

The addition of a Kossel camera for making divergent beam x-ray diffraction photographs enables an electron microprobe to be used to obtain diffraction information such as lattice parameters and orientations on a selected single grain with a volume down to $< 10^{-9}$ cm³, as pointed out by Castaing.[1]

Rutherford and Andrade[60] performed the classical experiment, involving diffraction of a divergent beam of gamma x-rays from radium. Kossel[61,62] showed that the same technique could be employed using a divergent beam of x-rays. Von Laue[63,64] was able to explain many of the finer details of Kossel patterns from dynamical x-ray theory. Lonsdale[65] applied the technique to practical studies on single crystals.

Characteristic x-rays are generated from what is effectively a point source, where the incident microprobe electron beam strikes the sample. In the pseudo-Kossel method [see Gielen et al.[66] for theory], the beam is allowed to strike an appropriate thin foil of a different metal overlaid on the specimen surface, thus permitting a free choice of the x-ray wavelength. Usually, the specimen is about 100 μ thick and the film is placed on the side opposite to the incident electron beam, a few centimeters

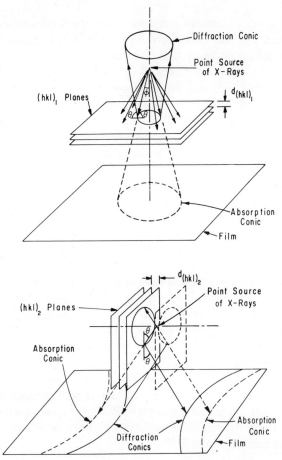

Figure 18   Schematic showing geometry of diffraction and absorption. Kossel line conics, formed from a single crystal. From Morris.[67]

from the sample, giving a transmission pattern. Back reflection patterns can be obtained with limited success in a similar manner.

The geometry of Kossel line formation is illustrated in Figure 18 after Morris.[67] Diffraction of characteristic x-rays occurs according to Bragg's law from, for example, the $(hkl)_1$ and $(hkl)_2$ sets of planes, giving diffracted cones of half-apex angle α:

$$\cos \alpha = \sin \theta = \frac{\lambda}{2d_{hkl}} \tag{39}$$

where λ is the x-ray wavelength, and d_{hkl} the interplanar spacing of the (hkl) planes.

A geometric pattern of conic sections results from the intersection of the cones from all diffracting planes with the film. There is background blackening, due to the continuous spectrum, and also from undiffracted characteristic radiation. The conic is deficient in intensity, relative to the background on its concave side, and shows a higher relative intensity on its convex side. Thus either diffraction or absorption cones are seen, depending on which effect is dominant. This depends on such factors as the wavelength of the x-rays, and on the nature and thickness t of the crystal. As a qualitative guide μt should be less than unity,[65,67] where μ is the linear absorption coefficient. Good Kossel patterns are obtained if the crystal is neither too perfect nor too imperfect.

Morris[67] has generalized the plotting of stereographic projections from Kossel patterns, so that it is no longer restricted to cubic crystals. He established procedures for determining orientations and crystal structures, and provided computer solutions based on the use of the reciprocal lattice and matrix algebra.

The specimen-film distance need not be accurately known nor the film be precisely parallel to the crystal in order to obtain precise lattice parameters by this technique.[68] Sources of error were discussed by Yakowitz.[6] The important factor is a ratio of distances in the pattern which can yield lattice parameters accurate to 1 part in 10^5, or better. The technique can thus be used to measure elastic strains in microvolumes and has been employed in such studies as irradiation damage, internal stress, deformation of single crystals,[69] and misfit strains at semiconductor junctions. Orientation relationships between precipitates and the matrix may be studied in the back reflection mode. Figure 19 is an example of a transmission conic obtained on nickel.[70]

F. Microradiography

The electron microprobe can be used to produce an intense point source for high resolution projection microradiography.[71]

G. Miscellaneous

The electron microprobe and its scanning facilities have found many important applications in the manufacture and dynamic testing of semi-

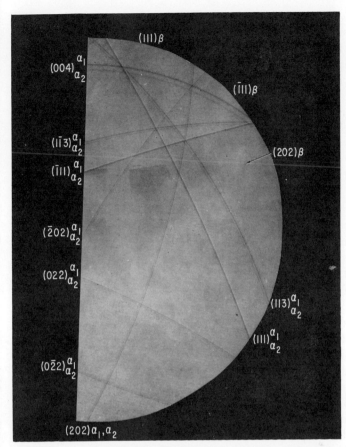

Figure 19 Enlarged region of transmission Kossel pattern of nickel single crystal. Note K_{α_1}-K_{α_2} doublet separation. From Hanneman.[70]

conductor devices and integrated circuits, as in Nealey et al.,[6] Ramsey and Weinstein,[6] and Everhart.[6] The examination of metal fractures for the presence of stress corrosion causative agents such as sulfur and chlorrine is another kind of application.

The electron microprobe is proving very useful in studies of nickel-base high temperature superalloys which often contain up to 10 or more elements distributed between the matrix, the gamma prime Ni_3Al-type precipitates, and other phases such as carbides. Understanding and prediction of phase changes during high temperature, long time service is dependent on knowledge, at present imperfect, of the way in which the elements are partitioned between the phases. Knowledge of the composition of the matrix and the various kinds of precipitates is important in understanding the resulting mechanical properties.

Applications to other fields such as mineralogy, biology (see Hall in Ref. 10), and medicine are outside the scope of the present discussion. Either the electron microprobe, or the scanning electron microscope, can in principle be used to measure the parameters on a planar section needed to determine such quantities as the volume fraction or size distribution of alloy phases (see discussion by G. Dörfler in Ref. 10).

XI. COMBINED ELECTRON MICROSCOPE AND MICROPROBE

Most electron microscopes have provision for selected area electron diffraction on small areas, typically down to approximately 0.2-μ diameter. While the information obtained often suffices for identification, there are many instances in which it would be valuable to supplement this information with the composition of a small area or particle. For example, in the case of a solid solution the parameter is often relatively insensitive to the change of composition. Many of the leading electron microscope makers have marketed microprobe attachments for this purpose; now microprobe manufacturers are beginning to market electron microscope attachments. Microprobe attachments for scanning electron microscopes employing solid state detectors have already been mentioned. Unfortunately, some compromises in performance are always necessary. It became apparent that better all-round performance might be obtained if a dual purpose instrument were specially designed; Nixon and Buchanan[37] and Duncumb[6] built such instruments. Duncumb's instrument, known as EMMA-1 (Electron Microscopy with Micro-Analysis), proved capable of combined transmission electron microscopy, electron diffraction, and microanalysis,[6] and has been jointly developed by Tube Investments Research Laboratories and AEI Scientific Apparatus into the EMMA-4, a commercial instrument (Figure 20).

Unlike the EMMA-1 prototype, which was basically an inverted electron microscope with 50-keV capability and a resolving power of $\leq$ 100 Å, EMMA-4 is based on a normal 100-keV electron microscope with about 10-Å resolution. A primary change is the use of a minilens based on a principle first demonstrated by J. B. LePoole at the Technical University, Delft, in 1967, as a probe-forming lens (Figure 21). The precision winding in this lens is supported on a copper former and the bulky iron pole-piece eliminated, thus reducing the lens diameter to $< \frac{1}{3}$ that of a conventional magnetic lens designed for 100-keV electrons. This permits more of the emitted x-rays to be deflected into the x-ray spec-

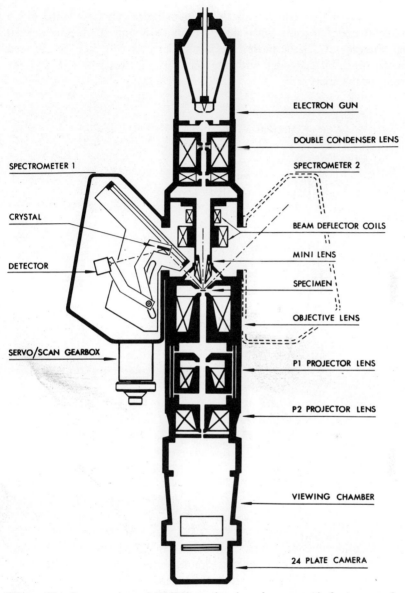

ELECTRON GUN

DOUBLE CONDENSER LENS

SPECTROMETER 1

SPECTROMETER 2

CRYSTAL

BEAM DEFLECTOR COILS

MINI LENS

DETECTOR

SPECIMEN

OBJECTIVE LENS

SERVO/SCAN GEARBOX

P1 PROJECTOR LENS

P2 PROJECTOR LENS

VIEWING CHAMBER

24 PLATE CAMERA

Figure 20 Cross section of EMMA-4, showing electron optical system and x-ray spectrometers. (Courtesy of AEI Scientific Apparatus, Harlow, England.)

trometers. With this instrument, thin film samples can be viewed at up to 160,000 magnification with a resolution of about 10 Å and normal beam diameter of 5 μ or more; the beam diameter can then be reduced to about 0.2 μ by focusing with the minilens, an area selected, and the emitted x-rays analyzed.

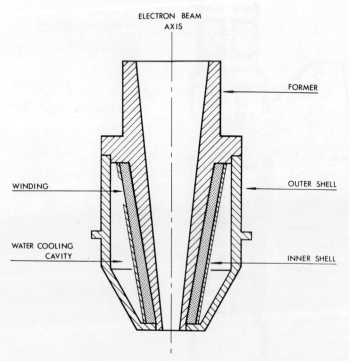

ELECTRON BEAM
AXIS

FORMER

WINDING

OUTER SHELL

WATER COOLING
CAVITY

INNER SHELL

Figure 21 Simplified section of a water-cooled minilens used in EMMA-4. (Courtesy of AEI Scientific Apparatus, Harlow, England.)

Two focusing crystal spectrometers and associated readout electronics enable two elements to be analyzed simultaneously and their ratios measured precisely, independent of drift. The elements from fluorine upward selected are indicated directly on large electroluminescent scales, located on either side of the column (Figure 22). The scales are element wave bands printed directly from a computer output. Each spectrometer has four crystals, covering different wavelength ranges, with a corresponding scale marked by the element name. Manual tuning is used to determine wavelength accurately, or a servosystem used, enabling any one of six

Figure 22 View of EMMA-4 in operation, showing electroluminescent element selection scales and readout x-ray electronics. (Courtesy of AEI Scientific Apparatus, Harlow, England.)

preset elements to be selected by push button. Fast or slow spectral scans may be made automatically with chart recorder-printout.

Six specimens up to 3-mm diameter are accommodated in the specimen stage and may be interchanged without breaking the vacuum. Electrical beam tilting permits dark field microscopy. An anticontamination device is provided, also a mechanized double-tilt specimen holder, permitting tilting a 2.3-mm grid about two orthogonal axes through $\pm20°$.

Selected area diffraction can be carried out on areas as small as 1μ,

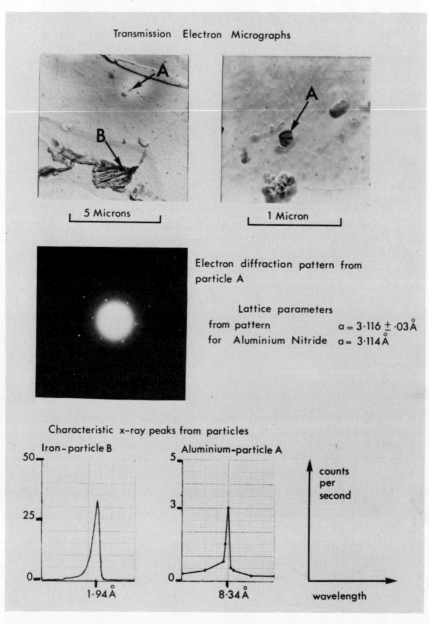

Transmission Electron Micrographs

5 Microns

1 Micron

Electron diffraction pattern from particle A

Lattice parameters

from pattern $a = 3 \cdot 116 \pm \cdot 03 \overset{\circ}{A}$

for Aluminium Nitride $a = 3 \cdot 114 \overset{\circ}{A}$

Characteristic x-ray peaks from particles

Iron - particle B

Aluminium - particle A

counts per second

50

25

0

$1 \cdot 94 \overset{\circ}{A}$

5

3

0

$8 \cdot 34 \overset{\circ}{A}$

wavelength

Figure 23 Metallurgical application of EMMA-4 to sheet containing aluminum nitride (*A*) and pearlite (*B*) particles. Electron micrographs of extracted particles on carbon film, selected area electron diffraction pattern, and spectrometer plots. (Courtesy of AEI Scientific Apparatus, Harlow, England.)

using conventional techniques, or the minilens used to reduce the area selected to about 0.2-μ diameter. An example of the results obtained on aluminum nitride particles (A) associated with surface cracks developed during hot working and pearlite particles (B) in steel tubing is shown in Figure 23.

This type of equipment has unique capabilities. However, it is difficult to obtain fully quantitative analyses because of the geometrical factors involved in making corrections on the raw x-ray data obtained from transmission samples; both the local foil thickness and the particle shape, if the particle is embedded in a matrix of different composition, are involved. If a particle is extracted, the x-ray intensity will be related to its size and shape, as well as composition. Further work is needed to develop more refined correction procedures.

XII. SECONDARY ION EMISSION MICROANALYSIS

The technique of secondary ion emission microanalysis was developed by Castaing and Slodzian about 1962 and has since undergone extensive study and development.[72-76] Secondary ions emitted from a sample bombarded by a beam of positive ions, usually argon or hydrogen, are collected and analyzed in a mass spectrometer. The technique is thus analogous to the conventional mass spectrometer analysis of solid samples but differs in that the ions emitted are focused with lenses to form an image of a sample surface. Using a magnetic filter, a single selected species may be used to form the image which then shows the distribution of the element or isotope.

The arrangement used by Castaing and Slodzian[72-75] is shown schematically in Figure 24. The incident ion beam, which is inclined to the specimen surface at 30°, typically has an energy of 6 to 7 kV and a diameter of about 0.5 mm². Using an accelerating voltage of 3.5 kV, secondary ions sputtered from the sample are focused by an electrostatic lens into the primary image plane. A magnetic prism analyzes the beam into ionic components, one of which is admitted into a projector lens by a slit placed at the point focus C'. After further acceleration, ions impinge on the cathode of an image converter, causing emission of secondary electrons which are focused onto a fluorescent screen. The back reflection image obtained is much brighter than if the ions themselves were allowed to impinge on the screen, and magnifications of about 250$\times$ are possible.

The ultimate resolution ϵ may be estimated[74] from the expression:

$$\epsilon = 10K^{-\frac{1}{4}}p^{-\frac{1}{2}} \exp\left(\frac{n}{m}\right)^{\frac{1}{4}} \tag{40}$$

where K is the number of atoms per cubic micron of sample, and p the percentage precision of analysis of a given element. Also, n/m is the ratio of the number of secondary ions, contributing to the formation of the image, to the number of neutral ions that must be detached to produce the ions required, the lens diaphragm being adjusted to give a resolving power of approximately 1 μ.

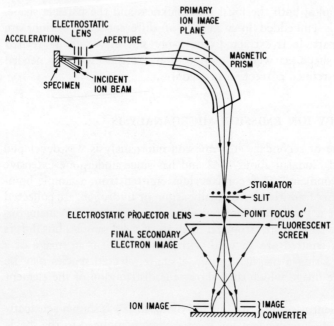

Figure 24 The Castaing-Slodzian secondary ion emission microanalyzer. After Castaing and Slodzian.[72-75]

The ionic yield (n/m) varies with the nature, concentration, and chemical form of the element analyzed. Thus, for pure aluminum, about 10^{-3} of the atoms are sputtered as ions which for $p = 10\%$ gives a resolving power limit of 350 Å. For pure copper, only 5×10^{-7} of the atoms are sputtered as ions; hence, for $p = 10\%$, the limiting resolving power is 0.25 μ. The actual resolving power was limited to about 1 μ by the optical characteristics of the Castaing and Slodzian equipment.[74]

The technique has a number of unique features. Since successive layers of the sample can be removed, the variation of concentration with depth may be explored. For example, the O+ peak from the natural oxide layer on magnesium a few tens of angstroms thick can be detected, and its disappearance followed as deeper layers are analyzed. The ionic yield increases as the atomic number decreases, so that the technique has great potential for the light elements which are the most difficult to analyze in the electron microprobe. Since isotopes can be distinguished, nonradioactive isotope tracers can be used for diffusion studies. The emission intensity depends on the chemical form of an element, thus Cu^+ ions emitted from copper oxide have about 10 times the intensity of those from pure copper.

The secondary ion emission is not unfortunately always proportional to the concentration of a given element. Thus Castaing and Slodzian[74] point out that the addition of 5% aluminum to copper gives a 10-fold increase in emission of Cu^+ ions over pure copper, but half the number of Al^+ ions compared to pure aluminum. Similarly, the addition of 2% beryllium to copper increases the emission of Cu^+ ions 60-fold over pure copper.

The heavier metals are difficult to image because of the small yield of ions. Even though initially smooth, the sample will tend to become etched and rough as sputtering continues. Although subject to these difficulties, the technique has considerable promise for certain applications.

Two instruments for secondary ion emission microanalysis are now commercially available. The CAMECA instrument is based on that of Castaing and Slodzian, whereas the ARL instrument is of the scanning type, where the incident ion beam is scanned over the sample surface and secondary ions picked up by a detector. The latter system is analogous to the scanning electron microscope.

XII. REFERENCES

1. R. Castaing, Thesis, University of Paris, France, ONERA No. 55, (1951), translated by P. Duwez and D. B. Wittry, Interim Technical Report No. 3, under contract, DA-04-495-Ord-463, 1955.
2. R. Castaing, "Electron Probe Microanalysis," in L. Marton, Ed., *Advances in Electronics and Electron Physics*, Vol. 13, Academic Press, New York, 1960, pp. 317–386.
3. L. S. Birks, *Electron Probe Microanalysis*, Chemical Analysis Series, Vol. 17, Interscience Publishers, New York, 1963.
4. *X-ray and Electron Probe Analysis*, 1963, ASTM Special Technical Publication No. 349, 1964.
5. H. A. Elion, "Instrument and Chemical Analysis Aspects of Electron Microanalysis and Macroanalysis," in *Analytical Chemistry*, Progress in Nuclear Energy, Series 9, 1st ed., Vol. 5, D. C. Stewart and H. A. Elion, Eds., Pergamon Press, New York, 1966.

6. T. D. McKinley, K. F. J. Heinrich, and D. B. Wittry, Eds., *The Electron Microprobe*, Symposium held in Washington, D.C., 1964, John Wiley & Sons, New York, 1966.

7. R. Castaing, P. Deschamps, and J. Philibert, Eds., *X-Ray Optics and Microanalysis*, 4th Int. Congress on Optics of X-rays and Microanalysis, held at Orsay, 1965, Hermann & Cie, Paris, 1966.

8. P. Duncumb, *Sci. Progr., Oxford*, **55**, 511 (1967).

9. R. E. Ogilvie, in *X-Ray and Electron Methods of Analysis*, Progress in Analytical Chemistry, Vol. 1, H. van Olphen and W. Parrish, Eds., Plenum Press, New York, 1968, pp. 55–75.

10. K. F. J. Heinrich, Ed., *Quantitative Electron Probe Microanalysis*, Proc. of seminar held at Natl. Bur. Standards, Gaithersburg, Md., June 1967, Dept. of Commerce, Natl. Bur. Standards Special Publication No. 298, 1968.

11. *Fifty Years of Progress in Metallographic Techniques*, Symposium, Atlantic City 1966, ASTM Special Technical Publication No. 430, American Society for Testing and Materials, Philadelphia, 1968.

12. B. L. Henke, *Advan. X-Ray Anal.*, **7**, 460, Plenum Press, New York, 1964.

13. V. D. Scott, *Metals and Materials*, **1**, (2), 39 (1967).

14. D. G. Brandon, *Modern Techniques in Metallography*, Van Nostrand Reinhold, New York, copyright Butterworth & Co. (Publishers) Ltd., London, 1966.

15. Proc. 11th Scintillation and Semiconductor Counter Symposium, Washington, D.C., IEEE Trans., **NS15**, (3), June 1968.

16. W. L. Brown, W. A. Higinbotham, G. L. Miller, and R. L. Chase, Eds., *Conference on Semiconductor Nuclear-Particle Detectors and Circuits*, Gatlinburg, 1968, Publication No. 1593, National Academy of Sciences, Washington, D.C., 1969.

17. E. Lifshin, in Proc. 3rd Annual Stereoscan Scanning Electron Microscope Colloquium, Chicago, 1970, Engis Equipment Co., Morton Grove, Ill., 1970, p. 33.

18. R. Fitzgerald, K. Keil and K. F. J. Heinrich, *Science*, **159**, 528 (1968).

19. J. Russ and A. Kabaya, in *Scanning Electron Microscopy, 1969*, Proc. 2nd Annual Scanning Electron Microscope Symposium, IIT Research Inst., Chicago, Ill., 1969, p. 57.

20. E. Lifshin, Symposium on Energy Dispersion Analysis, ASTM, Toronto, June 1970.

21. Lifshin and R. E. Hanneman, Electron Microbeam Probe Analysis, I and II, General Electric Research Laboratory Reports 65-RL-3944M, 1965; 66-C-250, 1966.

22. B. D. Cullity, *Elements of X-Ray Diffraction*, Addison Wesley, Reading, Mass., 1956.

23. A. H. Compton and S. K. Allison, *X-Rays in Theory and Experiment*, 2nd ed., Van Nostrand Reinhold, New York, 1935.

24. H. A. Liebhafsky, H. G. Pfeiffer, E. H. Winslow, and P. D. Zemany, *X-Ray Absorption and Emission and Analytical Chemistry*, John Wiley & Sons, New York, 1960.

25. A. Taylor, *X-Ray Metallography*, John Wiley & Sons, New York, 1961.

26. C. S. Barrett and T. B. Massalski, *Structure of Metals*, 3rd ed., McGraw-Hill Book Co., New York, 1966.

27. H. G. Moseley, *Phil. Mag.*, **26**, 1024 (1913).

28. A. Jönsson, *Z. Physik*, **43**, 845 (1927).

29. H. Kulenkampff, *Ann. Physik*, **69**, 548 (1922).

30. S. S. So and H. R. Potts, *Solid State Sci., J. Electrochem. Soc.*, **115**, 64 (1968).
31. T. O. Ziebold and R. E. Ogilvie, *Anal. Chem.*, **36**, 322 (1964).
32. C. R. Worthington and S. G. Tomlin, *Proc. Phys. Soc.* **A69**, 401 (1956).
33. G. D. Archard, *J. Appl. Phys.*, **32**, 1505 (1961).
34. M. Green and V. E. Cosslett, *Proc. Phys. Soc.*, **78**, 1206 (1961).
35. D. B. Brown and R. E. Ogilvie, *J. Appl. Phys.*, **37**, 4429 (1966).
36. D. M. Poole and P. M. Thomas, *J. Inst. Metals*, **90**, 228 (1962).
37. H. H. Pattee, Jr., V. E. Cosslett, and A. Engström, Eds., *X-Ray Optics and X-Ray Microanalysis*, 3rd Int. Symposium, Stanford University, 1962, Academic Press, New York, 1963.
38. S. J. B. Reed, *Brit. J. Appl. Phys.*, **16**, 913 (1965).
39. D. R. Beaman and J. I. Isasi, *Anal. Chem.*, **42**, 1540–68 (1970).
40. I. Adler and J. I. Goldstein, "Absorption Tables for Electron Probe Microanalysis," NASA Tech. Note, 1965.
41. R. Castaing and J. Deschamps, *J. Phys. Radium*, **16**, 304 (1955).
42. D. B. Wittry, *An Electron Probe for Local Analysis by Means of X-rays*, Ph.D. thesis, California Institute of Technology, Pasadena, Calif., 1957.
43. D. B. Wittry, "Fluorescence by Characteristic Radiation in Electron Probe Microanalysis," Rept. No. 84-204, University of Southern California Engineering Center, Los Angeles, July 1962.
44. H. A. Bethe, *Ann. Phys. (Leipzig)*, **5**, 325 (1930).
45. L. H. Pruden, E. J. Korda, D. P. Smith, and J. P. Williams, presented at 3rd Natl. Conf. Electron Microprobe Analysis, Chicago, 1968, abstract volume available from Electron Probe Analysis Society of America.
46. P. Duncumb and D. A. Melford, presented at 1st Natl. Conf. on Electron Probe Microanalysis, University of Maryland, College Park, Maryland, 1964.
47. E. J. Sternglass, *Phys. Rev.*, **95**, 345 (1954).
48. L. A. Harris, *J. Appl. Phys.*, **39**, 1419 (1968).
49. J. V. P. Long and S. O. Agrell, *Mineral. Mag.*, **34**, 318 (1965).
50. J. V. P. Long, in *Physical Methods in Determinative Mineralogy*, J. Zussman, Ed., Academic Press, New York, 1967, p. 215.
51. W. Crookes, *Phil. Trans. Roy. Soc.*, **170**, 641 (1879).
52. D. McMullan, Thesis, University of Cambridge, England, 1952.
53. K. C. A. Smith, Thesis, University of Cambridge, England, 1956.
54. R. Bernard, F. Davoine, and P. Pinard, *C. R. Acad. Sci., Paris*, **248**, 2564 (1960).
55. R. Bernard, F. Davoine, and P. Pinard, *Optik*, **17**, 129 (1960).
56. P. R. Thornton, *Scanning Electron Microscopy*, Chapman and Hall, London, 244 (1968).
57. E. Lifshin, General Electric Research and Development Center, Schenectady, N.Y., private communication.
58. E. Lifshin, "An Electron Microprobe Study of Diffusion in the Nickel-Gold System," Masters thesis, Rensselaer Polytechnic Institute, Troy, N.Y., 1966.
59. G. A. Hutchins, in *The Electron Microprobe*, T. D. McKinley, K. F. J. Heinrich, and D. B. Wittry, Eds., Proc. of Symposium held in Washington, D.C., 1964, sponsored by The Electrochemical Society, John Wiley & Sons, New York, 1966, p. 390.
60. E. Rutherford and E. Andrade, *Phil. Mag.*, **28**, 263 (1914).
61. W. Kossel and H. Voges, *Ann. Physik*, **23**, 677 (1935).
62. W. Kossel, *Ann. Physik*, **25**, 512 (1936).
63. M. von Laue, *Ann. Physik*, **23**, 705 (1935).

64. R. W. Jones, *The Optical Principles of the Diffraction of X-Rays*, G. Bell & Sons, London, 1950.

65. K. Lonsdale, *Phil. Trans. Roy. Soc.*, **240**, 219 (1947).

66. P. Gielen, H. Yakowitz, DeW. Ganow, and R. E. Ogilvie, *J. Appl. Phys.*, **36**, 773 (1965).

67. W. G. Morris, *J. Appl. Phys.*, **39**, 1813 (1968).

68. R. E. Hanneman, R. E. Ogilvie, and A. Modrzejewski, *J. Appl. Phys.*, **33**, 1429 (1962).

69. M. Umeno, H. Kawabe, and G. Shinoda, *Advan. X-Ray Anal.*, **9**, 23 (1966).

70. R. E. Hanneman, General Electric Research and Development Center, Schenectady, N.Y., private communication.

71. V. E. Cosslett and W. C. Nixon, *X-Ray Microscopy*, 1st ed., Cambridge University Press, London, 1960.

72. R. Castaing and G. Slodzian, *C. R. Acad. Sci., Paris*, **255**, 1893 (1962).

73. R. Castaing and G. Slodzian, *J. Microscopie*, **1**, (6) 395 (1963).

74. R. Castaing and G. Slodzian, Inst. Petroleum/ASTM Mass Spectrometry Symposium, Paris, 1964.

75. R. Castaing and J. F. Hennequin, *X-Ray Optics and Microanalysis*, 1965, Orsay, R. Castaing, P. Deschamps, and J. Philibert, Eds., Hermann & Cie, Paris, 1966.

76. G. Slodzian and J. F. Hennequin, *C. R. Acad. Sci., Paris*, **263**, 1246 (1966).

77. J. F. Hennequin, *Rev. Phys. Appl.*, **1**, 273 (1966).

78. J. F. Hennequin, *J. Phys. (Paris)*, **29**, 655 (1968).

79. P. Joyes, *J. Phys. (Paris)*, **29**, 774 (1968).

80. J. F. Hennequin, *J. Phys. (Paris)*, **29**, 957 (1968).

81. J. F. Hennequin, *J. Phys. (Paris)*, **29**, 1053 (1968).

82. G. Slodzian, *Rev. Phys. Appl.*, **3**, 360 (1968).

83. P. Joyes, *J. Phys. (Paris)*, **30**, 365 (1969).

84. P. Joyes, *J. Phys. (Paris)*, **30**, 243 (1969).

85. J. F. Hennequin, G. Blaise, and G. Slodzian, *C. R. Acad. Sci., Paris*, **268**, 1507 (1969).

86. G. Blaise and G. Slodzian, *J. Phys. (Paris)*, **31**, 93 (1970).

SUBJECT INDEX

Abrasion, study by reflection electron diffraction, 226
Absorption band, electron, 241
Absorption coefficient, for light, measurement, 21-22
for x-rays, equation for, 379
table of differences for iron alloys, 378
Absorption correction, 487-488; *see also* Electron microprobe analysis
Adsorption, study by, Auger electron analysis, 140
field emission microscope, 329
low energy electron diffraction, 129-130
reflection electron diffraction, 226
Alpha particle range in materials, equation for, 396
table of, 397
Alpha particles, use in autoradiography, 396-397
Alpha track autoradiography, 400-401
Aluminum, electron micrograph of tilt fringes, 242
ionic yield, 512-513
Aluminum alloys, inclusion identification in optical microscope, 21
x-ray imaging of defects in, 409
Aluminum nitride particles, study by electron microscopy with microanalysis, 510-511
Aluminum-tin alloy, microradiographs

of, 384
Amorphous solids, advantages of dark field electron microscopy, 228-229
formula for electron penetration of, 288
theory of electron scattering from, 227-229
Amplitude-phase diagram, 244-245
Anodized coatings, electron micrograph of ultramicrotomed section, 196
Anodizing, 8
Anomalous creep, study by hardness tests, 48-49
Antiphase domains, imaging by indirect lattice resolution, 270
Asbestos, lattice resolution of, 267
Astigmatism, *see* Optical lens and Electron lens
Atomic number correction, in electron microprobe analysis, 490-493
Atom imaging, *see* Field ion microscope
Atom probe field ion microscope, 338-341
Atoms, possibility of resolution in electron microscope, 230, 273
Auger electron analysis, 132-142
applications, absorption and desorption, 140
monolayer detection, 138-140
qualitative analysis of surfaces, 137-141
surface electron state, 141
surface segregation, 138-140

temper embrittlement, 139-140
combination with low energy electron diffraction, 136-137
electronic differentiation of signal, 134-135
equipment for, 133-137
Harris equipment, 133-135
Palmberg and Rhodin equipment, 135-136
principles of, 133
sample cleaning for, 135
use of reference spectra, 140-141
Autoradiography, 393-401
applications of, 394, 395, 400
factors affecting resolution, geometrical, 398
particle energy, 396-398
photographic grain size, 398-399
photographic exposures for, 398
resolution of, 396
technique, 393-401
emulsion types, 399
methods of adding activity, 394
use of alpha tracks, 400-401
use of electron microscope, 400
use of nuclear tracks, 400-401

Babinet compensator, 75
Backscatter coefficient, electron, 492
Backscattered electrons, use for analysis in electron microprobe, 494-495
Backscattered electron yield, effect of atomic number, 440
Barium titanate, mirror electron micrographs of, 354
Becke line, 22
Bend contours, 239-241; see also Electron microscope, extinction contours
Berg-Barrett x-ray method, 401-407; see also X-ray, imaging of defects
Beta brass, scanning electron diffractogram of, 365
Beta particles, table showing half-thickness of absorber, 397
use in autoradiography, 396-397
Biological materials, electron damage in high voltage microscope, 291
measurements possible with x-ray microradiography, 393
staining centers imaged by indirect lattice resolution, 270

study by, electron microprobe, 506
electron microscope, 230, 275-276
phase contrast light microscope, 78
see also Amorphous solids, and Macromolecules
Bireflection, 19
Birefringence, 23, 75
Bloch waves, 234-237
Bollman technique of electrothinning, 191-192
Bonse x-ray method, 407
Borrman divergent beam x-ray method, 402, 409, 411
Brewster's law, 72
Burgers vector, determination with electron microscope, 248-251

Cadmium sulfide, x-ray imaging of defects in, 409
Calcite, cathodoluminescence of, 495
x-ray imaging of defects in, 409
Calibration of electron microscope, 164-170
line resolution, test objects for, 166
magnification, 164-165
use of grating replica, 164
use of latex spheres, 165
use of lattice planes, 165
point resolution, test objects for, 165-166
see also Electron microscope
Cameca microprobe, 464-468; see also Electron microprobe analyzer
Capacitor, dielectric of, scanning electron micrograph, 442
Carbon fiber, lattice image of, 266-268
scanning electron micrograph of, 446
sectioning by ultra microtome for electron microscope, 196
Castaing and Slodzian microanalyzer, 511-513; see also Secondary ion emission microanalyzer
Castaing x-ray fluorescence correction, 489-490
Catalysis, study by low energy electron diffraction, 130
Cathodic etching, 9
Cathodoluminescence, 456
activator for, 495-496
study with scanning electron microscope, 495-496
use for analysis in electron microprobe,

495-496

Ceramics, cathodic etching of, 9
petrographic preparation for electron
microscopy, 202
Channeling, of protons, 342; see also Proton
scattering microscope
Channel plate, 339; see also Image intensifier
Channel plate intensifier, for field ion microscope, 334, 339
Characteristic electron losses, 298
Characteristic x-rays, see X-rays
Charge removal from surfaces, use of electron spray gun, 225
Chemical analysis, by Auger electrons, 132-
142
by electron microprobe, 461-511
by fluorescent x-ray emission, 390, 391
by photoelectron spectroscopy, 141-142
by secondary ion emission, 511-513
by x-ray absorption, 391-393
by x-ray emission in electron microscope,
179-180
low atomic number, in electron microprobe analyzer, 493-494
use of secondary ion emission, 512-513
use of x-ray absorption, 390-393
Chemical bonding, study by photoelectron
spectroscopy, 141-142
Chemical etching, 5
Chemical polishing, 2
Chemical structure, determination by photoelectron spectroscopy, 142
Chemical thinning, for electron microscopy,
197-200
Chemicomechanical polishing, 4
Chemisorption, study with thermionic
emission microscope, 324
Chromatic aberration, see Optical lens and
Electron lens
Chromatic error for electrons, 289-290
formula for, 289
Chromium sulfide scale, scanning electron
micrograph of surface, 447
Cleaning of surfaces, use of ion gun, 225
Cleavage, use to prepare samples for electron
microscopy, 200
Coates patterns, 446-467
Cobalt, slip bands, scanning electron micro-
graph of, 445
Coherency, study by electron micro-

scope, 274
Coherent precipitates, condition for visibility, 253-255
contrast at spherical particles, schematic
of, 254
distortion of lattice planes around, 251-
253
kinematical theory of diffraction contrast
at, 251-256
column approach, 252-255
for cuboidal inclusions, 255
for disk-shaped inclusions, 255-256
mismatch, determination using Moiré
patterns, 272
particle size, determination from diffraction contrast images, 254-255
sign of strain, determination from diffraction contrast image, 255
visibility limits, computed diagram of, 255
Column approximation for electrons, 233-
235
Coma, see Optical lens
Concentration gradients, electron microprobe determination of, 498-499,
501
Concentration mapping, 416; see also Scanning x-ray distribution microscopy
Contact microradiography, see X-ray microradiography
Contact potential, study with mirror electron microscope, 350, 351-353
Contact prints, 4
Contamination, sources of in vacuum systems, 493, 494
Continuous x-ray spectrum, see X-rays
Copper, filming etchants for, 6
ionic yield, 512, 513
proton blocking pattern on crystal of, 346
Copper-beryllium, diffraction contrast at
G. P. zones, 255-256
Copper-lead alloy, scanning electron micrographs of, 429, 443
x-ray distribution micrograph of, 429
Copper-silver alloy, x-ray imaging of defects
in, 409
Corrosion, study with field emission micro-
scope, 329
Cowley and Strojnik high voltage scanning
electron microscope, 358-362; see
also Scanning electron microscope

Creep, study by hardness test, 48-49
Crewe scanning electron microscope, 352-357
Crystal orientation, determined by, Kossel patterns, 504
  proton blocking patterns, 344-345
  Coates patterns, 446-447
  scanning electron microscopy, 446-447
  Weissmann x-ray technique, 403
  relative, determination using Moiré patterns, 269-273
  use of, electron diffraction, 217-218
  Kikuchi lines, 221-222
  polarizing light microscope, 77
Crystal structure, determination from, Kossel patterns, 504
  proton blocking patterns, 346

Dark field, 71, 103, 160-162; see also Electron contrast, and Optical microscope
Decoration, of dislocations in transparent crystals, 16
  of surfaces, 211-212
Defects, x-ray methods for direct imaging, 401-416
Defects in crystals, x-ray imaging of, 401-415
  x-ray Moiré fringe imaging of, 411
Deformation, study by etch pitting, 12-16
  study by electron microscope, 273
  study by hardness test, 48
  study by Kossel lines, 504
  techniques for topographical study of, 17
Deformation markings, study by, phase contrast light microscopy, 78
  scanning electron microscopy, 443, 445
  techniques for study, 17
Depth of field, electron lens, 153-154
Depth of focus, 155-156
  electron lens, 155-156
  see also Optical lens and Electron lens
Desorption, study by Auger electron analysis, 140
Deviation parameter, 223
  determined from Kikuchi patterns, 222
Diamond, controlled oxidation thinning for electron microscopy, 202
  x-ray imaging of defects in, 409
Diamond pyramid hardness, 29, 39-40

definition of, 29
  interpretation, 39-40
  relation to UTS, 40
  see also Microindentation hardness
Dielectric coatings, use in optical microscopy, 10-11
Dielectric films, study with mirror electron microscope, 350
Diffraction aberration, 431; see also Electron lens
Diffraction contrast, concept of the wave function, 232-233
  dynamical two-beam theory for perfect crystal, Bloch waves, 233-241
  column approximation, 233-235
  extinction contours, 236, 238-241
  extinction contours, 236, 238-241
  extinction distance, table of values, 237
  kinematical theory for imperfect crystal, coherent precipitates, 251-256
  dislocations, 248-251
  edge and mixed dislocations, 251
  general equations for, 245
  precipitates, 251-256
  screw dislocations, 249-250
  stacking faults, 246-248
  kinematical theory for perfect crystal, 241-245
  general equations for, 243
  intensity versus deviation parameter, 245
Diffraction gratings, for soft x-ray dispersion, 473
Diffusion, grain boundary, study by autoradiography, 394-395
  study by, electron emission microscope, 322
  secondary ion emission, 513
  surface, study with field emission microscope, 329
Diffusion gradient, electron microprobe determination of, 498, 499, 501
Direct lattice resolution, 231, 256-276
  applications, carbon fiber, 266-268
  dislocations, 261-265
  grain boundaries, 265-266
  twins, 265-267
  crossed fringe images, 263-264
  defocusing technique, 268
  relation between fringe and atom plane

positions, 257
resolution with axial illumination, 267-269
tilted illumination, 269
theory, complete and partial images, 230, 257
    conditions for fringe optimization, 260
    dynamical equations for, 259-261
    dynamical two-beam for perfect crystal, 258-261
    effect of dislocations on images, 261-265
    image regarded as interference pattern, 258
    sources of lattice fringe shifts, 260-261
Dislocation, field ion micrograph of, 341
Dislocation density, methods of determination, 16
    etch pits, 12-16
    x-ray microscopic determination, 16
Dislocation loops, in silicon, 202-203
    study by electron microscope, 272-274
Dislocations, Burgers vector, determination by electron microscopy, 248-251
    determination by x-ray imaging, 405-407
    crossed lattice image in silicon, 264
    decoration in transparent crystals, 16
    distinction from moiré fringes in electron micrographs, 272
    etch pit shape, significance of, 15
    etch pitting of, 12-16
    imaging by indirect lattice resolution, 272
    imaging in Moiré patterns, 272
    in gold platelet, 206, 208
    kinematical theory of diffraction contrast at, 248-251
    lattice imaging, by electron microscopy, 261-265
    in germanium, 263
    mobility, study by hardness test, 49
    pileup analysis from etch pits, 15
    sign determination using etch pits, 15
    study by, electron microscope, 272-273
        high voltage electron microscope, 291-293
    x-ray imaging of, 401-415
    theory of contrast, 403
    x-ray moiré fringe imaging of, 411
Distortion, see Optical lens

Double diffraction of electrons, 218-219
Double refraction of light, 72-75
Duncumb and Reed atomic number correction, 490-493; see also Electron microprobe analyzer

Elastic microstrain, determination from Kossel lines, 504
Elastic strain, diffraction effects from, 220
Electrical conductivity, study with mirror electron microscope, 350
Electrodeposits, use of contact prints, 4
Electroluminescent devices, study with scanning electron microscope, 456
Electrolytic, etching, 7-8
Electrolytic polishing, 2
Electromechanical effects, in microhardness testing, 49
Electromechanical polishing, 4
Electron beam detector, use of Zircaloy crystal, 379
Electron contrast, bright field, 160-162
    dark field, 160-162
        techniques for, 161-162
        multiple dark field imaging, 162
Electron damage, in high voltage electron microscope, 290-291
Electron diffraction, 212-226
    reflection, 223-226; see also High energy electron diffraction
    attachment for electron microscope, 180-181
    scanning high energy, 362-366
    see also High energy electron diffraction and Low energy electron diffraction
Electron emission microscope, 316-325
    applications, metallurgical, 322-325
    image contrast, origin of, 317, 321-322
    Balzers KE-3 Metioscope, 317-320
    modes of operation, 318-319
        photoemission, 318
        secondary emission, 319
        thermionic emission, 319
    resolution, 321
    schematic of, 318
    thermionic mode, activators, use of, 322
    activators for steel, 322
Electron image, deflection modulation of, 453
Electron image contrast, in transmission

scanning electron microscope, principle of reciprocity, 360-362
Electron images, dark field, in high voltage
scanning microscope, 359
Electron lens, aberrations, astigmatism,
150-151
astigmatism correction, 150
chromatic, 151
diffraction, 153
spherical, 149-150
chromatic aberration, effect on beam
diameter, 431
depth of, field, 153-154
focus, 155-156
diffraction aberration, effect on beam
diameter, 431
high voltage microscope objective,
aberrations in, 289
Minilens, 506-508
objective, immersion, 175
schematic ray diagram for, 257
split immersion, 175
phase shift due to aberrations, 260-261
resolution, formulae for effect on chromatic aberration, 151
formulae for effect on spherical
aberration, 149
resolving power, 153
optimum aperture, 153
rotation, 168-169
spherical aberration, effect on beam
diameter, 431
Electron loss mechanisms in solids, 297-298
Electron microprobe analysis, 462-516
backscatter coefficient, equations for, 492
characteristic x-ray emission, 478-483
energy level diagram of, 480
intensity of, 482
relation to atomic number, 481
see also X-rays
corrections for, absorption, 487-488
atomic number, 490-493
contamination buildup, 493-494
continuum background, 484-485
counter dead-time, 484
fluorescence, 488-490
instrumental drift, 484
secondary fluorescence, 488-490
continuous x-ray spectrum, 479-482
intensity of, 482

of light elements, 493-494
problem of standards, 483-484
qualitative, 478-483
quantitative, 483-496
empirical approach using matching
standards, 485-486
theoretical approach, 486-493
specimen preparation for, 469-470
conducting mounts, 469
stopping power, equation for, 491
theoretical limit of detectability, 483
theoretical maximum probe current,
482-483
use of other signals, backscattered electron,
494-495
cathodoluminescence, 495-496
specimen current, 495
x-ray counts, formula for standard
deviation, 483
Electron microprobe analyzer, 461-516
AEI EMMA-4, 506-511; see also Electron
microscopy with microanalysis
application to, concentration gradient
determination, 498-499, 501
film thickness determination, 500-502
particle identification, 496-498
phase diagram determination, 499-501
phase identification, 496-501
attachment for electron microscope, 178-
180
particle analysis using, 179
thickness measurement using, 178-179
basic features, 463-464
Cameca, 464-468
anticontamination device for, 468
coaxial light optics for, 465-466
mirror objective for, 465-466
x-ray crystals for, 468
x-ray scanning imaging system, 467
contamination, sources of, 493-494
energy dispersion of x-rays, 475-477
pulse-height analysis, 475
solid state detector, 476-477
Kossel lines, interpretation of patterns,
503-504
Kossel line technique for, sources of
error, 504
specimen contamination, sources of, 493-
494
theoretical maximum probe current,

482-483
wavelength dispersion of x-rays, combination with energy dispersion, 477-478
crystals for, 468, 471-473; see also X-rays, wavelength dispersion
spectrometers for, 470-475
x-ray counters for, 473-475; see also X-ray counter
Electron microscope, attachments, electron microprobe, 178-180
reflection electron diffraction, 180-181
scanning electron microscope, 181-182
solid state x-ray detector, 180
x-ray spectrometer, 178-180
calibration methods, 164-170
for electron diffraction, 166-168; see also High energy electron diffraction
for image rotation, 168-169
for magnification, 164-165
for resolution, 165-166
see also Calibration of electron microscope
chromatic error, formula for, 289
commercial instruments, 170-175
magnification of, 174
Norelco EM 300, 170-175
resolution of, 170
Siemens Elmiskop #, 101, 170-175
stability of, 175
devices, anticontamination, 175
cooling, 175
double tilt, 175
goniometer stage, 175
heating, 175
stretching, 175
television display, 174
tilt and rotation, 175
electron emission, 316-325; see also Electron emission microscope
energy-analyzing, 299-310; see also Energy-analyzing electron microscope
energy-selecting, 310-316; see also Energy selecting electron microscope
extinction contours, thickness type, 236, 239
tilt type 239-241
use to measure foil thickness, 236, 239
field emission microscope, 325-330; see also Field emission microscope
field ion microscope, 330-338; see also

Field ion microscope
Fresnel fringe check of performance, 159
high voltage, 282-292; see also High voltage electron microscope
illuminated area, size of, 162-164, 182
image contrast, 309-310
loss versus no-loss electrons, 309, 310, 315
image contrast at, dislocations 248-251
lattice planes, 256-269
precipitates, 251-256
sources of, 256
stacking faults, 246-248
image contrast theory, for amorphous solids, 227-229
amplitude contrast, 226, 228
bright field observations on amorphous samples, 229
for crystalline solids, 230-272
dark field observations on amorphous samples, 228-229
mass thickness contrast, 226-272
for quasiamorphous solids, 229-230
instrument components, 170-175
lens aberrations, 149-151; see also Electron lens
loss and no-loss images, 314-315
mirror, see Mirror electron microscope
moiré fringes, 271-272; see also Indirect lattice resolution
ray paths, 159-161
reflection, 292-297; see also Reflection electron microscope
resolution, 148
effect of lens aperture, 149-150
effect of high voltage ripple, 151
factors limiting, 152
resolution with devices, 175
resolving power, Fresnel fringe measurement of, 159
scanning, 352-362; see also Scanning electron microscope
selected area diffraction in, 162-164
specimen preparation for, 182-212; see also Specimen preparation for electron microscope
replica methods, 182-190, 211-212; see also Specimen preparation for electron microscope
stereoscopy, 169-170

thickness contours, use to measure foil
thickness, 236, 239
wobbler, 174
Electron microscopy, fields of application,
crystal defects, 272-274
fracture, 275
noncrystalline materials, 275-276
solid state transformations, 274
structure and properties of thin films,
275
with microanalysis (EMMA), cross-section
of instrument, 507
Minilens for, 506-508
probe size, 508
resolution, 508
steel, results on, 510-511
Electron penetration, of materials, formula
for, 288
Electron probe, theoretical maximum current, 482-483
Electrons, diffraction at, hole, 159-160
particles, 157-159
straight edge, 156-157
interaction with materials, schematic of,
478
stopping power of materials, equation for,
491
Electron scattering, in amorphous solids,
227-229
by atoms, 227
in crystalline solids, 230-272
in quasiamorphous solids, 229-230
radial intensity distribution, 229-230
Electron source, coherence of, 159
Langmuir equation for brightness, 432
Electron spectroscopy for chemical analysis
(ESCA), 141-142; see also Photoelectron spectroscopy
Electron spray gun, 225
Electron wavelength, relation to voltage, formulae for, 148
table of, 147-148
Electrothinning, 190-192; see also Specimen
preparation for electron microscope
Energy-analyzing microscope, 299-310, 353
applications to alloys, 302-310
conversion of Elmiskop I, 299-302
high voltage, 359-360
magnification, 302
Möllenstedt analyzer for, principles,

301-302
resolution, 302, 309
Energy level diagram for atom, 480
Energy loss peaks in solids, table of, 313
Energy-selecting microscope, 310-316, 353
applications, study of diffraction contrast,
316
energy spectrum display, 312-313
high voltage, 359-360
loss and no-loss images, 314-315
selected area diffraction, with selected
energy, 312, 314, 316
use for microanalysis, 315-316
Watanabe and Uyeda instrument, 310-313
applications of, 312-316
schematic of, 312
Epitaxial layers, double diffraction due to,
218-219
Epitaxy, study, by electron microscopy,
274
by low energy electron diffraction, 118-
130
by reflection electron diffraction, 226
using evaporated films, 204-207
Etchant, filming, 6-7
for copper, 6
for iron, 6
Etching, anodization, 8
cathodic, 9
cathodic vacuum oxidation, 9
chemical, 5
dielectric coatings, 10-11
double-etching, 7
electrolytic, 7-8
etchpitting, 5, 11-16
of lithium fluoride, 12-14
mechanism of, 13-14
heat tinting, 8-10
optical anisotropy produced by, 16-17
polish-attack, 4
potentiostatic, 8
ridging at grain boundaries, 209-210
selective, of monotectic alloy, 443
thermal, 9-10
use of laser, 10
types of, grain boundary, 5
grain contrast, 5-6
Etch pits, crystallographic, 11-12
for dislocation pile-up analysis, 15
dislocation sign determined from, 15

significance of shape, 12-13, 15
for sub-boundary misorientation determination, 15
tests of correspondence with dislocations, 14-15
Etch pitting, 5, 11-16
Gilman and Johnston technique, 12-14
Eutectic alloys, microradiographic study of, 383
study with energy-analyzing microscope, 305-308
Evaporated films, alloy, preparation of, 202-206
chromium on gold, mirror electron micrographs of, 352, 353
for study of, defects, 206-207
epitaxy, 204-207
nucleation and growth, 206
ordering, 206
metal, preparation of, 202-206
nucleation and growth, study by electron microscopy, 275
study by scanning electron diffraction, 362, 364
thickness measurement in interferometer, 85-86
Evaporation, field, see Field evaporation; Field emission microscope; Field ion microscope
Ewald reflecting sphere, 214-218, 222-223
Extinction contours, see Electron microscope
Extinction distance for electrons, formula for, 237
table of values, 237
Extinction of polarized light, 24
Extraction replica, 189-190, 211-212
Eyepieces, see Optical lens

Ferroelectric materials, study with mirror electron microscope, 350, 351, 354
Ferromagnetic materials, see Magnetic domains
Field emission microscope, 325-330
field evaporation in, 328-329
materials suitable for study, 328
resolution of, 327
factors limiting, 330
schematic of, 330
Field evaporation, 328-329

orientation dependence of, 329
table of fields, 335
Field ion microscope, 330-338
atom imaging, applications of, 336, 337
mechanism, 331-332
Brenner microscope, 330-331, 333-334
evaporation fields, table of, 335
field at tip, 332
field evaporation, 335-336
pulsed, 336
image intensifier, cascade type, 333-334, 338, 339
channel plate type, 334, 339
images, interpretation of, 332, 335, 337-338
imaging gas, choice of, 335
magnification, 332
micrographs of, dislocation in iridium, 341
grain boundary in tungsten, 340
inclusion in tungsten, 340
ion damage in platinum, 336
iridium, 341
platinum, 336, 337
tungsten, 333, 334, 340
vacancies, 337, 341
resolution, factors limiting, 330
Films, transparent, thickness by light section technique, 101-102
Film thickness, measurement, by taper sectioning, 17-19
Fluorescent x-ray emission, 478-482; see also X-rays, characteristic emission
Forbidden reflections, 216, 218
Fractography, use of scanning electron microscope, 444
Fractures, replication of, 188-189; see also Specimen preparation for electron microscope
study by electron microscopy, 275
Fracture surfaces, study by optical techniques, 105-109
Fresnel fringes, 150, 156-160
astigmatism, guide to, 150
at hole, 159-160
measure of resolving power of electron microscope, 159
at particles, 157-159
at straight edge, 156-157
test of electron source coherence, 159

Gallium arsenide, cathodoluminescence of, 495
instantaneous video display of x-ray topograph, 412, 414
study with scanning electron microscope, 456
Germanium, lattice image, of dislocation, 263
of low angle boundary, 266
x-ray imaging of defects, 409
Gilman and Johnston etch pitting technique, 12-14
Glass, ion thinning of, 202
scanning electron micrograph of, 444
Gold films, thickness of, determined in electron microprobe, 501-502
Gold-nickel, diffusion couple, concentration profile from microprobe, 499, 501
Gold-nickel alloys, characteristic x-ray intensity from, 486
Gold-nickel couples, study with electron microprobe, 498, 499, 501
Grain boundaries, lattice image in germanium, 266
ridging phenomenon on etching, 209-210
sectioning by ultra-microtome for electron microscope, 197
study by phase contrast light microscope, 78
thermal etching of, 9-10
in tungsten, field ion micrograph of, 340
Grain boundary segregation, study by hardness test, 35, 47
Grain growth, study by electron emission microscopy, 322
Graphite, gold decoration of dislocations in, 212
moiré images due to, 270
polarized light microscopy of, 104-106
scanning micrograph showing lattice, 357
Graphite fiber, lattice image of, 266-268
Guinier-Preston zones, diffraction effects from, 219
in copper-beryllium, diffraction contrast at, 255-256

Half-value thickness for x-ray absorption, 379-380
Hard metal alloy, study with photoemission microscope, 325

Hardness, 27-52; see also Microindentation hardness testing
Heat tinting, 8, 10
Hexagonal crystals, cleaved for electron microscopy, 200-201
High energy electron diffraction, 212-226
calibration, 166-168
camera length, 167
diffraction constant, 167
image rotation, 168-170
standards for, 167
use of internal standard, 167-168
reflection, 212-213, 223-226
accessory devices, 225
accuracy, 225
applications, 225-226
camera for, 223-224
pattern analysis, 225
rules for permitted reflections, 216
scanning, 362-366
history of equipment development, 362
intensity and reciprocal lattice vector display, 364-366
intensity profile display, 364-366
schematic of instrument, 363
use of no-loss electrons, 362, 365
selected area, in high voltage scanning microscope, 360
selected area technique, 162-164
selected energy, selected-area, 312
transmission, 212-223
crystal shape effects, 219
double diffraction, 218-219
effects due to elastic strain, 220
Ewald reflecting sphere, 214-218
intensity off Bragg position, 223
Kikuchi lines, 220-222; see also Kikuchi lines
permitted reflections, 216
reciprocal lattice concept, 213-214
structure factor, 218
at twins, 219
use for crystal orientation, 217-218
use of loss-electrons, 314-316
High temperature strength, study by hardness test, 44-46
High voltage electron microscope, 282-292
applications, 290-292
dislocation study, 291
electron damage study, 290-291
magnetic domain structure, 291

phase transformations, 291
commercial designs, 283-288
electron penetration, 288
energy loss in sample, 289-290
instrument design, accelerator, 284-285, 287-288
focal length, 283
gun, 283, 285-286
specimen stage, 286
objective lens, aberrations, 289
resolution of, 286, 289-290
High voltage microradiography, *see* X-ray microradiography
High voltage scanning electron microscope, 358-362; *see also* Scanning electron microscope
Hugo and Phillips jet technique of electro-thinning, 191-193

Ice, x-ray imaging of defects in, 409
Image intensifier, cascade, 333-334, 338-339
channel-plate, 339
Image transducer for x-rays, 412
Immersion oils, 22
temperature coefficient of refractive index, 23
Inclusions, identification using dielectric coatings, 10-11
identification using polarized light, 20-22
in tungsten, field ion micrograph of, 340
study by hardness test, 46
study by reflection electron diffraction, 226
Indirect lattice resolution, double diffraction, graphical representation of, 270-271
moiré fringes, crystal defects producing, 272
distinction from dislocations, 272
effect of dislocations, 272
general case, 271-272
in multilayer evaporated films, 206
parallel case, 271
rotational case, 271
tilt case, 271
Infrared microscopy, on silicon, 16
Inner potential, determined by low energy electron diffraction, 120-132
electron refraction due to, 225
Integrated circuit, Berg-Barrett topograph

of, 413
scanning electron micrograph of, 450
study with electron microprobe, 504-505
study with scanning electron microscope, 449-456
video x-ray topograph of, 413
Interference figures, 23
Interference light microscope, 86-102
Baker transmission, 93-94
image intensity in transmission, 91
Leitz transmission, 92
light section, 101-102
measurements possible with, 85-86
multiple beam, 96-101; *see also* Multiple beam interferometer
Nomarski, 88-90
reflection, 86-90
Reichert reflection, 88-90
transmission, 90-95
wave diagram representation of, 84
wave surface diagram representation of, 91
two-beam, 86-95
wedge interference filter, 95
Zeiss reflection, refractive index measurement with, 87-88
Zeiss transmission, 94-95
Interferometer, 83
light-cut, 101, 102
light profile, 101
multiple beam 96-101; *see also* Multiple beam interferometer
multiple profile, 101
x-ray, 411
see also Interference light microscope
Internally oxidized alloys, study by energy-analyzing microscope, 302-305
Internal stress, examined by polarized light, 20
Ion damage, in platinum, field ion micrograph of, 336; *see also* Proton scattering microscope
Ion gun, 225
Ionic yield, 512, 513
Ionization potential, determination by photoelectron spectroscopy, 142
Ionization state, study with field ion microscope, 341
Ion microscope, *see* Field ion microscope
Ion neutralization spectroscopy, 141

Ion-thinning for electron microscopy, 200-
202
Iridium, atom probe field ion micrograph of,
341
Iron, activators to increase electron emission,
322-323
filming etchants, 6
use to detect phosphorus segregation, 7
ridging at grain boundaries, 209-210
see also Steel
Iron-chromium alloy, contact microradiographs of, 381
Iron-silicon alloy, x-ray imaging of defects
in, 409
Isotopes, table of alpha-particle range in
materials, 397

Johann x-ray spectrometer, 470-472
Johansson x-ray spectrometer, 470-472

Kerr effect, 20-21
Kikuchi lines, 225
deviation parameter from, 222
in high voltage scanning microscope, 360
idealized patterns, 221-222
origin of, 220-221
pattern from silicon, 216-217
reciprocal lattice construction of, 222
use to orient crystal, 221-222
Kikuchi maps, 222
Knoop hardness, 30; see also Microindentation hardness testing
Köhler illuminating system, 68
Kossel lines, in electron microprobe, 502,
504
geometry of formation, 503, 504
in high voltage scanning microscope, 360
interpretation of patterns, 503-504
pseudo-Kossel method, 502-503
sources of error, 504
transmission conic on nickel, 505
x-ray source for, 389

Lambert's law, 376
Lang x-ray technique, 401, 402, 407-409
Laser, 56
use for thermal etching, 10
Lattice parameter, determined from Kossel
patterns, 504
differences from moiré patterns, 269-272

Lattice resolution in electron microscope,
256-272; see also Direct lattice
resolution and Indirect lattice resolution
Layer minerals, cleavage for electron microscope, 200-201
imaging by indirect lattice resolution, 272
Layer thickness, determined by hardness
tests, 47
Light, interference, at a wedge, 85
criteria for, 83-85
measurements possible, 85-86
laws of refraction and reflection, 57
measurement of optical path difference,
75, 83, 84
polarization by double refraction, 72-75
polarized, analyzer, 73
anisotropic materials, 77
applications of, 77-78
Babinet compensator, 75
birefringence, 75
Brace-Köhler compensator, 95
compensator types, 75
dichroism, 73
definition of, 71
Ehringhaus-type compensator, 95
identification of state, 76-77
isotropic materials, 77
Nicol prism, 72
polarizer, 73
quarter-wave plate, 74
by reflection, 72
sensitive tint plate, 76
sources of, 72-76
types of, 71-72, 75
Light contrast, see Optical microscope
Light microscope, see Optical microscope
Light sources, see Optical microscope
Linear absorption coefficient, for x-rays, 376
effect of wavelength, 379, 382
Lithium fluoride, dislocations in, etch pitting technique for, 12-14
x-ray imaging of, 405, 406, 409
Lorentz microscopy, 275
Loss-electron images, 314-315
Loss spectra for electrons, 298
Low angle boundaries, lattice images of,
265-266
misorientation determined by etch pits,
15

Low atomic number analysis, *see* Chemical analysis
Low energy electron diffraction, 111-132
  applications, 128-132
    absorption, 129-130
    catalysis, 130
    epitaxial structure determination, 128-129
    inner potential determination, 130-132
    oxidation, 124-126, 129-130
    surface atom displacement, 132
    surface diffusion, 130
    surface structure determination, 128-129
    surface topography determination, 128
  combined with Auger electron analysis, 136-137
  determination of surface atom location, 126-127
  diffraction pattern analysis, 118-127
    indexing of surface nets, 122-126
    types of surface nets, 122-126
  equipment for, 113-116
  intensity analysis, 126-127
  Mac Rae apparatus, 114-115
  sample preparation for, 117-118
  theory of, 118-126
  Tucker apparatus, 114-116

Macromolecules, study with electron microscope, 230, 275-276
Magnesium oxide, defects in, surface decoration of, 212
  x-ray imaging of, 409
Magnetic domains, study by, electron microscopy, 275
  high voltage-electron microscope, 291
  mirror electron microscopy, 350
  polarized light, 20
  scanning electron microscope, 436, 447-449
Mass absorption coefficient for x-rays, 376
Mass spectrometer, time of flight, use with field ion microscope, 338-339
Mechanical polishing, 2
  in relief, 2
Metallographic preparation, conducting mounts, 469
  for electron microprobe examination, 469-470

  see also Polishing
Metals, chemical thinning for electron microscopy, 198
  optical reflectivity of, 19
  study by phase contrast light microscope, 78
Meyer index, 40
Mica x-ray crystal, 471-473
Microindentation hardness, definition of, 29
  diamond pyramid, relation to U. T. S., 40
  grain size effect on, 39
  interpretation of, 39-40
  Knoop, 30
  load dependence, 38
  Meyer index, 40
  scale conversion, 38
Microindentation hardness testers, 40-46
  BNFMRA, 43-44
  GE high-temperature, 44-46
  Kentron, 41-43
  Zeiss, 41
Microindentation hardness testing, applications to study of, anisotropy of deformation, 48
  anomalous creep, 48-49
  bulk property determination, 46
  creep, 48-49
  crystallography of deformation, 47-48
  dislocation mobility, 49
  electromechanical effects, 49
  grain boundary segregation, 47
  grain size dependence of strength, 47
  hardness gradients, 47
  high temperature strength, 46
  inclusions, 46
  kinetics, 46-47
  photomechanical effects, 49
  polar surfaces, 48
  thickness of thin layers, 47
  error sources, 30-38
    adsorbed water on sample, 49
    anvil effect, 38
    elastic recovery of impression, 36
    impression shape, 37-38
    indenter shape, 35
    load inaccuracy, 31-32
    rate of loading, 32-33
    vibration, 33-34
  impression measurement, techniques for, 36-37

indenters for, diamond pyramid, 29-30
    Knoop, 30
load calibration, 32
load dwell time, 34
sources of vibration, 33-34
surface preparation for, 39
Microprobe, *see* Electron microprobe ana-
    lyzer
Microradiography, *see* X-ray microradio-
    graphy
Microtomy, 2; *see also* Ultramicrotomy and
    Specimen preparation for electron
    microscope
Minerals, cathodoluminescence of, 495
    petrographic preparation for electron
        microscope, 202
    study with photoemission microscope,
        326
Minilens for electrons, 506, 508
Mirror electron microscope, 347-355
    applications of, 350-355
    commercial instrument, 348-350
    image modes, contact potential, 352-353
        convergent mirror, 350-351
        divergent mirror, 350
    micrographs of, cleaved rock salt, 351
        evaporated chromium on gold film,
            352-353
        heated barium titanate crystal, 354
        P-N junction, 355
    principles of, 347-348
Moiré fringes, 271-272
    using x-rays, 411
    see also Indirect lattice resolution
Mollenstedt electron energy analyzer, 300-
    302
Molybdenite, dislocations in cleaved crystal,
    200-201
Monolayer, study by, Auger electron analy-
    sis, 138-140
    reflection electron diffraction, 224
Monotectic, scanning electron micrographs
    of, 429, 443
    x-ray distribution micrograph of, 429
Multiphase alloys, sectioning by ultramicro-
    tome for electron microscope, 196,
    199
Multiple beam interferometer, coating to
    optimize fringes, 100
    commercial instruments, 96

interpretation of fringe patterns, 98-99
principle, 97-99
sensitivity, 99
use for thickness measurement, 100-101

Neutrography, 374
Newtons rings, 99
Nickel, Kossel pattern of, 505
    surface segregation, study by Auger elec-
        tron analysis, 138-139
Nickel alloys, phase identification by heat
    tinting, 8
Nickel aluminide-chromium eutectic, micro-
    adiographs of, 383
Nickel-base superalloy, photoemission
    micrograph of, 327
    scanning electron micrograph of, 448
Nickel-chromium alloy scanning x-ray dis-
    tribution image of, 500
    specimen current scanning image of, 500
    study with electron microprobe, 496-498,
        500
    sulfided, microprobe x-ray line scan of,
        497
Nickel-silicon alloy, internal oxidation of,
    study by energy analyzing micro-
    scope, 302-305
Nomarski interference light microscope, 88-
    90
Nuclear track autoradiography, 400-401

Optical activity, due to etch pitting, 11-12,
    16-17
    tests for, 19-20
Optical anisotropy, use of etching to pro-
    duce, 16-17
Optical constants, determination by polar-
    izing microscope, 78
Optical contrast, amplitude features, 70
    differential interference, 90
    enhancement by filtered illumination,
        102-104
    interference, 90
    interference phase contrast, 83
    normal or absorption, 79-80
    phase, 78-83
    phase features, 70, 78-79
Optical lens, aberrations, astigmatism, 61-62
    barrel distortion, 62
    chromatic, 59-60

coma, 61
curvature of the field, 61-62
distortion, 62-63
pincushion
spherical, 60-61
angular aperture, 57
condenser, achromatic - aplanatic, 69
aplanatic, 69
effect of aperture, 68
curvature of the field, 61-62
depth of focus, 59
eyepieces, compensating, 63-66
Huygens type, 66
hyperplane, 66
negative-compensating, 66
periplane, 66
mirror objective, 465-466
numerical aperture, practical limits of, 58
objective, achromats, 63
anastigmats, 62
bloomed, 64
coated, 64
flat field, 62
fluorite, 64
immersion, 64
neofluor
parfocal, 64
reflecting, 64-65
semiapochromats, 64
working distance of, 64
resolution, relation to numerical aperture, 58
Optical microscope, color filters for, 102-104
components of, 63-70
contrast enhancement methods, 70-109
results by, 104-109
illumination, bright field, 71
dark field, 71, 103
oblique, 71
phase contrast, 78
use of color filters, 102-104
interference microscope see Interference light microscope
Köhler illuminating system, 68
lenses, see Optical lens
light sources, arc, 69-70
coherent, 56
color correction of, 69
critical illumination, 67-68

incoherent, 56
Köhler illuminating system, 68
laser, 56
monochromatic, 55
polarized, 72-75
requirements for interference of, 55
white, 55
phase contrast reflecting, schematic of, 81
polarizing reflecting, schematic of, 74
resolution, 58-59
vertical illuminators, 66-69
glass slip, 67
prism, 67
Optical properties, absorption coefficient, 21
extinction angle, 24
refractive index, 21
determination of, 22-23
Ordering, of bulk, study by electron micro-scopy, 274
of surfaces, study by low energy electron diffraction, 119-130
Ores, use of contact prints, 4
use of polarized light, 22
Orientation, see Crystal orientation
Orientation relationship studied by Kossel lines, 504
Oxidation, gradient by secondary ion emis-sion, 512-513
study by, Auger electron analysis, 135
electron emission microscope, 322
energy selecting microscope, 314
field emission microscope, 329
low energy electron diffraction, 124-126, 129-130
reflection electron diffraction, 226
Oxide coatings, sectioning by ultramicro-tome for electron microscopy, 196
thickness measurement, by taper section-ing, 17-19
use of polarized light, 20
Oxides, chemical thinning for electron microscopy, 198
study by light microscopy, bright field, 102-104
dark field, 102-104

Particle identification, use of electron micro-probe, 496-501
Pearlite, study by EMMA (Electron Micro-

scopy with Micro-Analysis), 510-511
Pendellösung fringes, 236
Phase contrast electron microscopy, 226,
    230-231, 256-276; see also Direct
    lattice resolution
Phase contrast light microscope, applica-
    tions of, 78
  nature of contrast, 79
  phase detail, 70, 78-79
  phase retardation plate, 81
  quantitative measurements with, 83
  schematic of, 81
  sensitivity, 83
Phase diagram, determination with electron
    microprobe analyzer, 499-501
Phase identification, use of, dielectric coat-
    ing, 10-11
  polarized light microscope, 20-22, 77
  potentiostatic etching, 8
Phase shift, due to aberrations in electron
    lens, 260-261
Phase transformations, study by, electron
    emission microscopy, 322-324
  hardness tests, 46-47
  high voltage electron microscopy, 291-
    293
Phosphorescence, study by scanning elec-
    tron microscopy, 456
Phosphorus prints, 3
Photoelectron emission microscope, see
    Electron emission microscope
Photoelectron spectroscopy, 141-142
  applications of, 142
  equipment for, 142
  sample types for, 142
Photoemission, ultraviolet light excitation
    of, 318
Photographic materials for autoradiography,
    liquid emulsions, 399-400
  table of, 399
Photomechanical effects on microhardness,
    49
Plasmon losses, 298
Platinum, field ion micrographs of, 336, 337
Pleochroism, 23
P-N junction, see Semiconducting junction
Point defect clusters, study by electron
    microscopy, 273
Point defects, field ion micrographs of, 340-
    341

Polarized light, application to, ferromag-
    netic domain study, 20
  internal stress analysis, 20
  oxide thickness determination, 20
  petrographic examination, 22
  phase identification, 20-22
  birefringence, 23
  elliptical, 19
  Kerr effect, 20-21
  see also Light
Polarizing microscope, 74
  analyzer, 73
  compensators for, 75
    Babinet, 75
    Brace-Köhler, 95
    Ehringhaus-type, 95
  light sources for, 72-75
    Nicol prism, 72
  polarizer, 73
  quarter-wave plate, 74
  reflecting, applications of, 77-78
  schematic of, 74
  sensitive tint plate, 76
  see also Optical microscope
Polish-attack etching, 4
Polishing, chemical, 2
  chemicomechanical, 4
  electrolytic, 2
  electromechanical, 4
  mechanical, 2
  polish-etch, 4
  relief, 2
Porous samples, ultramicrotomed for elec-
    tron microscopy, 195-196
Porter-Blum ultramicrotome, 193-196
Potentiostatic etching method, 8
Precipitation alloys, precipitates, diffraction
    effects from, 219
  sources of contrast in electron micro-
    scope, 256
  study by electron microscopy, 274
  study by energy-analyzing microscopy,
    305-310
  see also Coherent precipitates
Preferred orientation, determined by polar-
    izing light microscope, 77
Proton blocking camera, 342, 347; see also
    Proton scattering microscope
Proton scattering microscope, 342-347
  blocking patterns, analysis of, 342

on copper crystal, 346
  crystal orientation from, 343-346
  crystal structure from, 346
  on tungsten crystal, 345
  wave interpretation of, 342
  commercial equipment, 342-344
  ion damage in, 345
    annealing out of, 346
  proton channeling, 342
  resolution of, 347
  schematic of, 343
Pulse-height analysis of x-rays, 475-478

Quantitative analysis, by electron micro-
    probe, 483-494
  by photoelectron spectroscopy, 141-142
  see also Chemical analysis
Quantitative metallographic measurements,
    with electron microprobe, 506
  with scanning microscope, 445
Quarter-wave plate for light, 74
Quartz, x-ray imaging of defects in, 409
Quasiamorphous solids, definition of, 229
  theory of electron scattering from, 229-230

Radiation damage, by electrons, 290-291
  by ions, 345-346
  study by electron microscopy, 273
Radioisotopes, use in autoradiography, 394
Rebinder effect in microhardness testing,
    48-49
Reciprocal lattice, construction, 213-214
  definition of, 213-214
  disks, 219
  Ewald sphere construction, 214-218
  properties of, 214
  rules for permitted reflections, 216
  sheets, 219
  spikes, 219
  vector representation of double diffrac-
    tion, 270-271
Reciprocity principle, application to scan-
    ning electron microscope, 360-362
Recovery, study by electron microscopy,
    273
Recrystallization, study by electron emis-
    sion microscopy, 322
Reduction, study by electron emission
    microscopy, 322
Reflecting sphere construction, 214-218

Reflection electron diffraction, 223-226
  attachment for electron microscope, 180-
    181
  see also High energy electron diffraction
Reflection electron microscope, 292-297
  depth of field of, 297
  principle of, 293
  resolution of, 294-295
Reflectivity for light, of metals, 19
Refractive index, measurement by, inter-
    ferometer, 83-84
  light microscope, 21-23
  two-beam interferometer, 87-88
Refractometer, 22
Relativistic voltage of electrons, 147-148
Replica methods, 182-190, 211-212; *see*
    also Specimen preparation for elec-
    tron microscope
Resolution limit in electron microscope,
    155
Ridging, at grain boundaries in etched iron,
    209-210
Rocks, petrographic preparation for elec-
    tron microscopy, 202
Rocksalt, cleaved, decoration of monatomic
    steps, 211
  mirror electron micrograph of, 351
Rowland focusing circle, 470

Scanning electron microscope, 422-460
  applications, 436, 442-456
    cathodoluminescence, 456
    crystal orientation, 446-447
    deformation, 443, 445
    dynamic experiments, 445
    fractography, 444
    insulators, 441
    magnetic domains, 436, 447-449
    metals and alloys, 442-449
    observation of electrical phase in de-
      vice, 454-456
    quantitative metallographic measure-
      ments, 445
    semiconducting devices, 449-456
    surface potential mapping, 451
  attachment for electron microscope, 181-
    182
  probe size, 182
  resolution of, 182
  backscattered electron yield, 440

beam diameter, 426
    effect of aberrations on, 431-433
beam interactions with sample, 438-439
    schematic of, 439
Coates patterns, 446-447
Crewe high resolution transmission, 352-357
    cross-section of, 356
    lattice fringe imaging in, 357
    micrograph of graphite lattice, 357
    ray diagram of, 355
    resolution of, 356-357
high voltage transmission, 358-362
    cross-section of, 358
    imaging theory, 360-362
historical development, 424
image contrast in reflection, 436, 438-442
    depth sampled, 440-441
    effect of backscattered electron yield,
        439-441
    effect of beam voltage, 439-440
    effect of incident beam angle, 438-439
    effect of secondary electron yield, 439-
        440
image contrast in transmission, theory,
    360-362
image contrast types in reflection, com-
    positional, 436, 441
    induced conductivity, 453-454
    induced current, 451-453
    magnetic, 448-449
    surface field, 451
    surface voltage, 449-453
    topographical, 436, 441
image display, 427, 428, 433-435
    brightness modulation, 428, 434-435
    choice of scan time, 435
    deflection modulation, 453
    line number, 433-434
    shot noise, 434-435
    sources of noise, 437
    stroboscopic imaging, 454-456
instrumentation, 424-430
    secondary electron detector, 427
magnification of, 428, 434
micrographs of, aluminum oxide capacitor
        dielectric, 442
    carbon fiber, 446
    deformed cobalt whisker, 445
    glassy lunar rock, 444
    integrated circuit, 450

nickel-base superalloy, 448
nickel-chromium alloy, 500
phase in MOST ladder device, 455
planar transistor, 452
selectively etched monotectic, 443
sulfide scale on Ni-Cr alloy, 447
microprobe attachment for, 426, 428
minimum beam diameter, 433
operating modes, 435-437
operating signals, backscattered electron,
    436
    light, 437
    secondary electron, 435-436
    specimen current, 437
    subsurface current, 437
resolution, 424
    factors affecting, 428-429
schematic for, 425
stage for, 426-427
Stereoscan Mark II A instrument, 426-427
stereoscopy, 441
television display, 430
theoretical maximum probe current, 432
theory, 430-442
transmission, reciprocity principle, 360-
    362
transmission image contrast, applicability
    of transmission electron microscope
    theory, 360-362
use of light as a signal, 496
use of secondary ion emission signal, 513
Scanning high energy electron diffraction,
    362-366; see also High energy elec-
    tron diffraction
Scanning x-ray distribution microscopy,
    415-416
    computerized concentration mapping, 416
    resolution of, 416
    factors limiting analytical accuracy, 416
    magnification of, 415
    technique for, 415-416
Secondary electron emission, ion bombard-
    ment excitation of, 319
Secondary electron yield, effect of atomic
    number, 440
Secondary fluorescence correction, in
    microprobe analysis, 488-490
Secondary ion emission microanalyzer, 511-
    513
    applications, 512-513

light element analysis, 512-513
  study of oxygen gradient, 512-513
  Castaing and Slodzian instrument, 511-512
  commercial instruments, 513
  ionic yield, 512
  resolution, 511-512
  scanning concentration imaging, 513
Segregation, study by, autoradiography, 394-395
  contact prints, 4
  scanning x-ray distribution microscopy, 415
  x-ray imaging, 409
  surface, study by field emission microscopy, 329
Selected area electron diffraction, 162-164, 360
selected energy, 312
Selected energy electron diffraction, selected area, 312
Semiconductor, junction, mirror electron micrograph of, 355
  preparation for electron microscopy by, chemical thinning, 197-198
    cleavage, 200
  study of surface by, Auger electron analysis, 140
    mirror electron microscopy, 350, 351, 355
    scanning electron microscopy, 437, 449-456
Semiconductor device, study of phase with stroboscopic scanning microscope, 454-456
  study with electron microprobe, 504-505
Sensitive tint plate, 5, 76
Sensitized materials, see Photographic materials
Shadow casting of replicas, 189
Shot noise, in scanning electron image, 434-435
Silicon, crossed lattice image of dislocation, 264
  dislocations in, x-ray image of, 410
  infrared microscopy of, 16
  Kikuchi pattern from, 216-217
  lattice image of twin in, 265-267
  stacking fault in, x-ray topograph of, 410
  x-ray imaging of defects in, 409

Silicon carbide, x-ray imaging of defects in, 409
Silver, surface decoration of, 207, 211-212
Silver chloride, optical anisotropy due to deformation, 20
  x-ray imaging of defects in, 409
Slip steps, monatomic, decoration in rock-salt, 211
Solid solution alloys, study with energy-analyzing microscope, 306-308
Solid state devices, video display by x-ray topography, 411-415
Solid state x-ray detector, 476-477
  attachment for electron microscope, 180
Specimen current, use for analysis in electron microprobe, 495
Specimen preparation for electron microscope, chemical thinning, 197-200
  cleavage of layer-type crystals, 200-201
  controlled oxidation, 202
  electrothinning, 190-192
    Bollman technique, 191-192
    Hugo and Phillips technique, 191-193
  ion thinning, 200-202
  one-sided thinning technique, 209
  petrographic thinning, 202
  replica methods, 182-190, 211-212
  replicas, advantages of, 183
    carbon, 184-187, 188-190, 211
    extraction replicas, 189-190, 211-212
    fractography, 188-189
    oxide, 187-188
    plastic, 183-184
    positive and negative, definition of, 183
    shadow casting, 189
    surface decoration technique, 211-212
  ultramicrotomy, 192-197
    applications, 195-199
    deformation during, 195
    knives for, 195
    mounting specimens for, 194-195
  use of, cast films, 209
    crystals precipitated from solution, 207
    electrodeposited films, 207-208
    epitaxially grown evaporated films, 205-206
    etched foils, 209-210
    evaporated films, 202-206
    extracted precipitates, 207
    natural crystals, 206-207

ordered evaporated films, 206
splatt-cooled films, 209
sputtered films, 208
sublimed films, 209
vapor grown platelets, 206, 208
Spherical aberration, *see* Optical lens and
        Electron lens
Splatt-cooling, 209
Stacking fault, amplitude phase diagram
        for, 247-248
    schematic of diffraction contrast at, 246
    segregation to, study in etched foils, 210
    in silver, 204
    theory of diffraction contrast at, dynamic,
        248
        kinematic, 246-248
    x-ray imaging of, 409-410
Stacking fault energy, determination with
        electron microscope, 274
Stearate x-ray crystal, 468, 472-473
Steel, cracking in, study by EMMA (Elec-
        tron microscopy with microanalysis),
        510-511
    high voltage electron micrographs of, 292-
        293
    phase identification, by etching, 7
        by heat tinting, 8
    segregation in, study by contact prints, 3
    study by, emission microscopy, 322-324
        microradiography, 378, 380-382
        photoemission microscopy, 323-324
    temper embrittlement of, study by Auger
        electron analysis, 139-140
    see also Iron
Stereoscopy, in electron microscope, use for
        quantitative measurements, 168-169
Stigmator, 150
Stopping power for electrons, 491
Stroboscopic scanning microscopy, 454-456
Structure factor, 218
    rules for permitted reflections, 216
Subboundary, *see* Low angle boundaries
Sub-grain structure, study by, Weissmann
        x-ray technique, 403
        x-ray imaging, 403, 404, 407
Sulfidation, of nickel-chromium alloy, micro-
        probe x-ray line scan, 497
    study with, Auger electron analyzer, 138
        electron microprobe, 496-498, 500
Sulfide, nucleation at defects in silver,

206-207
Sulfur prints, 3
Superconducting domains, study with
        mirror electron microscope, 350
Superlattice, imaging by indirect lattice
        resolution, 270
Surface atom displacement, determined by
        low-energy electron diffraction, 132
Surface charge removal, by electron spray
        gun, 225
Surface cleaning, use of ion gun, 225
Surface composition, determination by,
        Auger analysis, 132-142; *see also*
        Auger electron analysis
        photoelectron spectroscopy, 141-142
Surface contamination, study by Auger
        electron analysis, 138
Surface decoration technique, 211-212
Surface diffusion, study by low energy
        electron diffraction, 130
Surface electrical conductivity, study with
        mirror electron microscope, 350
Surface electron state, study by Auger
        electron, 141
Surface films, thickness determination by
        hardness test, 47
Surface layers, study by reflection electron
        diffraction, 225-226
Surface potential, study by mirror electron
        microscope, 347
Surface profile, measurement with light
        section microscope, 101-102
Surface segregation, in nickel, study by
        Auger electron analysis, 138-139
Surface structure, study by low energy
        electron diffraction, 118-129
Surface topography, study by, decoration,
        211-212
    replication, 182-190, 211-212; *see also*
        Specimen preparation for electron
        microscope
    various optical microscope techniques,
        micrographs showing results, 105-109
    low energy electron diffraction, 128

Taper sectioning, 17-19
    use in transmission electron microscopy,
        19
Temper embrittlement, study by, Auger
        electron analysis, 139-140

phase contrast light microscope, 78
Thermal etching, 9-10
Thermionic emission, temperature dependence, formula for, 321
Thermionic emission microscope, *see* Electron emission microscope
Thickness fringes, *see* Electron microscope, extinction fringes
Thin films, thickness measurement, by electron microprobe, 178-179, 500-502
    by extinction contours in electron microscope, 236, 239
    by interferometer, 85-86
    by x-ray emission, 178-179, 500-502
Thoria, cathodoluminescence of, 495
Threshold energy for electron damage, 290
Tilt contours, *see* Electron microscope, extinction contours
Time of flight mass spectrometer, use with field ion microscope, 338-339
Tolansky multiple beam interferometer, 96-101; *see also* Multiple beam interferometer
Topography, study by interferometry, 83-84
Transistor, scanning electron micrograph of, 452
Transmission electron diffraction, 212-223; *see also* High energy electron diffraction
Tungsten, field ion micrographs of, 333, 334, 340
    proton blocking pattern of, 345
    study with, atom probe field ion microscope, 339-340
    field ion microscope, 332-334, 340
Tungsten carbide, thermionic emission micrograph of, 329
Twins, double diffraction due to, 218-219
    lattice image in silicon, 267
    study by electron microscope, 265-267, 273

Ultramicrotomy, 192-197
    deformation mechanism in cutting, 195
    see also Specimen preparation for electron microscope
Uranium, cathodic vacuum oxidation of, 9

Vacancies, field ion micrographs of, 337, 341

possibility of resolution in electron microscope, 273
Vertical illuminators, *see* Optical microscope

Wave function concept for electrons, 232-233
Wavelength of electrons, relation to voltage, 148
Wobbler device, for electron beam, 174
Work function, effect on photoemission, 321-322
    reduction by activator, 322-323
    study with field emission microscope, 325, 327, 328

X-ray absorption coefficient, *see* Absorption coefficient for x-rays
X-ray absorption micrography, *see* X-ray microradiography
X-ray counter, gas ionization, 473
    Geiger-Muller, 473-474
    ionization chambers, 474-475
    proportional counter, 473-474
    scintillation counters, 473, 475
    sensitivity as function of voltage, 473-474
X-ray imaging of defects, 401-415
    instantaneous video display, image tube for, 412
    resolution, 411
    techniques for, Berg-Barrett, 401-407
        Bonse reflection, 407
        Borrman transmission, 402, 409, 411
        Lang transmission, 401, 402, 407-410
        Weissmann variation of Berg-Barrett, 403
    theory of contrast in Berg-Barrett-technique, 403
    x-ray interferometer, 411
    x-ray moiré fringes, 411
X-ray interferometer, 411
X-ray microradiography, 374-393
    absorption coefficient difference, table for iron alloys, 378
    compositional absorption analysis, 391-393
        accuracy of, 392
        calibration for, 392
    compositional analysis, by fluorescent x-rays, 390-391
    contact, 375, 387, 388

camera for, 387-388
radiation for, 387-388
stereographic, 388
factors affecting contrast, absorption,
376-382
Fresnel diffraction, 386-387
geometrical blurring, 384-386
photographic, 382-384
type of radiation, 380-382
high voltage, 375
mass per unit area measurement, 393
mass-thickness measurement, 393
point projection, 373-393
commercial instruments for, 390
fluorescent analysis by, 389
magnification of, 389, 390
resolution of, 390
stereographic, 389
use of electron microprobe as source,
504
x-ray sources for, 389
resolution of, 382-384, 386
stereographic, 388, 389, 393
X-ray microscopy, 373-421; see also Scanning x-ray distribution microscopy;
X-ray microradiography; Autoradiography; X-ray imaging of defects
X-rays, characteristic emission, 478-482
correction of intensity for sample absorption, 487-488
correction of intensity for secondary
fluorescence, 488-490
energy level diagram of, 480

excitation potential for, 481
equations for intensity of, 482, 487,
492
chemical analysis by, see Chemical analysis and Electron microprobe analysis
continuous emission, 479-482
equation for intensity of, 482
energy dispersion, 475-477
use of pulse-height analysis, 475-478
use of solid state x-ray detector, 476-
477
rhodium, spectrum of, 479
spectral display, 477
wavelength dispersion, by bent mica crystal, 471
combined with energy dispersion,
477-478
crystals for, 468
crystal geometry for, 470-473
by stearate coated mica crystal, 468,
472, 473
X-ray spectrometer, attachment for electron
microscope, 178-180; see also
Electron microprobe analyzer
X-ray spectrometers, crystal, 470-473
X-ray topographs, see X-ray imaging of defects

Zeiss interference microscope, 86-88
Zernike phase contrast method, 78
Zinc, x-ray imaging of defects in, 409
Zirconium, sectioning by ultramicrotome
for electron microscopy, 193, 198